Future Directions in Polymer Colloids

NATO ASI Series

Advanced Science Institutes Series

A Series presenting the results of activities sponsored by the NATO Science Committee, which aims at the dissemination of advanced scientific and technological knowledge, with a view to strengthening links between scientific communities.

The Series is published by an international board of publishers in conjunction with the NATO Scientific Affairs Division

A	Life Sciences	Plenum Publishing Corporation
B	Physics	London and New York
C	Mathematical and Physical Sciences	D. Reidel Publishing Company Dordrecht, Boston, Lancaster and Tokyo
D	Behavioural and Social Sciences	Martinus Nijhoff Publishers Dordrecht, Boston and Lancaster
E	Applied Sciences	
F	Computer and Systems Sciences	Springer-Verlag Berlin, Heidelberg, New York
G	Ecological Sciences	London, Paris and Tokyo
H	Cell Biology	

Series E: Applied Sciences – No. 138

Future Directions in Polymer Colloids

edited by
Mohamed S. El-Aasser
Emulsion Polymers Institute
Department of Chemical Engineering
Lehigh University
Bethlehem, PA 18015
USA

Robert M. Fitch
S.C. Johnson & Son, Inc.
Racine, WI 53401
USA

1987 **Martinus Nijhoff Publishers**
Dordrecht / Boston / Lancaster
Published in cooperation with NATO Scientific Affairs Division

Proceedings of the NATO Advanced Research Workshop on 'Future Directions in Polymer Colloids', Racine, Wisconsin, USA, June 30–July 4, 1986

Library of Congress Cataloging in Publication Data

Future directions in polymer colloids.

(NATO ASI series. Series E, Applied sciences ; no. 138)
Proceedings of a NATO workshop held in Racine, Wis., June 30–July 4, 1986.
Includes index.
1. Colloids--Congresses. 2. Polymers and polymer-ization--Congresses. I. El-Aasser, Mohamed S. II. Fitch, Robert McLellan, 1928–   . III. North Atlantic Treaty Organization. Scientific Affairs Division. IV. Series.
QD549.F97 1987   668.9   87-24818

ISBN 978-94-010-8150-4   ISBN 978-94-009-3685-0 (eBook)
DOI 10.1007/978-94-009-3685-0

Distributors for the United States and Canada: Kluwer Academic Publishers, P.O. Box 358, Accord-Station, Hingham, MA 02018-0358, USA

Distributors for the UK and Ireland: Kluwer Academic Publishers, MTP Press Ltd, Falcon House, Queen Square, Lancaster LA1 1RN, UK

Distributors for all other countries: Kluwer Academic Publishers Group, Distribution Center, P.O. Box 322, 3300 AH Dordrecht, The Netherlands

IN MEMORIAM

ALAN REMBAUM (1916 - 1986)

A great scientist, inventor, and friend has left us. Alan Rembaum died of cancer just before the beginning of this conference. He would have contributed greatly to our discussions and the published proceedings.

He had the idea over twenty years ago of chemically binding antibody molecules to particles of hydrophilic polymer colloids in order to effect specific cell and cell-site identification, and to develop targeted drug delivery systems. In the process, he invented entirely novel methods for the synthesis of hydrophilic colloids using highly unconventional techniques, so characteristic of the way he approached any problem. He studied the science - physics, chemistry, biology, whatever was required - and applied them in imaginative and innovative ways to the solution of practical problems. In this instance, he polymerized water-soluble monomers in the presence of soluble polymer to form sterically stabilized, insoluble latex particles, using gamma radiation to initiate the reaction. The latex particles could be labeled with fluorescent molecules of various colors, or they could be made paramagnetic by incorporation of magnetite. Preparative cell sorting thus became possible, initially by means of a Rayleigh jet in which electrically charged liquid microdroplets containing a fluorescently labeled cell were detected in a laser beam, and then deflected by electrodes to drop into containers separate from those for unlabeled cells. Thus lymphocytes, which were otherwise identical to each other, could be readily separated for analysis on the basis of their surface immunological characteristics. Later, preparative separations of viable T-cells from B-cells were carried out by means of differences in electrophoretic mobilities brought about by immuno-labeling.

This ultimately led to the landmark work in collaboration with John Ugelstad of The University of Trondheim, Norway, and J.T. Kemshead of the Institute for Child Health in London, England, on the successful clinical treatment of neuroblastoma, a cancer of the bone marrow. A slurry of bone marrow is made with the magnetic latex particles targeted for the tumor cells, which are subsequently removed with magnets. The bone marrow remaining in the body is destroyed by radiation, and the "cleaned" cells are reintroduced. These multiply rapidly, so that in a relatively short time the child is able to be sent home.

Alan liked to refer to his microspheres designed for targeted drug delivery as "smart bombs". In collaboration with Zoltan Tokes at The University of Southern California Medical School, he demonstrated that it was possible to deliver the anti-cancer drug, Adriamycin, selectively to tumor cells and to destroy them. The activity of the latex particles was undiminished, so that they could go on to act on other cells. In some cases attachment of the drug to the particles resulted in a ten-fold enhancement in the drug's activity.

Alan Rembaum's scientific interests and activities extended far beyond the field of Polymer Colloids. He was a member of the team under the direction of Michael Szwarc that invented and developed the technique of "living polymerization". He demonstrated the quantitative relationship between the ratio of monomer/initiator and the degree of polymerization. He developed the chemistry of, and named the "Ionene" polymers, in which quaternized nitrogen heteroatoms were placed in the main chain of the molecules. This led to patents in heart pacemaker batteries and hollow fiber ion-exchangers used for water treatment. He patented photo-active catalysts for the photolysis of water into hydrogen gas, and for chemical reactions such as hydrosilylation, hydroformylation, carbonylation and polymerization.

There was much more than Science in Alan's life. He was a warm person, a constant friend, and a wonderful father and husband. And he could wildly play Polish folk songs on the violin to bring tears to one's eyes. His colleagues gladly pay him the highest tribute which ultimately can be given to any person: the world is a better place to live because of him.

Robert M. Fitch
Racine, Wisconsin

TABLE OF CONTENTS

Preface.. IX

EMULSION COPOLYMERIZATION AND PARTICLE MORPHOLOGY

1. Emulsion Copolymerization and Particle Morphology
    Position Paper... 3

2. Semi-Continuous Emulsion Polymerization.......................... 23
    J.W. Vanderhoff

3. Interpenetrating Polymer Network Latexes:
    Synthesis, Morphology, and Properties........................... 47
    D.I. Lee, T. Kawamura, E.F. Stevens

4. Emulsion Copolymerization:
    Simulation of Particle Morphology............................... 65
    J. Guillot

5. Monomer Distribution and Transport in Miniemulsion
    Copolymerization.. 79
    J. Delgado, M.S. El-Aasser, C.A. Silebi,
    J.W. Vanderhoff and J. Guillot

RHEOLOGY OF LATEX SYSTEMS AND CONCENTRATED DISPERSIONS

6. Rheology of Latex Systems and Concentrated Dispersions -
    Position Paper.. 107

7. Polymer Colloids as Model Systems for Studying Rheology
    of Dispersions... 119
    I.M. Krieger

8. Theoretical Approaches to the Rheology of Concentrated
    Dispersions.. 131
    W.B. Russel

9. Structure Formation in Flowing Suspensions..................... 151
    R.L. Hoffman

10. Viscoelastic Properties of Concentrated Sterically
    Stabilized Latices and the Effect of Addition of Free
    Polymer.. 167
    Th F. Tadros

POLYMER STABILIZED LATEXES

11. Polymer Stabilized Latices - Position Paper.................... 183

12. The Stabilization and Controlled Flocculation of Sterically-
    Stabilized Latices... 191
    B. Vincent

VIII

13. Control of Particle Size in the Dispersion Polymerization
    of Sterically Stabilized Polymer Colloids...................... 209
    M.D. Croucher and M.A. Winnik

14. Polymers at Interfaces: Adsorption and Disjoining
    Pressure Theories... 229
    M.A. Cohen Stuart

    NEW TECHNIQUES IN CHARACTERIZATION OF POLYMER COLLOIDS

15. New Techniques in Characterization of Polymer Colloids -
    Position Paper... 243

16. Characterization of Polymer Colloids......................... 253
    R.H. Ottewill

17. The Characterization of Polymer Colloids by Fluorescence
    Quenching Techniques... 277
    M.A. Winnik and M.D. Croucher

18. Dielectric Spectroscopy of Model Polystyrene
    Colloids... 289
    R.M. Fitch, L.S. Su and S.L. Tsaur

    POLYMER COLLOIDS IN BIOMEDICAL FIELD

19. Biomedical Applications of Polymer Colloids:
    Future Directions - Position Paper #1........................ 307

20. Future Directions in the Latex Agglutination Assays -
    Position Paper #2.. 315

21. Biomedical Applications of Polymer Particles
    with Emphasis on Cell Separation............................. 321
    C.D. Platsoucas

22. Biomedical Applications of Monodisperse Magnetic
    Polymer Particles.. 355
    J. Ugelstad, A. Berge, T. Ellingsen, J. Bjorgum,
    R.Schmid, P. Stenstad, O. Aune, T.N. Nilsen, S. Funderud
    and K. Nustad

23. The Use of Polystyrene Latexes in Medicine................... 371
    J.M. Singer

    Subject Index   395

PREFACE

Future Directions In Polymer Colloids

Mohamed S. El-Aasser, and Robert M. Fitch (editors)

It is appropriate that the first NATO-Advanced Research Workshop on "FUTURE DIRECTIONS IN POLYMER COLLOIDS" was held approximately fifty years after the first synthetic polymer latexes were made on a commercial scale during the mid-1930s. Since that time the field of what is now known as polymer colloids has been evolving rapidly, not only on the practical level, but also on the scientific and engineering levels.

Billions of pounds of copolymers are manufactured annually by means of the emulsion polymerization process. "Commodity" polymers as well "specialty" polymers are prepared today for use in a wide variety of applications: synthetic rubber, floor coatings, paints, adhesives, binders for non-woven fabrics, high-impact polymers latex foam, additives for construction materials such as cement and concrete, and rheological modifiers. They are also used in numerous biomedical applications: such as diagnostic tests, immunoassays, biological cell-labeling, (identification and separation), and drug delivery systems. Small quantities of monodisperse polymer colloids are used as size calibration standards and find extensive use as model colloids to test theories in colloids surface and rheological studies. Advances have been made in our understanding of the mechanism and kinetics of the emulsion polymerization process as well as the stability of polymer colloids. Equal advances were made in engineering areas related to polymer colloids, e.g. modeling of batch, semi-continuous and continuous emulsion polymerization and copolymerization processes.

However, several questions are still being debated, for example the site of particle initiation and particle growth, as well as particle morphology and its control. In the engineering area several questions are still outstanding, e.g. transport phenomena of various species in the polymerization process, automation and control of the process, reactor fouling and aggregate formation.

These previous developments in the field of polymer colloids have been nurtured by a variety of international meetings which have helped in exchanging ideas among the various scientists and engineers in the field. For the most part, these meetings were always concerned with accomplished and completed work. Therefore, it was felt that the time had come to pause and assess these past developments, as well as to develop a plan of research action for the future. Consequently, a NATO-Advanced Research Workshop was organized which brought together sixty-three scientists and engineers from both academia and industry. They came from Australia, Europe, Japan, and the U.S.A. The main objective of the workshop was to develop position papers outlining future direction of research in five selected areas of polymer colloids:

x

1. Emulsion Copolymerization and Particle Morphology
2. Rheology of Latex Systems and Concentrated Dispersions
3. Polymer-Stabilized Latexes
4. New Techniques of Polymer Colloids
5. Polymer Colloids in the Biomedical Field

These objectives were met with the active participation of all the attendees.

This edition includes the five position papers, as well as all invited papers presented at the Workshop. We hope that this volume will achieve the objective of indicating the direction of the most promising research in this active, multidisciplinary field for the next one or two decades.

The NATO Advanced Research Workshop was made possible by grants from The NATO Scientific Affairs Division, The National Science Foundation and several industrial corporations. In addition to financial support, The Johnson Foundation contributed the use of their elegant Wingspread Conference Center in Racine, Wisconsin, as well as the services of its operating staff. The Emulsion Polymers Institute of Lehigh University co-sponsored the conference and donated its resources for retyping and preparing these manuscripts for publication. We gratefully acknowledge the contributions from these organizations.

We would also like to thank the following individuals from the Emulsion Polymers Institute: Kathy Devlin and Gilda Mulicka for typing all the manuscripts more than once, Olga Shaffer for the photographic work, Debra Nyby for coordinating all the activities in the manuscript preparation process, and Karen Hicks for coordinating all the arrangements associated with the conference.

EMULSION COPOLYMERIZATION AND
PARTICLE MORPHOLOGY

EMULSION COPOLYMERIZATION AND PARTICLE MORPHOLOGY

POSITION PAPER

1. INTRODUCTION
1.1. Background
Free radical-initiated vinyl addition polymerization comprises four reactions:

1. Initiation, in which a primary radical formed by decomposition of the initiator adds a monomer molecule to form a monomer radical;
2. Propagation, in which the monomer radical adds monomer molecules successively to form a long polymer chain ending in a monomer radical;
3. Termination, in which two polymer radicals combine to form a polymer molecule or disproportionate to form two polymer molecules;
4. Transfer, in which a growing polymer radical abstracts a hydrogen or halogen atom from another compound to terminate the polymer radical and form a new radical, which adds monomer molecules to grow another polymer chain.

This free radical-initiated vinyl addition polymerization can be carried out using four different polymerization processes:

1. Mass or bulk polymerization, in which the monomer is polymerized neat to form bulk polymer;
2. Solution polymerization, in which the monomer is dissolved in a solvent and polymerized to form a solution (or a heterogeneous dispersion if the polymer is not soluble in the solvent);
3. Suspension polymerization, in which the water-immiscible monomer is dispersed in water as drops and polymerized to form polymer beads;
4. Emulsion polymerization, in which the water-immiscible monomer is emulsified in water and polymerized to form a latex (colloidal dispersion of polymer particles).

In emulsion polymerization, the four reactions of free radical-initiated vinyl addition polymerization are superimposed on the heterogeneous colloidal latex system. Typically, the average diameter of the original emulsion droplets is 1-10 μm and that of the final latex particles is 100-300 nm. Thus the latex particles produced are an order of magnitude smaller than the original emulsion droplets, so that the mechanism of polymerization must comprise a first stage in which the particles are nucleated and a second stage in which the nucleated particles grow to a larger size by polymerization.

Experimentally, the number of particles nucleated depends upon:

1. The type and concentration of emulsifier;
2. The rate of radical generation;
3. The type and concentration of electrolyte;
4. The temperature;
5. The type and intensity of agitation;
6. Other parameters which are not easily discernible.

Thus the nucleation stage is not always reproducible, and successive polymerizations using the same recipe may give different numbers of particles. In contrast, the growth stage is tractable and reproducible. As a first approximation, the rate of polymerization is proportional to the number of particles, and the polymer molecular weight is proportional to the number of particles relative to the rate of free radical generation. Thus an increase in the number of particles gives an increase in both the rate of polymerization and the polymer molecular weight. These unusual kinetics of emulsion polymerization are different from those of mass, solution, and suspension polymerization, in which the rate of polymerization varies inversely with the polymer molecular weight. The two criteria for emulsion polymerization kinetics are:

1. A segregated free radical system, in which radicals grow in proximity to one another without termination;
2. The number of loci available for segregation must be within a few orders of magnitude of the number of radicals existent in the system.

In emulsion polymerization, the radicals are segregated in the latex particles; termination cannot occur across the aqueous phase between two neighboring particles containing free radicals. Moreover, the number of particles is generally within a few orders of magnitude of the number of free radicals existent. Thus these unusual kinetics suit emulsion polymerization to the preparation of high-molecular-weight polymer at rapid polymerization rates.

Many hypotheses have been advanced for the mechanism of emulsion polymerization, but all can be divided into four main categories, according to the locus of particle nucleation:

1. The monomer-swollen emulsifier micelles;
2. The adsorbed emulsifier layer;
3. The continuous aqueous phase;
4. The monomer droplets.

The mechanism of nucleation in monomer-swollen emulsifier micelles was developed during the Office of Rubber Reserve program of World War II, an outstanding example of cooperation between industry, university, and government, which resulted in the development of the synthetic rubber process within two years. This mechanism, which was postulated by Harkins (1) and quantified by Smith and Ewart (2), is still applicable today, particularly to monomers such as styrene and butadiene which are only sparingly soluble in water. It comprises nucleation by entry of a radical into a monomer-swollen micelle and its rapid growth to a polymer particle; the radicals enter micelles preferentially because their surface area is orders of magnitude greater than that of the monomer droplets; the nucleation stage ends when all of the micelles have been transformed into polymer particles or have given up their monomer and emulsifier to neighboring particles which have captured radicals. The mechanism of nucleation in the adsorbed emulsifier layer, which was postulated by Medvedev (3), is conceptually similar to the nucleation-in-micelles mechanism, although the kinetic equations differ somewhat. The mechanism of nucleation in the aqueous phase comprises the initiation of a radical in the aqueous phase, followed by addition of monomer molecules until the oligomeric radical exceeds its solubility and precipitates from solution; the precipitating oligomeric radicals either nucleate a particle or flocculate with a particle already nucleated according to whether the particle surface area exceeds a critical value; the function of the emulsifier is to stabilize these particles as they are formed.

This mechanism, which was proposed independently by Priest (4) and Jacobi (5), and developed further by Stannett et al. (6) and Fitch et al. (7), is most applicable to monomers such as vinyl acetate and vinyl chloride which are more soluble in water. The mechanism of nucleation in monomer droplets is significant only if the droplets are sufficiently small. The typical droplets of size 1-10 μm have surface areas which are negligible compared to those of of micelles; however, the droplets of size 100-200 nm made using mixed emulsifier systems (8) can compete with emulsifier micelles and the aqueous phase as a locus for particle nucleation.

The difficult-to-reproduce nucleation stage can be obviated by seeded emulsion polymerization, in which a predetermined number of latex particles is added to the polymerization and grown to a larger size. The increase in particle volume is stoichiometric provided no new particles are nucleated, e.g., one batch of a 20-nm size seed latex would be sufficient to prepare 1000 batches of 200-nm size final latex. Then, if the next batch of seed latex were of 25-nm size, the stoichiometry could be adjusted so as to give the desired 200-nm final size; in this case, one batch of seed latex would be sufficient to prepare 256 batches of 200-nm size final latex. Thus seeded emulsion polymerization can give batch-to-batch reproducibility by obviating the particle nucleation stage.

There are three types of emulsion polymerization processes used to prepare latexes:

1. Batch polymerization, in which all ingredients are charged to the reactor and heated with agitation at the polymerization temperature;
2. Semi-continuous or semi-batch polymerization, in which the monomer is added to the reactor, continuously or in increments, neat or in emulsion;
3. Continuous polymerization, in which all ingredients are added continuously to one part of the reactor system and partially or completely converted latex is removed continuously from another part.

All three processes are used in industrial polymerization. Usually, the polymerizations are carried out in stirred-tank reactors, a single reactor for batch and semi-continuous polymerization, and a series or cascade of stirred-tank reactors for continuous polymerization.

Latexes made by emulsion polymerization have found wide industrial application, and each of the following latex families represents a large business area, with production in many parts of the world:

1. Butadiene-styrene copolymer or polybutadiene latexes for synthetic rubber;
2. Styrene-butadiene copolymer latexes for paper coatings, carpet backing, nonwoven fabrics, reinforcement of concrete, and other coatings;
3. Polyvinyl acetate and vinyl acetate copolymer latexes for adhesives, paints, and other coatings;
4. Acrylate ester copolymer latexes for exterior paints, nonwoven fabrics, and other coatings;
5. Polyvinyl chloride and vinyl chloride copolymer latexes for plastisol resins and coatings;
6. Polychloroprene latexes for synthetic rubber;
7. Polytetrafluoroethylene and other fluorinated polymer latexes for varied uses;
8. Inverse polyacrylamide latexes for flocculation, papermaking,

sludge treatment, and enhanced oil recovery.

1.2. State of the Art

The polymerization of a mixture of two or more monomers gives a copolymer with properties intermediate between those of the corresponding homopolymers. The properties depend, not only on the overall composition of the polymer molecules, but also on the compositional distribution of monomer molecules in the polymer chains and the distribution of polymer molecular weights. The reaction conditions which govern the initiation, propagation, termination, and transfer reactions determine the polymer molecular weight distribution but not the copolymer compositional or sequence distribution.

Some industrial latexes such as polyvinyl acetate are made by the polymerization of a single monomer. Others such as styrene-butadiene copolymers are made by the polymerization of two or more monomers, selected to give specific polymer properties and performance in a given application. These copolymers can be divided into two general categories:

1. Copolymers comprised of significant proportions of two or more monomers, which determine the polymer properties, particle morphology, and application performance;
2. Copolymers comprised primarily of one monomer, with a small proportion of functional monomer(s), which give specific properties such as improved colloidal stability or enhanced adhesion to a given substrate;

The overall composition of a copolymer formed from a given monomer mixture can be predicted by copolymerization theory and verified by experiment. In the copolymerization of two monomers, there are four propagation reactions. The copolymerization reactivity ratio r1 is the ratio of the propagation rate constant for a polymer chain ending in radical 1 to add monomer 1 relative to that to add monomer 2; the reactivity ratio r2 is the ratio for a polymer chain ending in radical 2 to add monomer 2 relative to that to add monomer 1. These copolymerization reactivity ratios can be determined experimentally for a given monomer pair. Moreover, the reactivity of a given monomer in copolymerization can be expressed by its reactivity Q and polarity e; these parameters can be calculated from the r1 and r2 values of a given monomer pair and then used to calculate the r1 and r2 values for monomer pairs which have not been measured experimentally. One problem that remains, however, is to adapt this copolymerization theory developed for homogeneous polymerization to the heterogeneous emulsion polymerization.

The reaction engineering of copolymers made by bulk, solution, and suspension polymerization is concerned with the prediction and control of such parameters as polymer molecular weight and molecular weight distribution, molecular architecture, copolymer compositional and sequential distribution, and residual concentrations of monomer and other nonpolymeric components. The emulsion polymerization process, however, because of the colloidal size of the latex particles, offers additional opportunities for systematic variation of polymer properties and application performance. The additional parameters of the latex that are important are the number of particles, particle size and size distribution type and concentration of surface groups, particle morphology, and nature of the continuous phase.

In emulsion polymerization, the monomer-polymer phase comprises colloidal particles, and polymerization can occur in both the monomer-polymer and the aqueous phases, especially where the monomers such as vinyl acetate are relatively soluble in water. In an extreme case, the polymer molecule could be formed almost entirely in the aqueous phase and then precipitate just before termination. Thus the transport of oligomeric radicals from the aqueous phase to the monomer-polymer phase as well as the polymerization reactions determine both the polymeric and colloidal parameters of the system.

The industrial latexes prepared for practical applications must have excellent colloidal stability combined with high solids content. Generally, the higher the concentration of surface functional groups, the better is the colloidal stability. Moreover, the higher the solids content, the easier it is to formulate the latex in a given practical application. The industrial latexes which have been developed by extensive research over the past forty years can be divided into three main generations:

1. First-generation latexes, which are prepared using conventional emulsifiers for the polymerization, steam-stripping, and poststabilization for a given application, typically have excellent stability to added electrolyte, mechanical shear, and freezing-and-thawing; however, the high concentrations of emulsifier required give low surface tensions (and hence much foaming) and poor adhesion to some substrates;

2. In second-generation latexes, all or part of the conventional emulsifier is replaced by a functional monomer such as acrylic acid, sodium styrene sulfonate, or acrylamide; typically, these latexes have excellent stability to added electrolyte, mechanical shear, and freezing-and-thawing with lower concentrations of functional groups; moreover, their surface tensions are higher (and hence their foaming is less), and the functional groups, particularly carboxyls, enhance the adhesion to certain substrates;

3. To prepare a third-generation latex, a good second-generation latex is characterized to determine the composition and molecular weight of the functional copolymer adsorbed on the particle surface; this copolymer is then prepared in a separate reaction and used as emulsifier in the polymerization; typically, these latexes have excellent stability to added electrolyte, mechanical shear, and freezing-and-thawing; however, their surface tensions are high, sometimes approaching that of water itself (hence their foaming is minimal), and the concentration of functional groups required for stability is minimal.

Thus an emerging technology is the use of polymeric emulsifiers tailored to the specific application, which give excellent colloidal stability in lower concentration and interfere less with end-use properties than conventional emulsifiers. Moreover, some polymeric emulsifiers are prone to grafting and thus can be chemically bound to the latex particles.

The emulsion polymerization processes used today couple the polymeric and colloidal parameters; one type cannot be changed significantly without affecting the other. Often, industrial developments concentrate on achieving the requisite colloidal stability, with little consideration given to the effect of the colloidal parameters on the polymer properties. A major challenge for researchers is to learn how to uncouple these parameters, thereby allowing their independent optimization. Over the past thirty years, much work has been directed implicitly towards this objective. A powerful tool here is the use of copolymerization, especially that of water-soluble functional monomers to improve colloidal stability and control latex viscosity. Thus far the approach has been largely empirical; the research has been carried out mostly in industrial laboratories, and, except for the patent literature the published results have been sparse. Consequently, progress in this area has been slow despite its technical importance and great potential.

Emulsion polymers are prepared by batch, semi-continuous, and continuous polymerization. In batch polymerization, the decomposition of initiator forms free radicals, which initiate the polymerization and nucleate the particles; nucleation proceeds concurrently with particle growth until it ceases and only particle growth occurs. There is little control over the polymerization other than cooling to remove the heat of polymerization and variation of the recipe to alter the number of particles. The particle nucleation stage is difficult to reproduce; thus seed latex is often added to the initial charge to control the number of particles.

In semi-continuous polymerization, the particles are nucleated in two ways: a small proportion of the monomer is charged initially and polymerized in batch to prepare a seed latex in situ or the continuous monomer addition and the polymerization are started at the same time. In the second case, nucleation proceeds concurrently with particle growth until it ceases and only particle growth occurs. In both cases, the number of particles nucleated may vary from batch to batch. This variation may be obviated by addition of a seed latex, which gives rigorous control of the number of particles and stoichiometric particle growth. The monomer is added either continuously or in increments, either neat or in emulsion. These different modes of addition give different results: the addition of neat monomer generally results in the growth of the particles nucleated early in the reaction; the addition of monomer in emulsion may give continual nucleation throughout the polymerization. The mode and rate of monomer addition control the rate of polymerization rigorously; moreover, they also control the copolymer compositional distribution and particle morphology, and furnish the means to minimize the formation of coagulum and achieve the requisite latex properties for the practical application. Semicontinuous polymerization is the preferred process for rigorous process control, and it is used to prepare many industrial latexes; however, our understanding of its fundamentals is still primitive.

Finally, in continuous emulsion polymerization, the polymerization is started as soon as the monomer emulsion is heated to the polymerization temperature, and particle nucleation occurs concurrently with particle growth. The number of particles nucleated (and hence the rate of polymerization and conversion of the exit stream) varies cyclically with mean residence time according to the local conditions in the reactor system. This variation can be obviated by the continuous addition of a seed latex or the use of a short tube reactor ahead of the continuous reactor, in which a seed latex is formed continuously in situ, to furnish the requisite number of particles to the system, so that the conversion of the exit stream becomes constant after a few mean residence times. Continuous polymerizations are run in series or cascades of stirred-tank reactors, a single stirred-tank reactor with an outside loop, tubular reactors, and other types. The process is economical and gives latexes of constant quality; however, without a detailed understanding of its fundamentals, it is difficult to alter the polymeric and colloidal properties of the latex. In the laboratory, it is often used for the study of fundamental reaction and transport phenomena.

The development of seeded emulsion polymerization to control the number of particles and the particle size distribution was a significant advance in emulsion polymerization technology. (This development resulted from the fundamental work of Willson et al. (10) and Smith (11), which was directed towards understanding the polymerization mechanism.) Typically, the seed latex is colloidally stable and of very small particle size. It is prepared in a separate polymerization or a separate process step. Problems with seeded emulsion polymerization concern the reproducibility of the seed latex preparation and stability, and new approaches to the control of particle size and size distribution are still needed.

Despite new process improvements, many variables of the emulsion polymerization are still poorly controlled, which results in serious batch-to-batch variation in molecular weight and molecular weight distribution, and degree of branching and crosslinking. In emulsion copolymerization, this poor control is complicated further by compositional drift and phase separation in the particles. The compositional distribution can be controlled by variation of the rate of addition of the monomer mixture; the phase separation can be controlled by the formation of the appropriate graft copolymer of the two polymers.

Semi-continuous emulsion polymerization can be used to make particles with a variation in composition from center to surface, by adding different monomer compositions in sequence at controlled rates during the polymerization. These latexes are called "structured" or "multiphase" latexes; those with a well-defined concentric compositional gradient are called "core-shell" latexes. Structured particle and core-shell latexes are a new class of polymer blends of unique morphology which can be prepared only by emulsion polymerization.

1.3 Current Applications
Industrial latexes are used in practical applications in three different ways:

1. The latexes are coagulated to recover the solid polymer;
2. Latexes are formulated into coatings which form continuous films upon drying,;
3. The latexes are used as latexes.

The science of emulsion polymerization began with the develop ment of butadiene-styrene and polybutadiene latexes for synthetic rubber during World War II; this application is still the largest single use for emulsion polymers. The latexes are made by continuous emulsion polymerization in a series or cascade of stirred-tank reactors, then coagulated and processed in the same way as natural rubber latex. Another application is ABS resins; polybutadiene or acrylonitrile-butadiene copolymer latexes prepared by emulsion polymerization are used as seed latexes for the grafting of styrene-acrylonitrile copolymer to give a core-shell latex, followed by coagulation and dispersion in a solid styrene-acrylonitrile copolymer matrix, to give a colloidal dispersion of stable particles in a high-viscosity polymeric matrix.

The drying of latexes at temperatures above the apparent second-order transition temperature T_g of the polymer gives continuous films; the water-air and polymer-water interfacial tension forces act to cause coalescence of the particles. The film-forming latexes are used unpigmented, e.g., vinylidene chloride copolymer latexes for barrier coatings on polyolefin films, or formulated with aqueous dispersions of pigments or fillers, e.g., the many different latexes used for latex paints, paper coatings, adhesives, carpet backing, and many other applications; the composite dispersions dry to form a dispersion of pigment or filler in a continuous matrix of polymer. The use of film-forming latexes began with the development of poly(vinyl acetate) latexes for adhesives in the 1930's and styrenebutadiene copolymer latexes for latex paints in the 1940's, and has continued up to the present time in a bewildering variety of copolymer compositions for many different film-forming applications. Moreover, the need to tailor the latex properties to the specific application stimulated the development of semicontinuous emulsion polymerization as an improvement over batch polymerization.

Other applications use latexes as latexes, e.g., polystyrene latexes used as pigments and opacifiers for detergents, where the particles of high refractive index (1.59) and proper size relative to the wavelength of visible light confer opacity on solid films or liquids. Moreover, the development of monodisperse polystyrene latex calibration standards has enhanced scientific investigations in many different fields all out of proportion to the relatively small quantities of latexes used, and the development of these latexes as substrates for immunological diagnostic tests and other immunological reactions has stimulated much research in the biomedical field.

2. GOALS FOR THE NEXT DECADE

Emulsion copolymerization processes have developed greatly during the past four decades to give a wide variety of sophisticated latexes used for many different applications. For this growth to continue over the next decade requires that certain goals be met in the following areas:

1. Copolymerization in emulsion;
2. Characterization of latexes;
3. Particle morphology;
4. Physical and colloidal properties of latex polymers;
5. Mathematical models of copolymerization;
6. On-line measurements and control of reactors.

2.1. Emulsion Copolymerization

Most of the known copolymerization parameters were derived from experiments carried out in homogeneous systems. To apply these parameters to the heterogeneous emulsion polymerization system requires that other parameters be taken into account: the different solubilities of the monomers in the aqueous phase; the diffusion coefficients of the monomers from the aqueous phase into the particle phase; the capacity of the monomers to swell the particles, and the phase-separation of incompatible monomer-swollen copolymers. Therefore, the goals for the next decade include the following basic experimental investigations of heterogeneous emulsion polymerization systems:

1. Determination of all of the parameters that affect copoly-merization;
2. Accurate determination of copolymerization reactivity ratios in emulsion polymerization;
3. Consideration of the effect of such thermodynamic parameters as the polymer-solvent interaction parameter and interfacial tension;
4. Determination of the diffusion coefficients of the monomers;

5. Systematic studies of the emulsion copolymerization of three or more monomers;
6. Simulation and modeling of the emulsion copolymerization as a function of monomer composition, copolymerization reactivity ratios, monomer water solubilities, monomer diffusion coefficients, and monomer addition rate;
7. Simulation and modeling of particle morphologies prepared by variations in the copolymerization process;
8. Understanding of the phase-separation of incompatible polymers during polymerization, after aging in latex form, and in the latex film.

For the simulation and modeling of emulsion copolymerization to be relevant and reliable, the variation of polymer microstructure, particle morphology, and latex colloidal stability with conversion must be taken into account.

In semi-continuous polymerization, the mode of monomer addition (continuously or in increments, neat or in emulsion) is important. The addition in increments gives a slightly different distribution of copolymer compositions than continuous addition; the addition of a monomer emulsion gives a different particle size distribution (and hence a different number of particles) than addition neat. The rate of polymerization and the breadth of the copolymer compositional distribution are controlled by the rate of addition; generally, the slower the rate of monomer addition, the slower the polymerization rate and the narrower the distribution of copolymer compositions. "Starved" semi-continuous copolymerization can be used to prepare a narrow distribution of copolymer compositions for systems with disparate copolymerization reactivity ratios, e.g., vinyl acetate-butyl acrylate. Further improvements are the systematic variation of the composition of the monomer mixture added so as to make a constant-composition copolymer. Other variations include:

1. Initial addition of all of the slower-polymerizing monomer and some of the faster-polymerizing monomer, to give a constant-composition copolymer with minimum long-chain branching and crosslinking;
2. Initial addition of part of both monomers to give the desired copolymer composition, followed by addition of the monomer mixture with a time-variant feed rate so as to keep the monomer ratio constant, to give a copolymer with long-chain branching and crosslinking, according to the rate of monomer addition.

2.2. Characterization of Latexes

Latexes can be characterized as to the loci of functional groups as well as the properties of the latex copolymer. The functional groups are incorporated by the initiator or a functional monomer. Persulfate ion initiator gives sulfate or hydroxyl endgroups, according to the pH of the polymerization, and the presence of adventitious oxidizing agents gives carboxyl endgroups. These functional groups enhance the colloidal stability of the latex, improve the adhesion of the latex film to the substrate, control the rheology of the latex, and give active sites for further reactions. The copolymerization of a small proportion of functional monomer gives five possibilities:

1. The monomer copolymerizes in the aqueous phase to form a copolymer that remains there;

2. The monomer copolymerizes in the aqueous phase to form a copolymer that adsorbs on the latex particle surface;
3. The monomer copolymerizes at the particle surface;
4. The monomer copolymerizes inside the particle;
5. The monomer does not polymerize.

The proportion of the functional monomer that is found in each of these five categories depends upon its concentration, partitioning between the aqueous and monomer-polymer phases, the copolymerization reactivity ratios, the type of polymerization process, and the time and mode of addition of the functional monomer. The loci of the functional monomer can be determined as follows:

1. Separation of the colloidal components of the latex by serum replacement or ion exchange (dialysis is ineffective);
2. Determination of the concentration of acidic or basic surface or buried groups by conductometric or potentiometric titration;
3. Determination of surface groups by spectroscopic methods, e.g., nuclear magnetic resonance (NMR), X-ray photoelectron (ESCA), or reflective infrared spectroscopy.

The relative hydrophobicity of the latex particle surface can be determined by the adsorption of emulsifier.
The goals for the next decade in the determination of the loci of functional groups include further development of the foregoing techniques as well as the development of the following new techniques:

1. Determination of emulsifier adsorption by ion-selective electrodes and more rapid surface tension measurements;
2. Determination of the surface functional groups by X-ray photoelectron spectroscopy (ESCA);
3. Determination of functional groups by nuclear magnetic resonance spectroscopy (NMR) or Fourier transform infrared spectroscopy (FTIR) after reaction with other functional molecules to enhance the resolution;
4. Determination of the free radical concentration by electron spin resonance (ESR) spectroscopy.

Emulsion copolymerization gives, not only a distribution of molecular weights, but also a distribution of copolymer compositions and sequences as well as significant degrees of branching and crosslinking by transfer. The molecular weight can be determined by gel permeation chromatography (GPC) as well as such classical methods as intrinsic viscosity, light scattering, and ultracentrifugation. Of these methods, only gel permeation chromatography and ultracentrifugation give the molecular weight distribution. One significant advance is the use of thin-layer chromatography combined with flame ionization detection (TLC-FID) for characterization of copolymer composition, graft copolymers molecular weight and molecular weight distribution, degree of crosslinking, and presence of minor constituents. The goals in this area also include the complete characterization of the copolymer composition in the aqueous phase (solute polymer), monomer-polymer phase (buried polymer), and latex particle surface (adsorbed water-soluble surface-active polymer).
Gel permeation chromatography fractionates polymers according to their hydrodynamic volume, so that the same element of the fractionated copolymer would contain molecules of the same hydrodynamic volume but different molecular weights and compositions. To characterize the

molecular weight distribution of a heterogeneous copolymer with a distribution of copolymer compositions requires a detector with improved response. The response of the present ultraviolet or infrared detectors depends, not only on the number of absorbing units, but also on the local microstructure of the copolymer molecule about these units. Except in a few special cases, the concentration and composition of the copolymer molecules cannot be measured quantitatively; however, the ratio of detector response to retention time can be used to determine molecular size qualitatively. Fractionation techniques based on composition as well as molecular size must be developed, so that the detector can measure copolymers of uniform composition.

Another goal is the characterization of the insoluble or crosslinked fraction of the polymer, including:

1. Measurement of the swelling of latex particles by solvents;
2. Measurement of copolymer composition, nature of crosslinking groups, and crosslink density by solid-state NMR spectrosopy;
3. Characterization of the rheology of the solid latex polymer;
4. Measurement of the mechanical properties of the solid polymer by dynamic mechanical spectroscopy;
5. Degradation of the polymer and analysis by hydrolysis and pyrolysis.
6. Characterization of the rheology and surface-active properties of the water-soluble polymer fractions adsorbed on the latex particle surface.

The goals of the characterization should also include better techniques for the fractionation of the solid polymer:

1. Extension of thin-layer chromatographic separation methods to other chromatographic methods, including high-performance liquid chromatography (HPLC);
2. Improvements in NMR techniques for measuring sequence length and degree of blockiness;
3. Determination of the factors that control the formation of surface-active water-soluble polymers and their effects on film formation, water sensitivity of latex films, and latex rheology;
4. Development of a better understanding of the factors that control the electrophoretic mobility and zeta potential of latex particles;
5. Electrophoretic or acoustophoretic fractionation of the water-soluble polymer formed in the aqueous phase;
6. Fractionation of the polymer by field flow fractionation (FFF) in electric and mechanical shear fields.

Following the development of improved fractionation methods, improved methods for molecular weight determination should be developed, including:

1. Improved columns and software for gel permeation chromatography;
2. Field flow fractionation of polymer solutions in a thermal gradient.

The average particle size and particle size distribution of latexes can be determined by transmission electron microscopy, various light scattering functions, photon correlation spectroscopy (quasi-elastic light scattering), ultracentrifugation, small-angle X-ray scattering, flow ultramicroscopy, and hydrodynamic chromatography. These methods are generally satisfactory, although each measures a slightly different diameter and each has its drawbacks. The goals for improved methods of particle size measurement include:

1. Evaluation of field flow fractionation for particle size measurement;
2. Development of better software for quasi-elastic light scattering for resolution of multimodal distributions;
3. Development of better columns and software for hydrodynamic chromatography.

2.3. Particle Morphology

Latexes prepared in the laboratory and in production have a broad range of particle morphologies, i.e., homogeneous particles and multiphase particles such as core-shell particles, particles with microdomains ranging from fine to coarse dispersions, interpenetrating polymer networks, and combinations of the above. The latexes comprising multiphase particles often combine the desirable properties of both components. The recipes used to prepare them are often empirical and complex, and there is presently no unified understanding of the mechanism by which these morphologies are prepared. Although there is some understanding of the relative importance of the variables that control the particle morphology, this knowledge is insufficient to realize the full potential of these multiphase particles. The need for better understanding of the control of particle morphology affects, not only product development, but also research in polymerization reaction kinetics, control of grafting reactions, and polymerization methods.

The overall goal of future research in particle morphology is:

1. The development of a basic understanding of the mechanisms of particle structure development;
2. The identification of the role of each component of the recipe and the process type (i.e., batch, semi-continuous, continuous) in controlling particle morphology.

It is likely that advances in this area can be achieved by assuming that the formation of particle structure is a rate process involving mass transfer of polymer molecules to different locations within the particle. Thus there are driving forces for this transport, which may be described in terms of thermodynamic potentials (e.g., interfacial energies), as well as resistances, which may be described in terms of polymer chain mobility.

To achieve this understanding of particle structure requires a combined experimental and theoretical approach. The treatment of the thermodynamics (interfacial energy) should consider the polymer-polymer compatibility, and the polymer-water, polymer-polymer, and water-air interfacial tensions. The mobility of the polymer molecules depends upon the molecular weight, degree of crosslinking, monomer-polymer ratio during polymerization, and anchoring of the polymer chains at the surface by polar endgroups. The particle structure will also be affected by the rate at which the viscosity of the particle increases during polymerization. The proposed mechanisms and models must take these parameters into account.

The traditional examination of dried latex dispersions by transmission electron microscopy is often ineffective in delineating the morphology of multiphase particles. Newer techniques such as selective staining of individual particles or microtomed sections, examination of frozen dispersions, and freeze-fracture of frozen dispersions give better results. Selective staining of individual multiphase particles works best with a negative stain to give a dark background and a selective stain for the different domains of the particle. Examination of frozen dispersions on the cold stage shows the particle morphology without the distortion introduced by drying. Examination of freeze-fractured surfaces of frozen dispersions also shows the morphology of dispersions which would deform upon drying. Even these techniques need improvement, however, and one goal for the next decade is their further development and use to amass data for various copolymerization systems. Preliminary work has shown that the morphology of multiphase particles is far more complex than thought earlier, and therefore another goal is to re-examine systems studied earlier using these new techniques.

An indirect method for determining the multiphase particle morphology is dynamic mechanical spectroscopy, which determines the continuous and dispersed phases of the particles as well as the presence of graft copolymers. A goal for the next decade is to further develop this method for the characterization of particle morphology.

Other goals include the development of the following new methods for studying particle morphology:

1. Nuclear magnetic resonance spectroscopy tuned to a resonance specific to one monomer unit;
2. Fluorescence decay methods to determine the locus of a monomer containing a fluorescent group in a structured particle;
3. Better thermal and mechanical methods;
4. Better selective staining methods for electron microscopy.

2.4. Physical and Colloidal Properties

A correlation between the performance of the latex in a given application and the particle structure and latex colloidal properties must be developed using well-defined physical and chemical methods. Latexes are complex in that they contain emulsifiers, water-soluble polymers, functional monomers in various loci, decomposition products of initiators and buffers, coalescing aids, and plasticizers in addition to the latex polymer. The factors that are important in this correlation are:

1. The control of the polymer composition and particle morphology during polymerization;
2. The characterization of the polymer composition and particle morphology of the final latex;
3. The determination of the effect of particle size and serum composition on rheological behavior of the latex;

The mechanical properties of the latex polymer can be characterized by:

1. Dynamic mechanical spectroscopy for small deformations of the polymer;
2. Stress-strain measurements at different temperatures for large deformations of the polymer;
3. Fracture-mechanics methods to investigate the toughening and energy dissipation of the polymer;

4. Gel permeation chromatography to determine the molecular weight distribution of the polymer;

5. Nuclear magnetic resonance spectroscopy to determine the copolymer sequence distribution.

The colloidal properties of the latex can be characterized by:

1. Transmission electron microscopy to determine the particle size distribution of the latex;

2. Viscometry using such instruments as the Weissenberg Rheogoniometer or Haake Rotovisco to determine the rheological properties of the latex;

3. Mechanical stability using such instruments as the Hamilton-Beach ice cream soda mixer to determine latex stability as a function of shear rate and temperature;

4. Electrolyte stability using visual or spectrophotometric methods to determine latex stability as a function of electrolyte valence and concentration;

5. Electrophoretic or acoustophoretic mobility to determine the electrokinetic properties of the latex particles.

2.5. Mathematical Models

One important goal is the development of predictive mathematical models of copolymerization for industrial research and computer control of industrial polymerization processes. Our present models of batch, semi-continuous, and continuous polymerization comprise material and population balances for:

1. The number of particles, including the nucleation, flocculation, and coagulation processes which determine the final value in homogeneous nucleation;

2. The average particle size and particle size distribution;

3. The monomer concentration in the latex particles based on thermodynamics;

4. The average number of free radicals per particle;

5. The instantaneous and overall conversions;

6. The copolymer composition and microstructure;

7. The average molecular weight and molecular weight distribution;

8. The degree of branching and crosslinking.

For homopolymers, an important point in the development of a mathematical model is the elucidation of the mechanism and prediction of the rate parameters. Although considerable progress has been made, it has been shown that the fitting of limited data by global modeling can lead to false conclusions. Therefore, it is recommended that rate constants be measured directly using methods that are not based on model assumptions, e.g., the direct measurement of k_p by electron spin resonance. Mechanisms and models have been proposed for the desorption of radicals from the particles, and for the various aqueous phase processes; however, the radical distributions and the other parameters in these models are not known precisely. Mechanisms and models are being developed for particle nucleation and entry of radicals into particles. The demonstration of the importance of coagulation in homogeneous nucleation has given a quantitative understanding of the effects of ionic strength, and emulsifier type and concentration, on the number, size, and size distribution of the particles.

These homopolymerization models can be applied directly to emulsion copolymerization. Equations also exist for the time evolution of the particle size distribution (rather than just the mean size). Moreover, the full molecular weight distributions can be calculated from the basic kinetic parameters. A major lack for all free radical-initiated vinyl addition polymerization systems is reliable mechanistically-based predictive models for kp, kt, and ktr, particularly the dependence of these parameters on the weight fraction of polymer; moreover, only a few model-free measurements of these parameters are available. One goal for the next decade is the development of a theory for these rate parameters and its verification by experiment. The use of reliable rate parameters will lead to equally reliable quantification of copolymer systems, based on global modeling. Additional complications for copolymer systems include multiphase particles, phase-separation of incompatible polymers, and diffusional (mass transfer) limitations.

Another important goal is a better understanding of the thermodynamics of copolymer systems, e.g., surface tensions, swelling parameters for monomer mixtures, mutual solubilities in monomer-copolymer systems, and polymer phase diagrams. Inclusion of such thermodynamic effects (e.g., using the Flory-Huggins equation) is essential in mathematical modeling. Many of the overriding characteristics of copolymer systems result from the monomer partitioning between the two phases and the copolymerization reactivity ratios.

Earlier work on physical understanding and mathematical modeling has treated thermodynamic and kinetic effects separately, whereas in fact they are coupled. These effects should be coupled through the fundamental equations of irreversible thermodynamics, which specify the spatial and temporal evolution of entropy and chemical potential, subject to both thermodynamic and kinetic constraints. This treatment could in principle define parameters such as particle morphology, but formidable problems must be solved to obtain quantitative results.

The proper combination of thermodynamics and kinetics should give reliable, predictive models which do not require inordinate computer resources for practical application. Therefore, the development should concentrate on precise models with all necessary refinements, which can be used to predict the behavior of new systems, and more approximate models which can be used to control and modify industrial processes.

Copolymer molecules rarely have identical compositions. There fore, simultaneous specification of both the molecular weight distribution and the copolymer compositional distribution requires the use of a bivariate distribution. Three different approaches can be used for modeling copolymerization systems:

1. Statistical methods;
2. "Black box" methods;
3. Detailed kinetic methods.

The most powerful approach by far is that based on the detailed kinetic analysis. According to this approach, an infinite number series of differential conservation equations is written to describe the time variation of all polymer molecules in the reactor. This approach assumes an understanding of the elementary kinetic mechanism of polymerization and a knowledge of the rates of the individual reactions. In practice, such detailed information on the joint copolymer molecular weight distribution-copolymer compositional distribution can be obtained from solution of a detailed mathematical model; however, much research is required to develop analytical methods that can be used to measure experimentally the joint copolymer molecular weight distribution-copolymer compositional distribution. This analysis can be extended to a simple core-shell latex particle using the foregoing treatment and mass balances for the aqueous phase (monomer 1, monomer 2, initiator, radicals, emulsifier), a population balance for the number density of polymer particles, and appropriate mass balances within each particle (monomers, radicals) including a conservation equation for the leading moments of the total molecular weight distribution-copolymer compositional distribution.

Concomitant with the development of mathematical models is the assurance that these are mechanistically correct and employ valid kinetic and thermodynamic parameters. A major problem in the past has been the divergence of different schools of thought in this context, leading to duplication of effort and promulgation of errors. It is therefore recommended that NATO/ASI workshops be held biennially to review and refine model development, to compare practical experience with predictions, and to unify results from precise studies on laboratory model systems with the requirements of industry. These workshops should have as their goal the development and publication of accepted mechanisms, and tables of parameters and descriptive equations, rather than the mere presentation of papers.

2.6. On-Line Measurement and Control of Reactors

In industry, there is a great incentive to develop on-line sensors and optimal control policies to improve product quality, increase production rate, and to eliminate or minimize unacceptable product. The term "product quality" includes all molecular microproperties (molecular weight distribution, copolymer compositional and sequence distributions, degree of branching, stereoregularity) and morphological macroproperties (particle size distribution, pore size distribution, surface area, bulk density) that characterize the morphological structure of the polymer.

One of the main difficulties in this control is the relation of polymer quality to nonmolecular properties such as tensile strength, clarity, swellability, and melting point. Although optimal control theory has been applied successfully to other chemical processes, there have been only a few successful attempts to apply advanced control techniques to emulsion polymerizations. The reasons are as follows:

1. The emulsion polymerization reactors are highly nonlinear systems with time-varying parameters, which are difficult to model and control;
2. The formulation of a meaningful objective function is difficult;
3. The on-line measurement of important copolymerization process parameters such as conversion, composition, molecular weight distribution, particle size distribution, and copolymer sequence distribution is not easy; indeed, these difficult measurements are the weakest link in any closed-loop control system.

In spite of these difficulties, recent improvements in modeling and on-line polymer characterization methods have increased the scope for closed-loop control of emulsion copolymerizations. The monitoring may be accomplished by using some combination of the following sensors:

1. On-line densitometry (overall conversion);
2. Gas chromatography (conversion, overall composition);
3. On-line viscometry (latex viscosity);

4. Gel permeation chromatography (molecular weight distribution);
5. Hydrodynamic chromatography (particle size distribution);
6. Surface tensiometry (concentration of free emulsifier);
7. Turbidity (latex particle size).

For batch polymerization reactors, the control problems are as follows:

1. The determination of the time-optimal open-loop policies (temperature, initiator addition, monomer addition) needed to obtain the desired product properties;
2. The determination of the closed-loop control policies which minimize the effects of process disturbances such that the stated parameters remain on the optimal trajectory.

For continuous polymerization reactors, the control problems are as follows:

1. The selection of the optimal steady-state controls needed to produce product of the desired quality;
2. The determination of the time-optimal policies needed for moving from one steady state to another;
3. The regulatory controls needed to maintain the states of the system at, or near, their optimal values in the presence of process disturbances.

A general approach to the control of polymerization reactors should include the following three steps:

1. Mathematical modeling (see above);
2. The determination of the optimal policies to move the system from its initial to its final state while minimizing a quality-economic objective function;
3. The determination of the most efficient regulatory controls to maintain the process along its optimal trajectory in the presence of process disturbances and the evaluation of the optimal adaptive model-predictive and internal model-control policies.

For these new reactor designs to be ready for the new mathematical models and characterization methods, work must be started soon. Therefore, it is proposed to hold a NATO/ASI Workshop on Reactor Design.

3. CONCLUSIONS
3.1. Qualifications
This workshop comprised representatives from most of the important academic, industrial, and government laboratories working on emulsion polymers in the United States, western Europe, and Japan. Thus its review of the state-of-the-art and its development of goals for the next decade is authoritative. The workshop atmosphere was conducive to the systematic well-planned discussion of the problems of emulsion polymers, the arrival at a group consensus, and the codification of the ideas into a position paper.

3.2. Education
An important byproduct of the work on emulsion polymers is the training of graduate students who have backgrounds in polymer, colloid,

and surface chemistry, and chemical engineering, and who are experienced in working with emulsion polymers. The American academic workshop members represent the most important source of graduates experienced in this field in the United States, e.g., the Emulsion Polymers Institute of Lehigh University with 25-45 graduate students graduates an average of eight Ph.D. candidates each year.

3.3. Need for Innovation

To attain the foregoing goals requires, not only much theoretical and experimental work, but also much innovation in the theoretical and experimental approaches, and new ideas for the solution of the problems. Innovation cannot be programmed; it is a creative act that springs from the mind, often without warning. Innovation, however, can be enhanced by the the proper atmosphere and environment. The atmosphere for innovation for the workshop members varies according to their individual organizations; however, the workshop provided an excellent atmosphere for innovation -- uninhibited discussion, frank exchange of viewpoints, and cooperative organization of the agreed-upon problems to be solved. And the comfortable environment was eminently satisfactory. Therefore, one result expected from the workshop is the development of innovative ideas to solve these problems.

3.4. Technical

The history of emulsion polymerization and the development of the many sophisticated latexes produced today for many different applications was reviewed, and the importance of copolymerization was emphasized. The state-of-the-art of emulsion copolymerization was reviewed, and the goals for the next decade were outlined in terms of emulsion copolymerization, characterization of latexes, physical and colloidal properties, mathematical models, and on-line measurements and control of reactors. Two new NATO/ASI workshops were proposed, one on the correlation of mathematical models with experiment and the other on the reactor design and engineering. Both workshops should be held in two years time.

4. REFERENCES

1. D. Harkins, *J. Am. Chem. Soc.* **69**, 1428 (1947).
2. W. V. Smith and R. H. Ewart. *J. Chem. Phys.* **16**, 592 (1948).
3. S. S. Medvedev, *Ric. Sci. Suppl.* **25**, 897 (1955); ibid., International Symposium on Macromolecular Chemistry, Prague, Pergamon, New York, 1957, p. 174.
4. W. J. Priest, *J. Phys. Chem.* **56**, 1077 (1952).
5. B. Jacobi, *Angew. Chem.* **64**, 539 (1952).
6. R. Patsiga, M. Litt, and V. Stannett, *J. Phys. Chem.* **64**, 801 (1960).
7. R. M. Fitch, *Polymer Colloids*, R. M. Fitch, ed., Plenum, New York, 1971, p. 73.
8. J. Ugelstad, M.S. El-Aasser, and J.W. Vanderhoff, *J. Polym. Sci., Polym. Letters Ed.* **11** 503 (1973).
9. E. A. Willson, J. R. Miller, and E. H. Rowe, *J. Phys. Colloid Chem.* **53**, 357 (1949).
10. W. V. Smith, *J. Am. Chem. Soc.* **70**, 3695 (1948); ibid. **71**, 4077 (1949).

5. **PANEL**

Discussion Leader:
    J. W. Vanderhoff (Lehigh University, USA)

Recording Secretary:
    D. C. Sundberg (University of New Hampshire, USA)

Participants:
    D. I. Lee (Dow Chemical, USA)
    J. Guillot (CNRS, France)
    R. G. Gilbert (The University of Sydney, Australia)
    H. Kast (BASF, Germany)
    G. W. Poehlein (Georgia Institute of Technology, USA)
    R. A. Wessling (Dow Chemical, USA)
    J. Asua (Quimica, Spain)
    J. S. Dodge (BFGoodrich, USA)
    M. S. El-Aasser (Lehigh University, USA)
    A. E. Hamielec (McMaster University, Canada)
    C. Kiparissides (Aristotle University of Thessaloniki, Greece)
    D. Lorah (Rohm & Haas, USA)
    L. Morgan (S.C. Johnson, USA)
    M. Nomura (Fukui National University, Japan)
    W. Pavelich (Borg-Warner, USA)
    J. Schork (Georgia Institute of Technology, USA)

SEMI-CONTINUOUS EMULSION POLYMERIZATION

J. W. Vanderhoff

Emulsion Polymers Institute and Department of Chemistry
Lehigh University
Bethlehem, Pennsylvania 18015

1. Introduction

There are three types of emulsion polymerization processes: 1. batch polymerization in which all ingredients are added at the beginning of the reaction; 2. semi-continuous or semi-batch polymerization in which the monomer is added continuously or in increments, neat or in emulsion; 3. continuous polymerization in which all of the ingredients are added continuously to one part of the reactor system and partially or completely converted latex is removed continuously from another part. All three processes comprise particle nucleation and particle growth stages, which may occur sequentially or concurrently. All three processes benefit by the use of a seed latex to obviate the particle nucleation stage. The seed latex may be prepared in a separate reaction or by polymerization in situ.

Batch polymerization is the simplest process, but the polymerization exotherm is difficult to control for monomer systems which polymerize rapidly, e.g., acrylate ester copolymers; the number of particles is usually constant after a short particle nucleation stage; layered or core-shell particles may be prepared in successive stages. Semi-continuous polymerization is more complex in that the rate of monomer addition is used to control the rate of polymerization; moreover, it is extremely versatile in that layered or core-shell particles of complex morphology can be prepared with minimal coagulum; the number of particles is usually constant after a short particle nucleation stage; the seed latex may be prepared by batch polymerization of a small part of the monomer mixture or by polymerization in situ when the monomer addition is started. Both batch and semi-continuous polymerizations are usually carried out in stirred-tank reactors. Continuous polymerization is even more complex in that the rate of polymerization is also controlled by the rate of monomer addition, but the particle nucleation occurs throughout the reaction, so that the number of particles varies continually and thus the particle size distribution is more difficult to control; moreover, it requires more complex equipment, e.g., a train or cascade of stirred-tank reactors or a stirred-tank reactor with a loop reactor. Most industrial latexes are prepared by batch or semicontinuous polymerization, but some are prepared by continuous polymerization.

In semi-continuous polymerization, the rate of polymerization is controlled by the rate of monomer addition. Earlier studies[1,2] postulated the existence of "flooded" (the monomer concentration increases with increasing monomer addition time) and starved" (the monomer concentration remains low and essentially constant) systems. More recent studies of the 80:20 vinyl acetate-butyl acrylate and 60:40 butyl acrylate-methyl methacrylate systems[3], however, showed that the monomer concentration decreased with decreasing rate and increasing time of monomer addition; moreover, above a critical rate of monomer addition, the rate of polymerization was the same as that of batch polymerization, and slower rates of monomer addition gave decreasing rates of polymerization (Figure 1). Thus these semi-continuous polymerizations were never flooded in the sense of the monomer concentration in the particles increasing with

24

monomer addition time, but were of varying degrees of starvation, according to the rate of monomer addition.

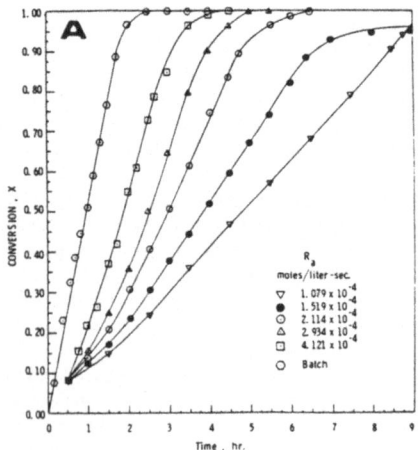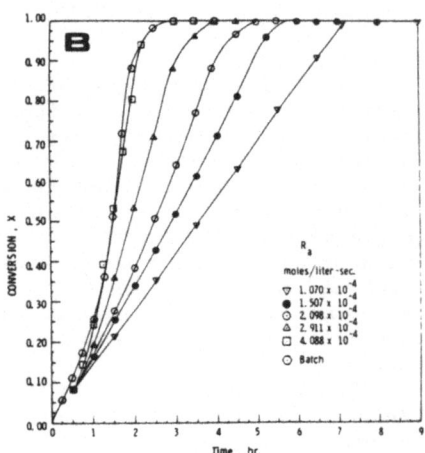

Figure 1: Conversion-time curves for semi-continuous polymerization: A. 80:20 vinyl acetate-butyl acrylate; B. 60:40 butyl acrylate-methyl methacrylate[3].

Systematic variation of the polymerization process and recipe gives a variation in the composition of the latex particle from center to surface. Thus the latex particle may have a core of one polymer and a shell of another, or it may have a continuously varying compositional gradient from center to surface. Both types have been referred to as "core-shell" or "structured" latexes, although the latter type may not have a well-defined boundary between core and shell. These core-shell latexes may be prepared by batch or semi-continuous polymerization (and in principle by continuous polymerization).

These core-shell latexes pose an interesting problem in polymer compatibility. Most polymers are incompatible with other polymers of different composition. Thus two incompatible polymers formed in a submicroscopic latex particle tend to separate or demix, which may result in the formation of spherical concentric or eccentric core-shell particles or nonspherical particles; however, in the latter case, the separation must overcome the restoring force arising from the polymer-water interfacial tension, which tends to keep the particles spherical.

2. EARLIER WORK

The term "core-shell" is relatively new, but core-shell latexes have been prepared in industry for many years. One of the earliest examples is the Dow Styralloy resins of the mid-to-late 1940's prepared by batch polymerization[4]. In one type, butadiene was polymerized in a polystyrene seed latex to give a copolymer with the low-temperature flexibility of polybutadiene (T_g -78°) filled with rigid polystyrene particles (T_g -

80°)[5]. In another type, styrene was polymerized in a polybutadiene seed latex to give the reverse morphology[6]. The core-shell morphology of both types was confirmed by electron microscopy.

Also, Monsanto Lytron 680 latex for exterior paints was prepared by semi-continuous polymerization of a styrene-acrylonitrile mixture to form a seed latex followed by semi-continuous polymerization of a styrene-2-ethylhexyl acrylate-methacrylic acid mixture[7]; electron microscopy showed that the particles had a well-defined core-shell morphology. This latex had the film-forming characteristics of the shell copolymer but with much lower film tack; moreover, paints made with this latex showed much less dirt pickup than those made with a homogeneous copolymer. Thus this latex behaved as if the hard cores reinforced the soft copolymer shell, so that the films did not soften and become tacky in the midsummer sun.

2.1. Compatible Core-Shell Systems

The foregoing systems comprised a copolymer core which was in compatible with the copolymer shell, so it was not unexpected that there was a well-defined demarcation between the core and the shell. However, core-shell systems have also been postulated for systems in which the shell polymer was compatible with the core polymer, e.g., it was postulated that, in the batch seeded emulsion polymerization of styrene in polystyrene latex, the styrene-swollen particles comprised a monomer-rich shell around a polymer-rich core and that the polystyrene molecules formed by polymerization were located predominantly in the shell[8-10]. This hypothesis was supported by:

1. The linear or near-linear conversion-time curves over a range in which the monomer concentration in the particles varied two- to threefold;
2. The addition of a small amount of butadiene to the core or shell monomer, followed by electron microscopy of osmium tetroxide-stained microtomed sections of embedded particles, which gave a butadiene-containing shell around a polystyrene core (Figure 2A) or a butadiene-containing shell around a polystyrene shell around a butadiene-containing core (Figure 2B);
3. The addition of tritiated styrene to the core or shell monomer, followed by coating of individual particles with a photographic emulsion, development after 90 days in the dark, and examination by electron microscopy, which gave silver traces for the tritiated polystyrene core (Figure 3A) but no silver traces for the nontritiated polystyrene shell around the tritiated polystyrene core (Figure 3B);
4. A thermodynamic argument.

This hypothesis was supported by the characterization of monodisperse polystyrene latexes prepared by batch conventional and seeded polymerization using persulfate ion initiator, which showed that at least half of the polymer endgroups remained on the particle surface[11,12].

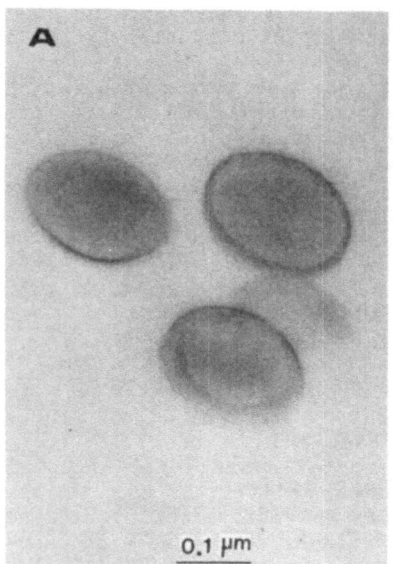
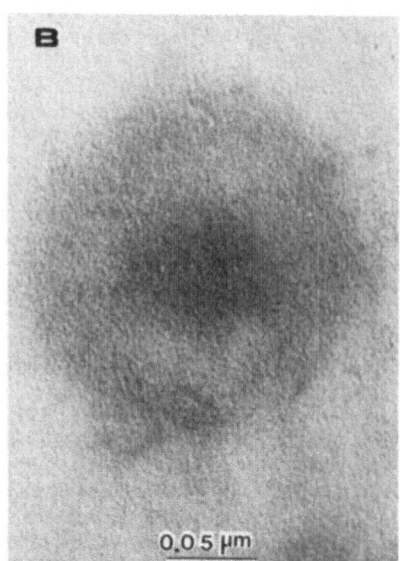

Figure 2: Transmission electron micrographs of microtomed osmium tetroxide-stained polystyrene particles containing butadiene in: A. shell; B. core and outer shell[10].

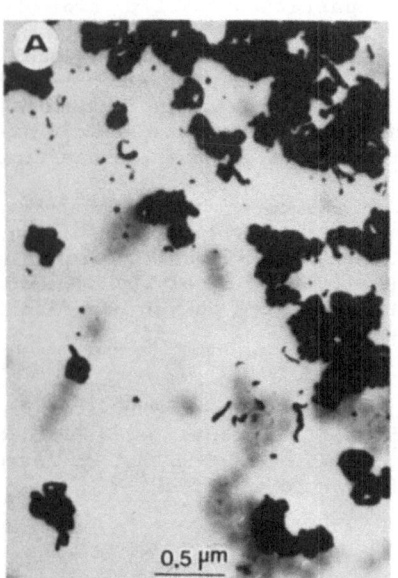
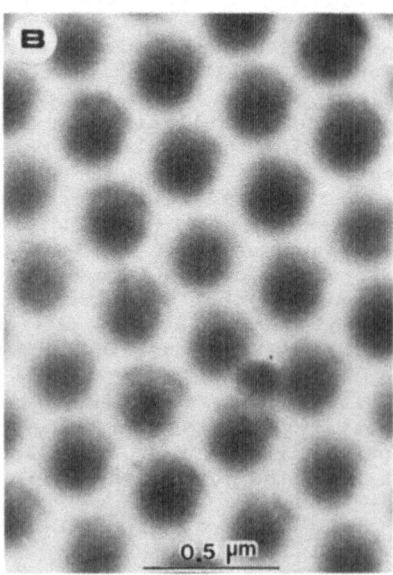

Figure 3: Transmission electron micrographs of polystyrene particles with photographic emulsion coating after three months in the dark: A. seed particles containing tritium; B. seed particles covered with shell of polystyrene without tritium[10].

Furthermore, in the seeded emulsion polymerization of lightly crosslinked particles, the time allowed for swelling with monomer is important. In one series of successive seeded batch polymerizations[13], spherical particles swollen with monomer for only one hour were transformed to spheroids, which were then transformed back to spherical particles again by swelling for three hours (Figure 4). More recent work[14] showed that seeded batch emulsion polymerization of lightly crosslinked particles gave uniform nonspherical particles such as symmetric and asymmetric doublets, ellipsoids, and egg-shaped particles (Figure 5).

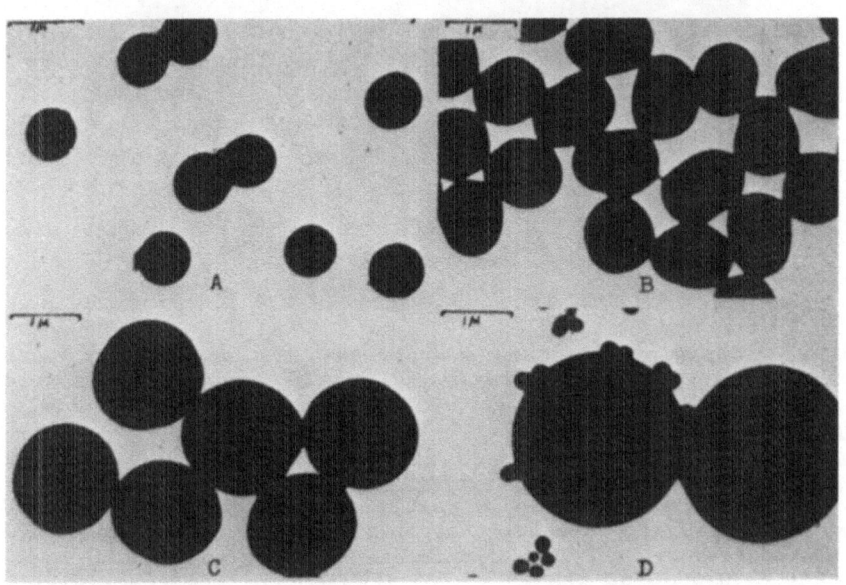

Figure 4: Transmission electron micrographs of polystyrene particles prepared by successive seeded polymerization (0.06% divinylbenzene; M/P 2.0-2.4; 70°C) with monomer swelling times of: A. 660nm seed; B. one hour; C. three hours; D. three hours[13].

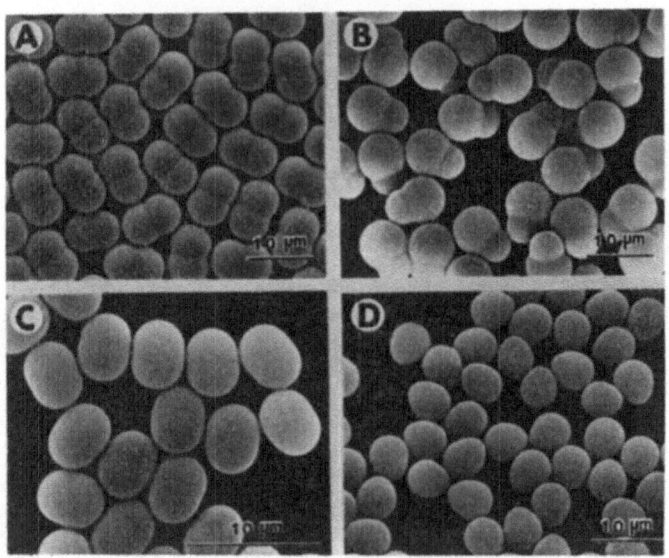

Figure 5: Scanning electron micrographs of nonspherical particles prepared by seeded polymerization of lightly crosslinked polystyrene particles: A. symmetric doublets; B. asymmetric doublets; C. ellipsoids; D. egg-shaped[14].

2.2. Batch vs. Semi-Continuous Polymerization

Core-shell morphology can result from batch copolymerization of monomer mixtures with disparate copolymerization reactivity ratios. The vinyl acetate-butyl acrylate system has the copolymerization reactivity ratios $r_1 = 0.00-0.04$ and $r_2 = 3-8$. These disparate reactivity ratios mean that chains ending in either a vinyl acetate or a butyl acrylate radical will both add butyl acrylate preferentially to form a butyl acrylate-rich copolymer until the butyl acrylate is depleted, after which a vinyl acetate-rich copolymer or polyvinyl acetate will be produced. Thus, at all vinyl acetate-butyl acrylate ratios, batch polymerization will produce a distribution of copolymer compositions comprising a butyl acrylate-rich fraction and a vinyl acetate-rich or polyvinyl acetate fraction, and semi-continuous polymerization will produce a single-peaked distribution of copolymer compositions of the same ratio as the monomer mixture. These distributions were confirmed by dynamic mechanical spectroscopy (Figure 6, Table 1)[3], which showed a single compositional peak for the semi-continuous copolymer and two peaks for the batch copolymer. The Tg values by dynamic mechanical spectroscopy, which vary with frequency, were confirmed by differential scanning calorimetry (Table 1).

TABLE 1. Tg Values of Vinyl Acetate-Butyl Acrylate Copolymers[3]

VAc/BA	T_g/deg C				
Molar		Batch		Semi-Continuous	
Ratio	DSC	DMS	Difference	DSC	DMS
100/0	34	60	--	27	54
89/11	-15 & 21	0 & 40	40	21	38
71/29	-33 & 17	-17 & 28	45	4	18
49/51	-45 & 31	-32 & 43	75	-20	5
0/100	-53	-40	--	-73	-58

DSC - differential scanning calorimetry
DMS - dynamic mechanical spectroscopy

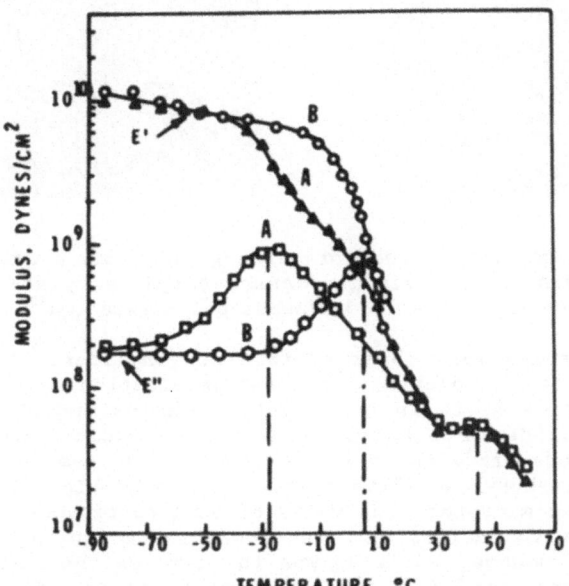

Figure 6: Variation of log storage modulus and loss modulus with temperature for 49/51 vinyl acetate-butyl acrylate copolymers prepared by A. batch; B. semi-continuous polymerization[3].

The emulsion copolymerization of the vinyl acetate-butyl acrylate mixtures showed that batch polymerization gave butyl acrylate-rich particle cores that comprised as much as 50% of the particle, whereas semi-continuous polymerization gave much smaller butyl acrylate-rich cores which comprised only about 10% of the particle (Figure 7)[15]. These small cores in the semicontinuous copolymer were attributed to the batch polymerization of the first 8% of the monomer mixture to give a seed latex for the semi-continuous polymerization of the remaining 92%.

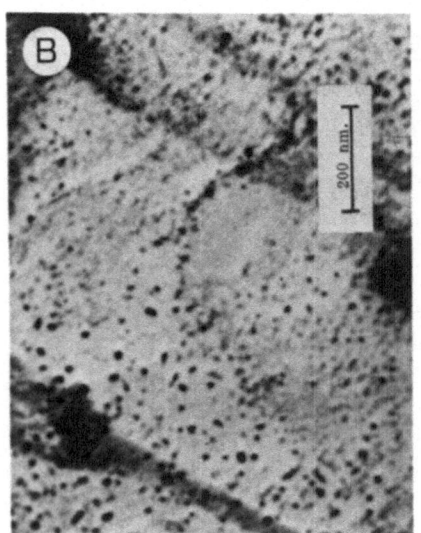
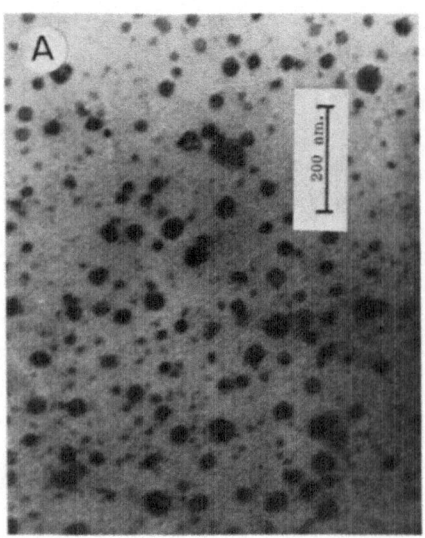

Figure 7: Transmission electron micrographs of hydrazine-osmium tetroxide-stained 63/37 vinyl acetate-butyl acrylate copolymers prepared by: A. batch; B. semi-continuous polymerization[15].

Similar results were observed for the vinylidene chloride-butyl methacrylate system[16], which has the copolymerization reactivity ratios $r_1 = 0.22$ and $r_2 = 2.41$. In this system, chains ending in either a vinylidene chloride or a butyl methacrylate radical will both add butyl methacrylate preferentially, but not to the same extent as with the vinyl acetate-butyl acrylate system. For 83/17 vinylidene chloride-butyl methcrylate mixtures, the rates of semi-continuous copolymerization increased with increasing rate of monomer addition; that for the fastest rate of monomer addition was the same as the identical rates of the batch and seeded batch copolymerizations (Figure 8). The monomer concentration in the particles varied strongly with monomer addition rate (Figure 9); only the slowest rate of monomer addition gave a near-constant concentration. Batch and seeded batch copolymerization gave insoluble copolymers, which showed crystallinity (determined by infrared spectroscopy and X-ray diffraction) upon heating to 120° or aging for three months at ambient temperature; semi-continuous polymerization gave soluble copolymers which showed no crystallinity. The insolubility of the batch copolymers was attributed to crystallization of the longer vinylidene chloride sequences formed late in the polymerization after most of the butyl methcrylate had polymerized. These differences between the batch and semi-continuous copolymers were confirmed by measurements of their physical properties, as given in Table 2. The semi-continuous copolymers showed lower Young's moduli and ultimate tensile strengths, and higher

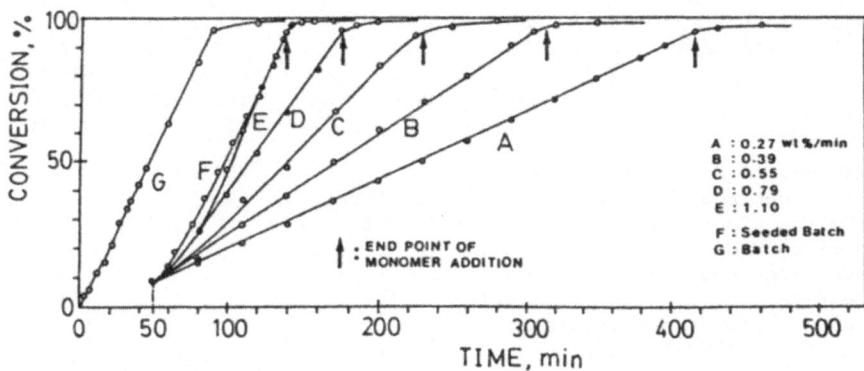

Figure 8: Variation of conversion with time for 83/17 vinylidene chloride-butyl methacrylate copolymers prepared by semi-continuous polymerization with feed rates of: A. 0.27; B. 0.39 C. 0.55; D. 0.79; E. 1.10 wt%/min; F. seeded batch polymerization; G. batch polymerization[16].

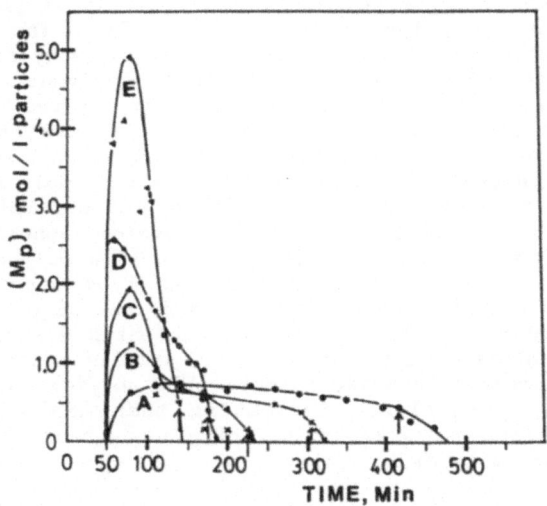

Figure 9: Variation of monomer concentration with time for 83/17 vinylidene chloride-butyl methacrylate copolymers prepared by semi-continuous polymerization with feed rates of: A. 0.27; B. 0.39; C. 0.55; D. 0.79; E. 1.10 wt%/min[16].

TABLE 2. Vinylidene Chloride-Butyl Methacrylate Copolymers

Sample	A	B	C	D	E	F	G
Young's modulus/MPa	3.4	6.2	7.3	12.7	12.7	230	296
Ultimate strength/MPa	1.3	2.7	2.7	2.9	2.9	7.6	8.4
Elongation to break/%	588	483	480	480	468	74	87
Energy to break/MJ/m^3	15.6	11.4	9.9	8.3	7.7	5.4	6.1
T_g/°C	16.3	16.3	15.8	15.7	16.9	23.2	24.2
T_m^g/°C	none	none	none	none	none	184	187
MFFT*/°C	<5	<5	<5	<5	<5	<5	<5

* - minimum film formation temperature
A - semi-continuous copolymerization (0.27%/min)
B - semi-continuous copolymerization (0.39%/min)
C - semi-continuous copolymerization (0.55%/min)
D - semi-continuous copolymerization (0.79%/min)
E - semi-continuous copolymerization (1.10%/min)
F - seeded batch copolymerization
G - batch copolymerization

energies to break and percents elongation, than the batch copolymers. Also, they showed higher T_g values and the presence of T_m values at the same minimum temperature for film formation.

2.3. Polystyrene-Polybutyl Acrylate System

Batch polymerization of styrene in a polybutyl acrylate seed latex (50:50 ratio) gave 25.5% graft copolymer prepared in situ; batch with-equilibrium-swelling polymerization gave 16.0% graft copolymer, and semi-continuous polymerization only 7.7%[17]. The percent graft copolymer was determined by thin-layer chromatography-flame ionization detection and the composition of the graft copolymers was determined by nuclear magnetic resonance spectroscopy after separation by preparative thin-layer chromatography. Transmission electron microscopy showed that the particles prepared by semi-continuous polymerization changed in shape from spheres to doublets upon aging in the latex form for six months (Figure 10). The particles prepared by batch or batch-with-equilibrium-swelling remained spherical upon aging (Figure 11). Thus the graft copolymer acted as a polymeric emulsifier to prevent the two incompatible polymers from separating. In the sample containing only 7.7% graft copolymer, the polystyrene separated from the polybutyl acrylate despite their high viscosities and the polymer-water interfacial tension. In the samples containing 16.0% and 25.5% graft copolymer, the two incompatible polymers did not separate under the same conditions, indicating that the amount of graft copolymer required to prevent demixing was greater than 7.7% and perhaps as great as 16.0%.

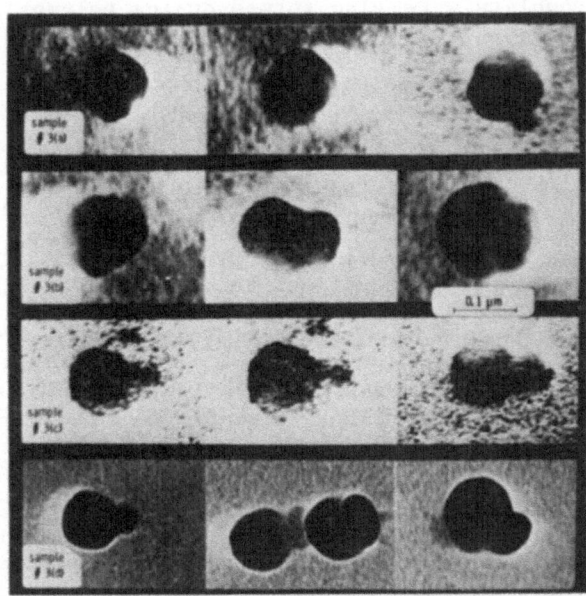

Figure 10: Transmission electron micrographs of 50/50 polybutyl acrylate core-polystyrene shell particles prepared by semi-continuous polymerization (after aging in the latex form for six months)[17].

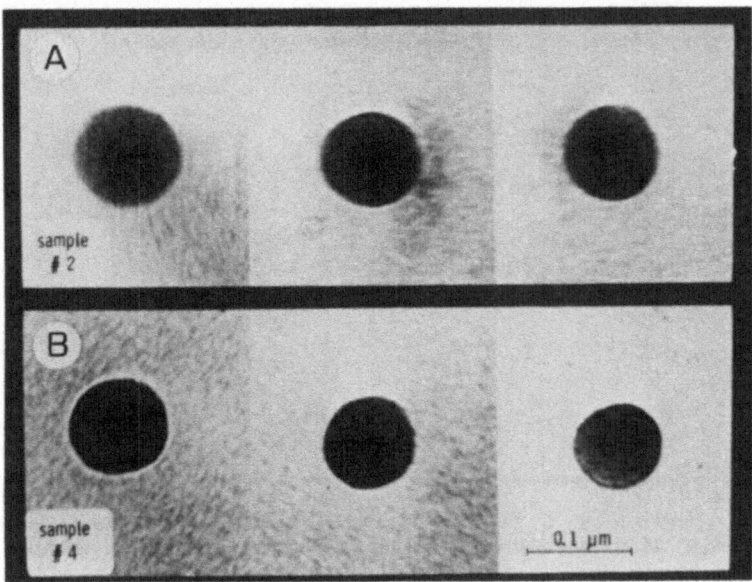

Figure 11: Transmission electron micrographs of 50/50 polybutyl acrylate core-polystyrene shell particles prepared by: A. batch polymerization with equilibrium swelling; B. batch polymerization (after aging in the latex form for six months)[17].

Similar doublets were also formed during polymerization under different conditions[18]. They were attributed to the aqueous-phase polymerization of styrene followed by precipitation of polystyrene oligomers which flocculated with the polybutyl acrylate seed latex particles to give polystyrene surface domains, which then grew to 60-90 nm diameter and formed a concentric shell.

2.4. Grafting of Styrene-Acrylonitrile Mixtures

Dimonie et al.[19] studied the copolymerization of the azeotropic styrene-acrylonitrile mixture in 190- and 300-nm monodisperse polystyrene seed latexes by batch, batch-with-equilibrium-swelling, and semi-continuous polymerization. Because of the high solubility (7%) of acrylonitrile in water, the copolymerization of this mixture sometimes formed a new crop of particles and sometimes not. The critical factor was the surface area of the seed latex; at ca. 200 m^2/dl latex or above, no new particles were formed; below ca. 180 m^2/dl, new particles were formed in every case, the number increasing with decreasing surface area (Table 3). This formation of a new crop of particles was attributed to the aqueous-phase polymerization of styrene-acrylonitrile mixture to form oligomers which grew until they precipitated and flocculated with particles already existing; if the surface area of the particles was below the critical value, the oligomeric particles did not flocculate immediately but persisted, absorbed monomer, and grew to mature latex particles.

TABLE 3. Formation of New Styrene-Acrylonitrile Copolymer Particles

Seed/nm Grafted PA	M/P Ratio	N/1014	Seed Surface Area/m^2/dl	New Particles	Percent
300	4.2	4.0	113	many	18
300	2.6	6.0	170	few	--
190	4.2	15.7	179	very few	--
190	3.2	19.9	226	none	22
300	2.0	7.9	226	none	24
190	2.6	24.8	281	none	29
300	1.6	9.8	280	none	31
190	2.6	27.5	312	none	49
300	1.5	11.0	311	none	40
190	2.1	29.9	339	none	45
300	1.4	12.0	339	none	43

The azeotropic styrene-acrylonitrile copolymer swollen to equilibrium with the azeotropic monomer mixture showed an unexpected T_g of 65° (Figure 12) as compared to 120° for the copolymer; obviously, the swelling was limited, so that the T_g was in the usual range for polymerization. The rate of polymerization, the activation energy for polymerization, and the swelling of the copolymer were different above and below this critical temperature. The degree of grafting to the polystyrene substrates as well as the grafted styrene-acrylonitrile copolymer was determined by thin-layer chromatography-flame ionization detection. The degree of grafting was greatest for the semi-continuous, less for batch-with-equilibrium-swelling, and least for batch polymerization (the reverse order to the polystyrene-polybutyl acrylate system) (Table 4). Latex particles with a grafted styrene-acrylonitrile copolymer layer of 46 nm thickness or greater formed a stable colloidal dispersion in acetone (good solvent for styrene-acrylonitrile copolymer; nonsolvent for polystyrene) whereas those with thinner layers coagulated (Table 5). This stability in acetone was proposed as an indication of the dispersibility of these particles in a styrene-acrylonitrile copolymer matrix.

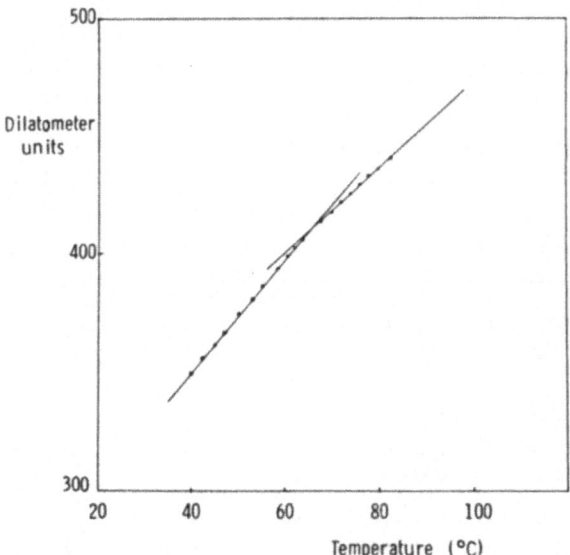

Figure 12: Variation of specific volume with temperature of azeotropic 72/28 styrene-acrylonitrile copolymer swollen with azeotropic 72/28 monomer mixture[19].

TABLE 4. Degree of Grafting for Different Polymerization Methods

Seed/nm	M/P Ratio	Seed Surface Area/m²/dl	Method	Percent Grafted PS	Particle Size/nm exptl	calcd
300	4.2	113	batch	18	425	503
300	4.2	113	batch-swell	27	469	493
300	4.2	113	semi-cont	53	470	480
300	1.5	311	batch	40	365	367
300	1.5	311	batch-swell	54	370	367
300	1.5	311	semi-cont	54	382	382
190	4.2	172	batch	37	275	271
190	1.5	488	batch	39	275	270
190	1.5	488	semi-cont	48	223	221
190	0.95	640	batch	53	210	202

TABLE 5. Colloidal Stability and Thickness of Grafted Layer

Seed/nm	Seed Surface Area/m²/dl	Polym Method	Percent Acetone Soluble	Percent PS Grafted	Shell Thick-ness/nm	Dispersion in Acetone
190	control	control	3.1	none	0	agglomerated
190	111	batch	83	26	70	fine dispersion
190	172	batch	79	37	55	fine dispersion
190	312	batch	66	49	45	agglomerated
190	488	batch	57	39	30	agglomerated
190	640	batch	53	53	22	agglomerated
190	488	semi-cont	58	48	29	agglomerated
300	113	batch	77	18	60	fine dispersion
300	312	batch	54	40	42	agglomerated
300	111	batch-swell	79	25	87	fine dispersion
300	312	batch-swell	59	54	40	agglomerated
300	111	semi-cont	80	53	95	fine dispersion
300	312	semi-cont	59	54	46	dispersion

The morphology of thin copolymer films cast from toluene dispersion and stained with ruthenium tetroxide (preferential stain for polystyrene) comprised dispersions of styrene-acrylonitrile copolymer in polystyrene, dispersions of polystyrene in styrene-acrylonitrile copolymer, or two interpenetrating polymer networks (Figure 13), in good agreement with the T_g values determined by dynamic mechanical spectroscopy (Figure 14), close to the 104° value for polystyrene for (PS-AS-5), close to the 120° value for styrene-acrylonitrile copolymer for (PS-AS-19), and an intermediate value for (PS-AS-21). The physical mixture of polystyrene and styrene-acryloni-

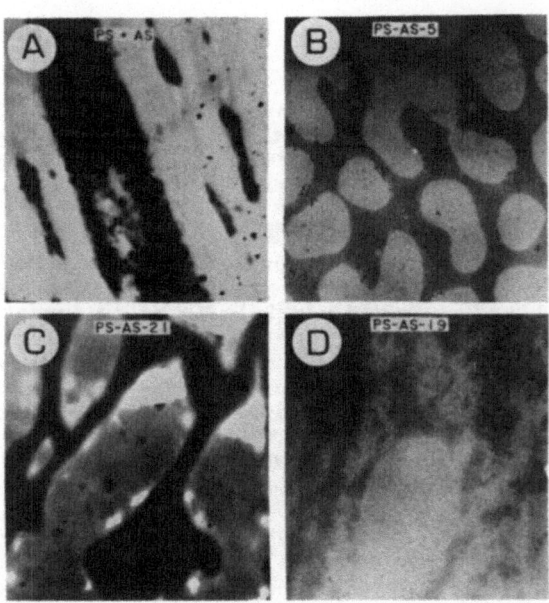

Figure 13: Transmission electron micrographs of thin films of: A. physical mixture of polystyrene and 72/28 styrene-acrylonitrile copolymer (PS + AS); B. 72/28 styrene-acrylonitrile mixture grafted onto polystyrene seed latex particles (PS-AS-5); C. ditto (PS-AS-21); D. ditto (AS-PS-19)[19].

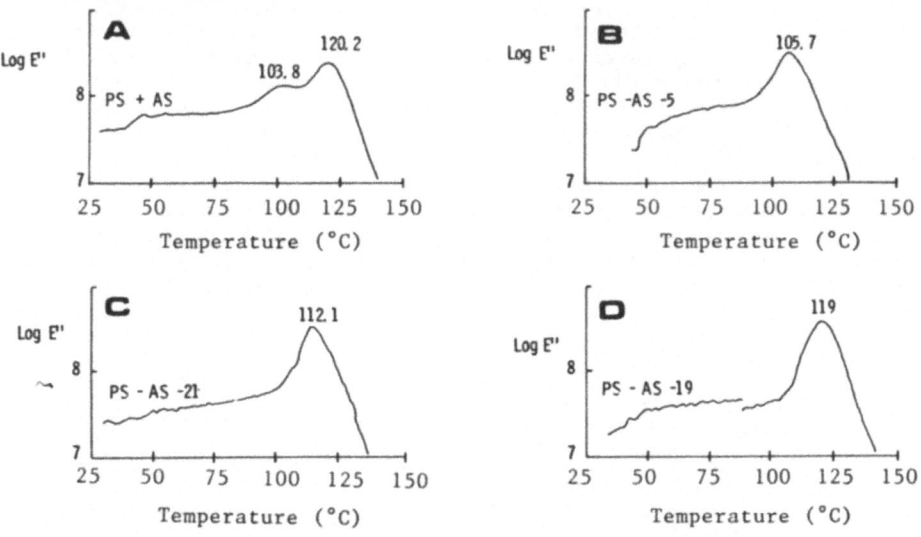

Figure 14: Variation of log loss modulus with temperature: A. physical mixture of polystyrene and 72/28 styrene-acrylonitrile copolymer (PS + AS); B. 72/28 styrene-acrylonitrile mixture grafted onto polystyrene seed latex particles (PS-AS-5); C. ditto (PS-AS-21); D. ditto (AS-PS-19)[19].

trile copolymer comprised a mixture of both phases and showed the two T_g values.

These results showed the importance of the seed latex surface area during polymerization, and the effect of the grafted layer thickness as well as the depth of grafting in the polystyrene seed particles on the morphology and properties of the final polymer.

2.5. Grafting of Methyl Methacrylate

Polymethyl methacrylate is compatible with polyvinyl chloride and some nylon and epoxy resins. Thus polymethyl methacrylate chains grafted to a rubbery or rigid particle forms an impact modifier for these polymers. The grafting of methyl methacrylate to nonionic polybutadiene or styrene-butadiene copolymer latexes was studied as a function of monomer-polymer ratio and other polymerization parameters[20]). Although most of the experiments used batch polymerization, the results are of interest in semi-continuous polymerization. The morphology changed with increasing monomer-polymer ratio, from spheroidal particles with a lumpy surface to smooth ellipsoidal particles (Figure 15), and the degree of grafting decreased. Characterization by transmission electron microscopy, thin-layer chromatography-flame ionization detection, dynamic mechanical spectroscopy, infrared spectroscopy, and nuclear magnetic resonance spectroscopy showed that the degree of grafting decreased with increasing crosslinking of the rubber seed latex (Figure 16); moreover, the T_g of the polymethyl methacrylate particle shell decreased with increasing degree of grafting. The degree of grafting decreased with increasing initiator concentration (Figure 17), increasing monomer-polymer ratio (Figure 18), decreasing seed latex surface area (Figure 19), and increasing conversion (Figure 20). The grafting was postulated to occur on the particle surface by abstraction of a hydrogen atom from the polymer substrate and addition of monomer to the radical. Comparison with theory[21] showed that the degree of grafting decreased with decreasing seed latex surface area (Figure 21) and decreasing initiator concentration (Figure 22) as expected.

The grafting of methyl methacrylate to 190- and 300-nm diameter monodisperse polystyrene particles gave spherical particles, independent of the monomer-polymer ratio; the degree of grafting was determined by the monomer-polymer ratio, initiator concentration, conversion, degree of crosslinking of seed polymer, and seed surface area.

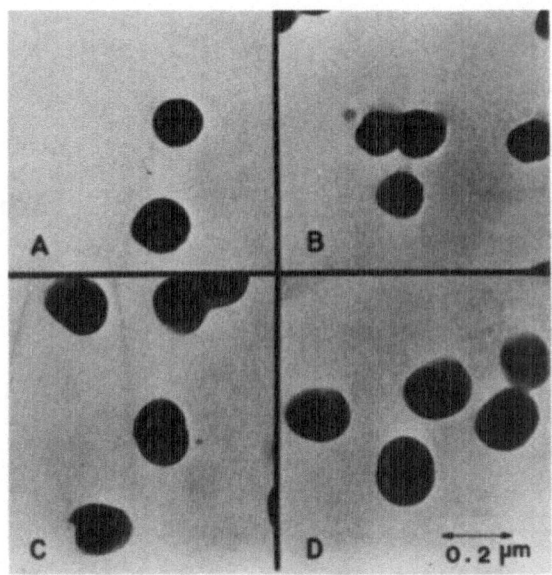

Figure 15: Transmission electron micrographs of polybutadiene core-polymethyl methacrylate shell particles prepared using different monomer-polymer ratios: A. 0.45; B. 0.89; C. 1.38; D. 2.03[20].

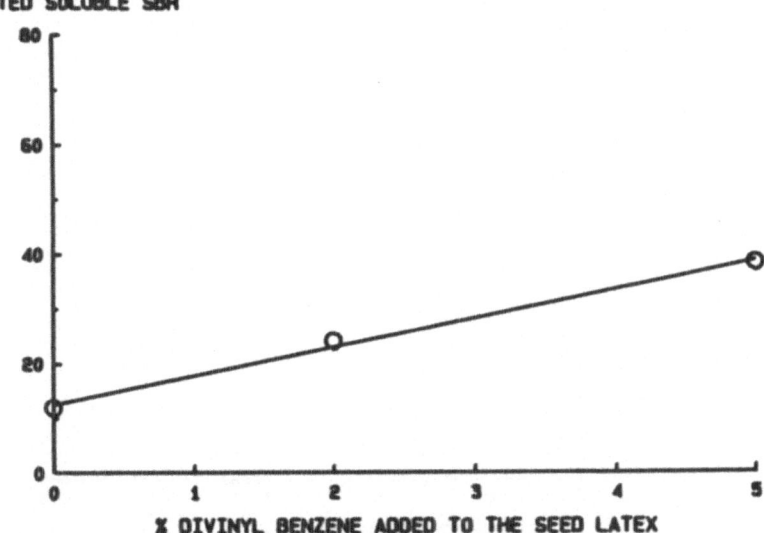

Figure 16: Variation of the degree of grafting of polymethyl methacrylate with divinylbenzene concentration in the rubber seed particles[20].

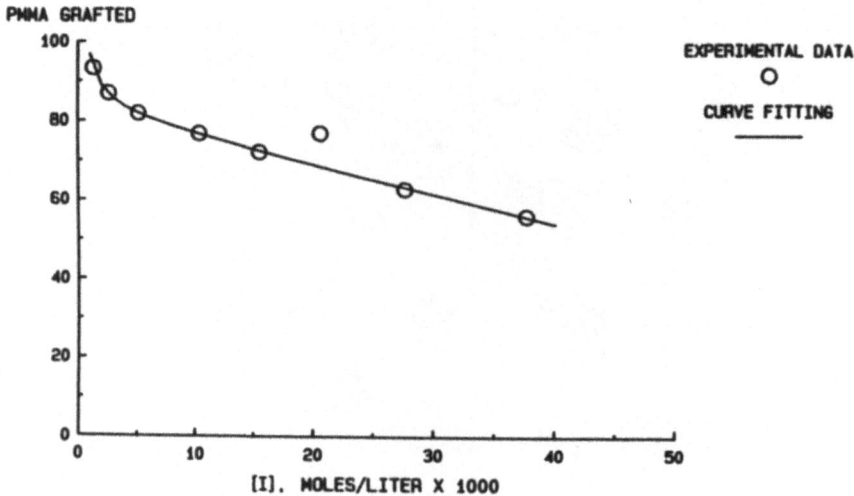

Figure 17: *Variation of the degree of grafting of polymethyl meth-acrylate on rubber seed latex with 2,2'-azobis(2,4-dimethyl-valeronitrile) initiator concentration*[20].

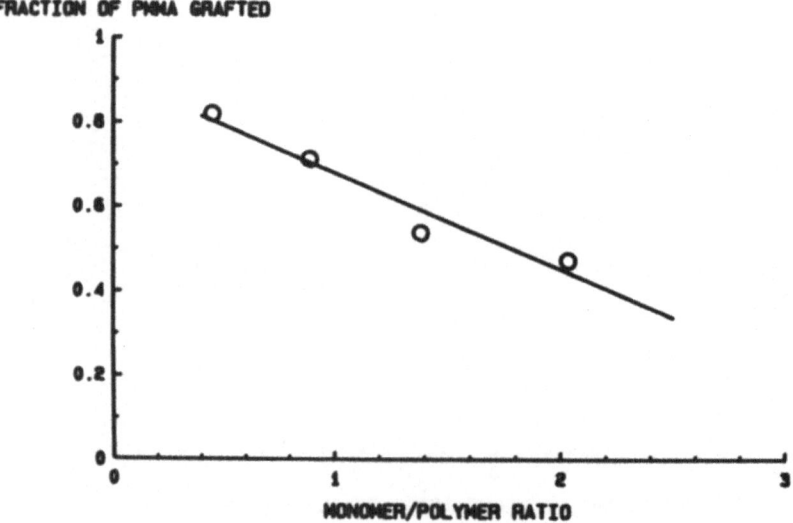

Figure 18: *Variation of the degree of grafting of polymethyl meth-acrylate on rubber seed latex with monomer-polymer ratio*[20].

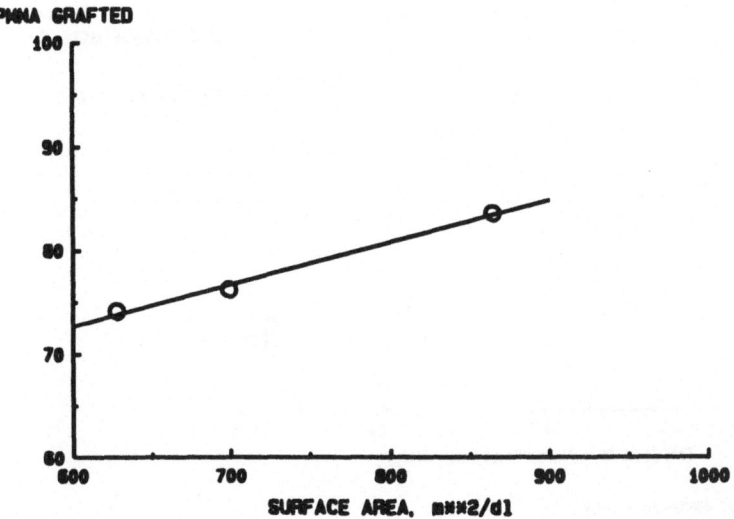

Figure 19: Variation of the degree of grafting of polymethyl methacrylate with surface area of the rubber seed latex[20].

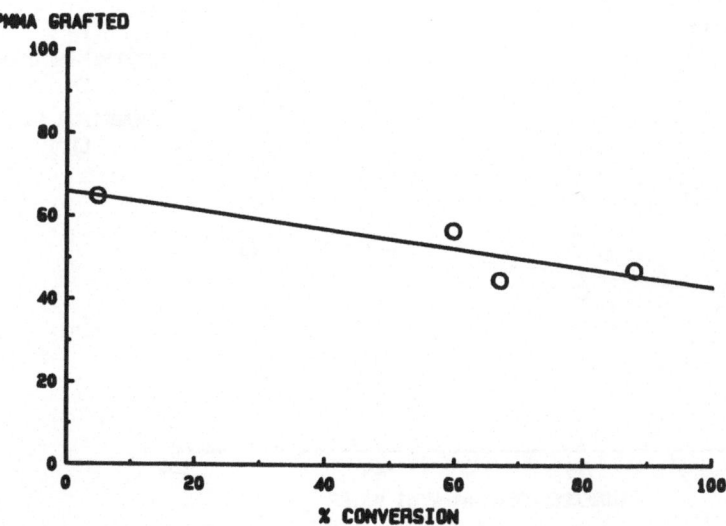

Figure 20: Variation of the degree of grafting of polymethyl methacrylate with conversion[20].

42

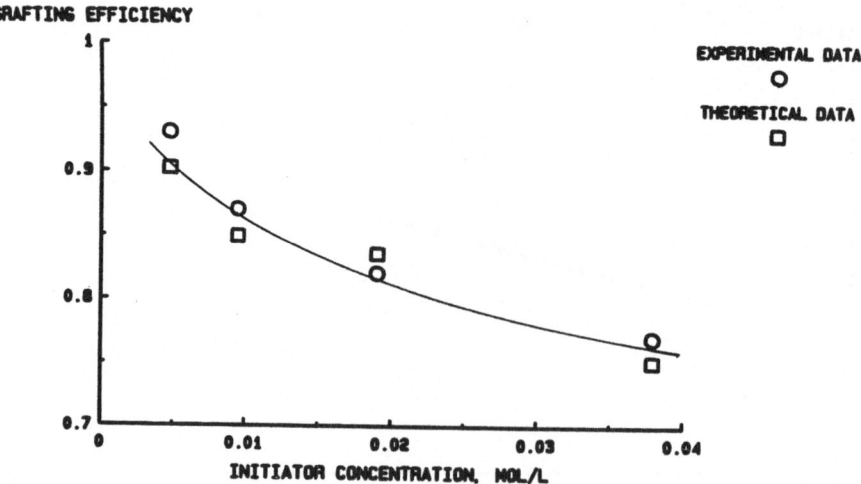

GRAFTING EFFICIENCY

EXPERIMENTAL DATA
THEORETICAL DATA

INITIATOR CONCENTRATION, MOL/L

INITIATOR CONCENTRATION (ORGANIC PHASE)

Figure 21: *Variation of theoretical and experimental degree of grafting of polymethyl methacrylate with rubber seed surface area*[20].

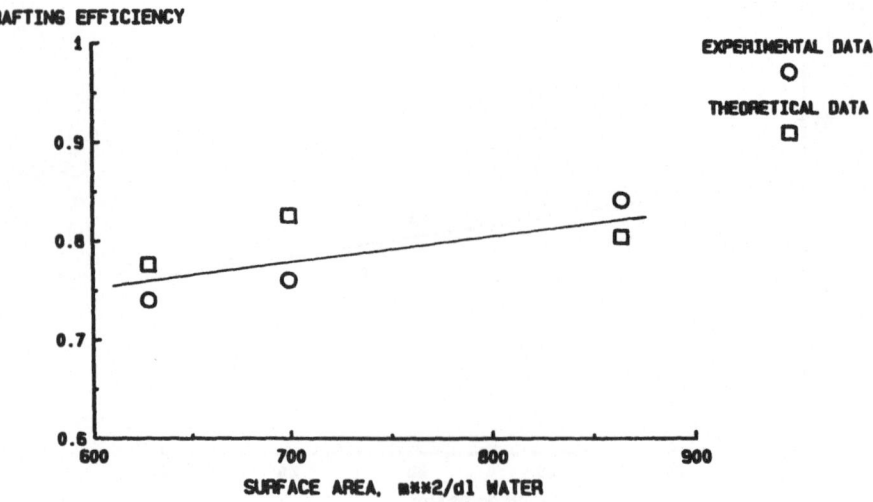

GRAFTING EFFICIENCY

EXPERIMENTAL DATA
THEORETICAL DATA

SURFACE AREA, m**2/dl WATER

Figure 22: *Variation of theoretical and experimental degree of grafting of polymethyl methacrylate on rubber seed latex with initiator concentration*[20].

2.6. Morphology of Core-Shell Particles

The different morphologies of the grafted particles were shown by transmission electron microscopy, with or without staining of one component. The staining of one component defined its morphology, but often the other component was not defined satisfactorily. Negative staining with phosphotungstic acid showed the overall morphology of the particles but did not show the individual components. The combination of negative and positive staining, however, defined both the overall morphology of the particles as well as their components.

Polyethyl methacrylate was polymerized in 190- and 300-nm diameter monodisperse polystyrene seed latexes. The final latexes were stained with phosphotungstic acid negative stain and ruthenium tetroxide positive stain for the polystyrene methacrylate[22]. For the 190-nm diameter seed, the particles were prolate spheroids with a dark center section of polystyrene and symmetrical end sections of light polyethyl methacrylate (Figure 23), indicating that the two polymers were incompatible and separated during polymerization. The corresponding electron micrographs of the particles stained with phosphotungstic acid or ruthenium tetroxide alone gave an inadequate definition of their morphology.

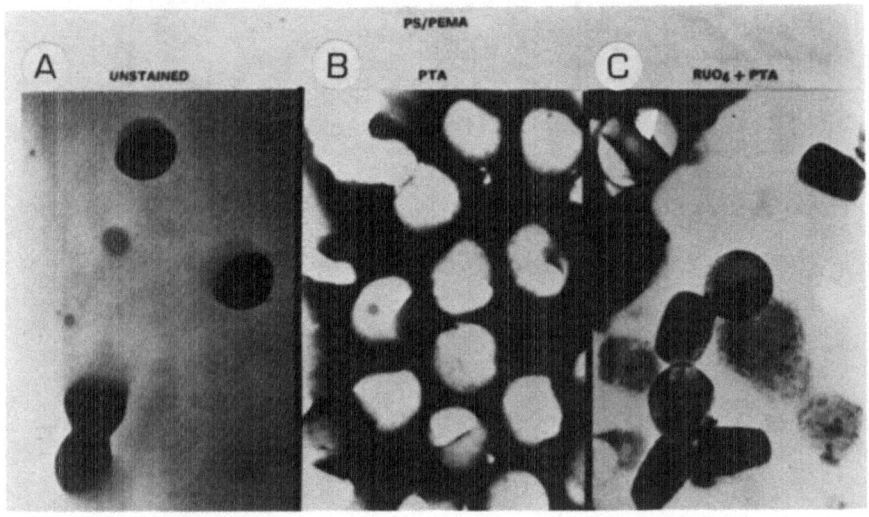

Figure 23: Transmission electron micrographs of 33:67 190-nm diameter polystyrene core-polyethyl methacrylate shell particles: A. unstained control; B. phosphotungstic acid negative stain; C. phosphotungstic acid negative stain and ruthenium tetroxide positive stain[22].

This technique was used to show the different morphologies of particles prepared by polymerizing ethyl methacrylate in 99-, 190- and 300-nm diameter monodisperse polystyrene seed latexes (Figure 24). The particles prepared using the 99- and 190-nm diameter seeds were prolate spheroids comprising near-hemispherical polystyrene and polyethyl methacrylate domains, whereas those prepared using the 300-nm diameter seed contained small circular patches of polyethyl methacrylate on the spherical polystyrene matrix. This unusual variation of morphology with particle size was unexpected.

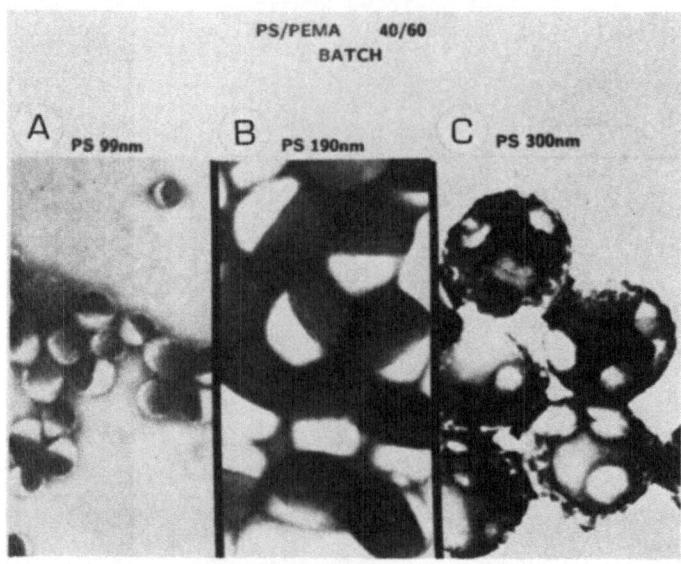

Figure 24: Transmission electron micrographs of 33:67: A. 99; B. 190; C. 300-nm polystyrene core-polyethyl methacrylate shell particles with phosphotungstic acid negative stain and ruthenium tetroxide positive stain[22].

REFERENCES

1. Wessling RA: *J. Appl. Polym. Sci.* 12, 309 (1958).
2. Gerrens H: *Z. Electrochem.* 67, 741 (1963).
3. Makgawinata T; El-Aasser MS; Klein A and Vanderhoff JW:
   J. Dispersion Sci. Technol. 5, 301 (1984).
4. Boundy RH and Boyer RF(eds): in *Styrene, Its Polymers and
   Copolymers*, Reinhold, NY, p.285, (1952).
5. Guss C and Amidon R: U.S. Patent 2,388,685, Nov. 13, (1945).
6. LeFevre WJ and Harding KG: U.S. Patent 2,460,300, Feb. 1, (1949).
7. Vanderhoff JW: unpublished research, Dow Chemical Co., (1959-
   1960).
8. Williams DJ and Grancio MR: *J. Polym. Sci.* A-1 8, 2617 (1970).
9. *ibid.* 2733 (1970).
10. Williams DJ and Keusch P: *J. Polym. Sci., Polym. Chem. Ed.*
    11, 143 (1973).
11. van den Hul HJ and Vanderhoff JW: *J. Colloid Interface Sci.* 28,
    336 (1968).
12. Vanderhoff JW; van den Hul HJ; Tausk RJM and Overbeek J.Th.G.:
    *Clean Surfaces: Their Preparation and Characterization for
    Interfacial Studies*; Goldfinger G(ed), Marcel Dekker, NY, p.15
    (1970).
13. Bradford EB and Vanderhoff JW: *J. Polym. Sci.* C3, 41 (1963).
14. Sheu HR; Tseng CM; El-Aasser MS; Micale FJ and Vanderhoff JW:
    Grad. Res. Progr. Reps. 24, 6 (1985); *ibid.* 25, 7 (1986).
15. Misra SC; Pichot C; El-Aasser MS and Vanderhoff JW: *J. Polym.
    Sci., Polym. Lett. Ed.* 17, 567 (1979).
16. Lee KC; El-Aasser MS and Vanderhoff JW: *Grad. Res. Progr. Reps.*
    23, 30 (1985); *ibid.* 24, 28 (1985); *ibid.* 25, 86 (1986).
17. Min TI; Klein A; El-Aasser MS and Vanderhoff JW: *J. Appl. Polym.
    Sci.* 21, 2845 (1983).
18. Stutman DR; Klein A; El-Aasser MS and Vanderhoff JW: *I&EC Prod.
    Res. Dev.* 24, 404 (1985).
19. Vanderhoff JW; Dimonie VL; El-Aasser MS and Klein A: *Makromol.
    Chem. Suppl.* 10/11 391 (1985).
20. Merkel MP: *Grad. Res. Progr. Reps.* 21, 39 (1984); *ibid.* 22, 13
    (1984); *ibid.* 23, 10 (1984); *ibid.* 24, 13 (1985); *ibid.* 25, 29
    (1986); *Ph.D. dissertation*, Lehigh Univ., (1986).
21. Chern SC: *Thesis Proposal*, Georgia Inst. Technol., Atlanta,
    (1985); Sundberg DC; Arndt J and Tang M: *J. Dispersion Sci.
    Technol.* 5, 433 (1984); Nelson D and Sundberg DC: paper,
    AICHE Meeting, Houston, March (1983); Nelson D: *M.S. Thesis*,
    University of New Hampshire, Durham, (1983).
22. Dimonie VL; El-Aasser MS; Vanderhoff JW and Klein A: *Grad. Res.
    Progr. Reps.* 25, 34 (1986); *ibid.* 26, 33 (1986).

INTERPENETRATING POLYMER NETWORK LATEXES: SYNTHESIS, MORPHOLOGY, AND PROPERTIES

D.I. Lee, T. Kawamura, E. F. Stevens

Dow Chemical U.S.A.
1604 Building
Midland, MI 48674

1. INTRODUCTION

Structured latexes are known to exhibit a variety of unique properties characteristic of their particle morphologies. For example, a hard core/soft shell latex is a low-temperature film-forming, soft binder or reinforced elastomer, whereas a soft core/hard shell latex is a high-temperature film-forming, stiff binder or toughened plastic. Non-core/shell latexes may exhibit a wide range of properties between those of the above-described core/shell latexes, depending on which polymer phase is more continuous.

This study is concerned with a special class of structured latexes, interpenetrating polymer network (IPN) latexes. An IPN can be defined as a blend of two polymers in a network form[1] Although two miscible polymers can form true IPN's as defined, two immiscible pairs cannot form molecular interpenetration due to phase separation. Instead, they may form one polymer phase more continuous and the other phase dispersed in microdomains, whose sizes depend on both the crosslinking density of the continuous phase and the extent of miscibility between the two polymers. Because of these inter-relationships, IPN's of immiscible polymer pairs may approach the morphology of IPN's of miscible polymer pairs, as the crosslinking increases and the microdomain size decreases.

This paper will describe the synthesis, morphology, and properties of IPN latexes based on crosslinked, carboxylated acrylate polymers as polymer I's and polystyrene and styrene-butadiene (S/B) copolymers as polymer II's.

2. LITERATURE SURVEY

Since the early 1970's, Sperling and his co-workers[2-6] have extensively studied various IPN latex systems. IPN latexes were synthesized by polymerizing monomer II with crosslinker in the presence of a crosslinked polymer I seed latex, without further addition of emulsifiers. They found that the structure of their IPN latex particles was predominantly a core-shell morphology with cellular structures within the core. They explained the core-shell formation of their IPN latex particles on the basis of polymerization in a monomer-rich shell surrounding a monomer-swollen core. Their explanation was consistent with the core-shell formation of styrene-polystyrene latex systems studied by Williams and his co-workers[7-11]. Vanderhoff[12] proposed another explanation for the experimentally-observed core-shell morphology without invoking a monomer concentration gradient within the particle.

Recently, Narkis, Talmon, and Silverstein[13] studied the properties and structure of elastomeric two-stage emulsion interpenetrating networks. They reported that the morphology of their IPN latex particles

48

was unique in that very small polystyrene microdomains were formed by phase separation during the polymerization of the second stage styrene monomer in the presence of a flexible polyacrylate seed latex. Lee et al[14] also studied the morphology and properties of IPN latexes and found that the morphology of their IPN latex particles was not a core-shell structure, but a cellular structure containing microdomains of polymer II within polymer I network latex particles, without the shell formation of polymer II.

3. PROPOSED SYNTHESIS SCHEMES FOR IPN LATEXES

Matsumoto et al[15,16], Eliseeva et al[17], Lee et al[18,19], and Muroi et al[20] have shown that the emulsion polymerization of a monomer or a mixture of monomers II in the presence of polymer latex I as a seed may either form polymer II microdomains within polymer I particles or result in an inverted core-shell morphology, if polymer I is more polar or hydrophilic than polymer II. Muroi[21] also has shown that carboxylated latex particles have acid-rich surface layers. Recently, Cho et al[22] have reported that sulfate end-groups can play an important role in the morphology of structured latex particles. These findings have allowed us to propose the following synthesis method for IPN latexes:

"IPN latexes can be synthesized by emulsion polymerization of a monomer or a mixture of monomers II in the presence of crosslinked polymer I latexes as seeds, whose overall or interfacial polymer molecules are more hydrophilic than polymer II".

Figure 1 shows a schematic representation of our IPN latex synthesis. As shown in the Figure, miscible polymer pair systems can form truly interpenetrations on the molecular level while immiscible systems result in microdomains by phase separation within the polymer I network.

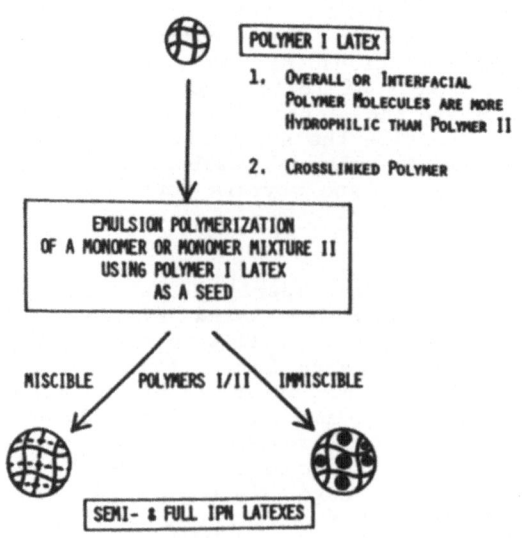

Figure 1: Schematic Representation of IPN Latex Synthesis.

4. EXPERIMENTAL
4.1. Materials

All monomers used in the present study were polymerization-grade, and were used as supplied without further purification. Sodium dodecyldiphenyl oxide sulfonate (DOWFAX 2A1, The Dow Chemical Co.) and sodium persulfate were used as an anionic surfactant and an initiator, respectively.

4.2. Preparation of Latex Samples

All two-stage latex samples were prepared by emulsion polymerization of the second-stage monomer mixture in the presence of the first-stage polymer seed latexes without any new particle generation. The mode of emulsion polymerization was a semi-continuous process for both stages, and polymerizations were carried out at 90°C.

For a series of semi-IPN latexes, polymer I seed latexes were prepared by polymerizing a mixture of n-butyl acrylate (nBA), styrene (S), and acrylic acid (AA) at the monomer ratios of 65, 31, and 4 (by weight), with varying amounts of a crosslinker at 0, 0.2, 0.4, 0.6, and 1.0 part per 100 parts monomers. The crosslinker used in the present study was allyl methacrylate (AMA), but other crosslinkers such as divinyl benzene and ethylene glycol dimethacrylate can be used for the preparation of crosslinked polymer I seed latexes. Subsequently, a series of Semi-IPN latexes were synthesized by emulsion polymerizing styrene in the presence of these polymer I seed latexes at a stage ratio of 50/50. For a series of full IPN latexes, three polymer I seed latexes containing the crosslinker levels of 0.2, 0.4 and 0.6 were selected and subsequently converted into IPN latexes. Polymer II's were crosslinked styrene (S) - butadiene (B) copolymers at varying ratios of S/B: 60/40, 52.5/47.5, 50/50 and 45/55. The stage ratios varied from 50/50 to 10/90 at the interval of 10 units.

5. RESULTS AND DISCUSSIONS
5.1. Electron Microscopy

The morphology of semi- and full IPN latex particles was studied by electron microscopy in conjunction with two staining methods: 1) staining butadiene-containing polymers with osmium tetroxide vapors[23] and 2) reacting acrylate ester-containing polymers with hydrazine, then reducing the hydrazides with osmium tetroxide vapors[24,25]. Latexes were dispersed in a 5% aqueous solution of methylcellulose, then drop-dried onto a copper slug for cryo-sectioning. Samples were microtomed at -55°C, and sections of about 1000 Å thickness were obtained, then stained according to the above-mentioned staining methods.

Figures 2 and 3 are transmission electron micrographs (TEM) showing the hydrazine/OsO$_4$-stained cross-sections of semi-IPN latex particles synthesized with and without crosslinker in polymer I, respectively. Dark and light regions are polymer I (nBA/S/AA: 65/31/4)-rich and polymer II (polystyrene)-rich phases, respectively. As can be seen from these micrographs, inverted core-shell latex particles were formed without crosslinker in the polymer I and interperetrating polymer networks were formed when 1.0 part crosslinker per 100 parts polymer I was used.

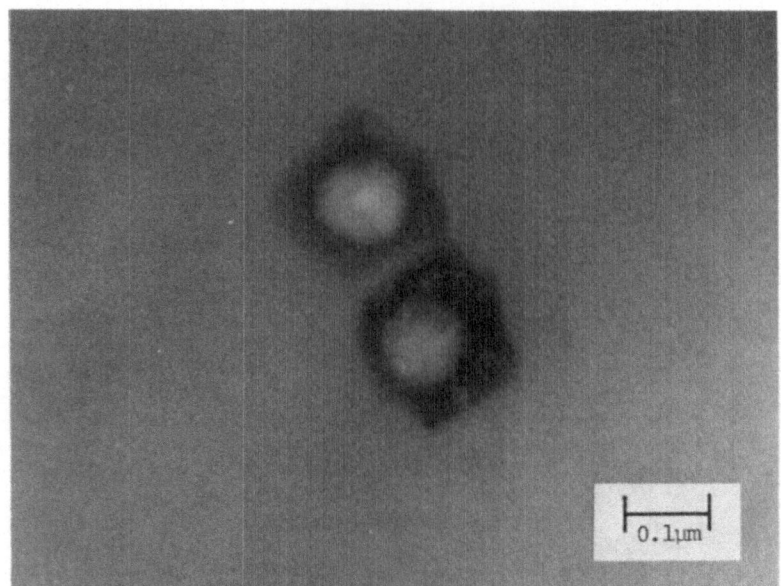

Figure 2: *Hydrazine/OsO₄-Stained Cross-Sections of Semi-IPN Latex Particles Synthesized Without Crosslinker In Polymer I: 50 Polymer I (nBA/S/AA: 65/31/4)//50 Polymer II (PS), Polymer I - Rich (dark) and Polymer II - Rich (light).*

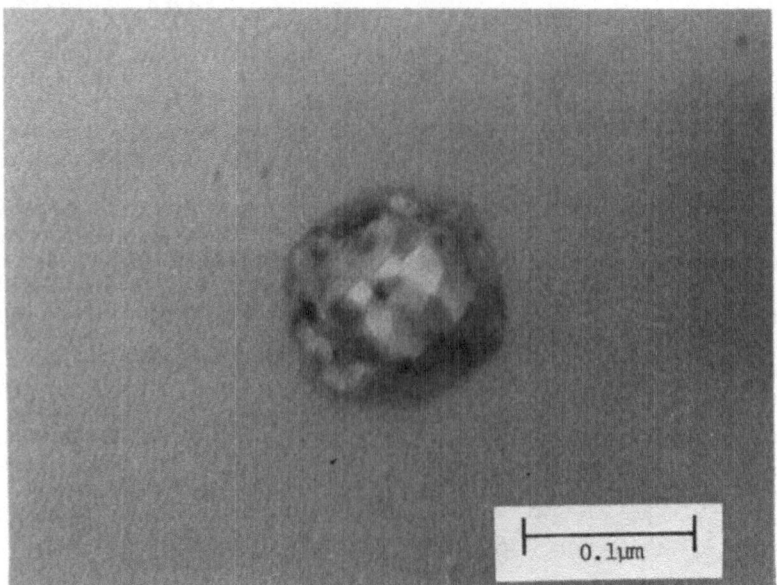

Figure 3: *Hydrazine/OsO₄-Stained Cross-Sections of Semi-IPN Latex Particles Synthesized With 1.0 Part Crosslinker Per 100 Parts Polymer I: 50 Polymer I (nBA/S/AA/X-Linker: 65/31/4/1)//50 Polymer II (PS), Polymer I - Rich (dark) and Polymer II - Rich (light).*

Figures 4, 5 and 6 show TEM's of the OsO$_4$-stained cross-sections of full IPN [50(nBA/S/AA/X-linker: 65/31/4/0.4)//50(S/B: 52.5/47.5)], reverse IPN [50(S/B: 52.5/47.5)//50(nBA/S/AA/X-linker: 65/31/4/0.4)], and homogeneous (nBA/S/B/AA/X-linker: 32.5/41.75/23.75/2/0.2) latex particles, respectively. From these three micrographs, it is quite obvious that the order of polymerization sequence has a profound effect on the morphology of latex particles containing the same overall composition. They represent the morphological features of IPN, core-shell, and homogeneous latex particles, respectively.

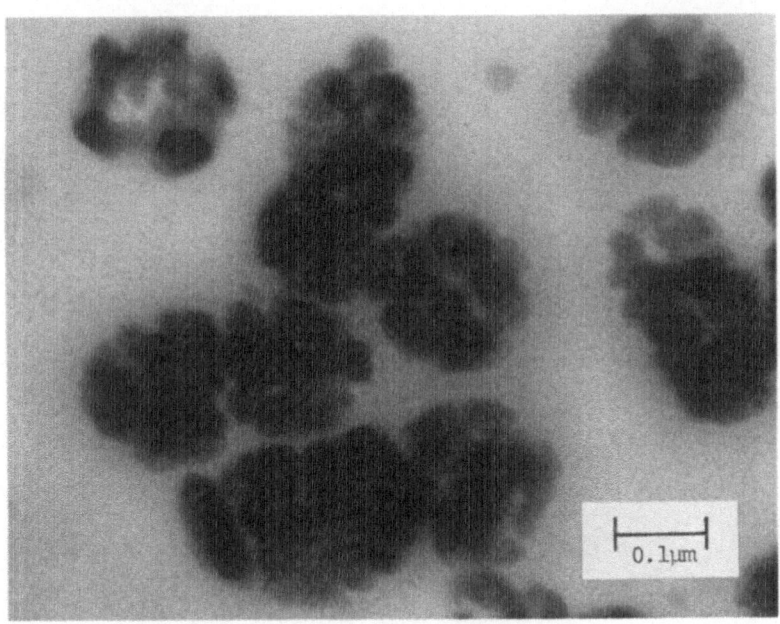

Figure 4: OsO$_4$-Stained Cross-Sections of Full IPN Latex Particles: 50 Polymer I (nBA/S/AA/X-Linker: 65/31/4/0.4)//50 Polymer II (S/B: 52.5/47.5), Polymer I - Rich (light) and Polymer II - Rich (dark).

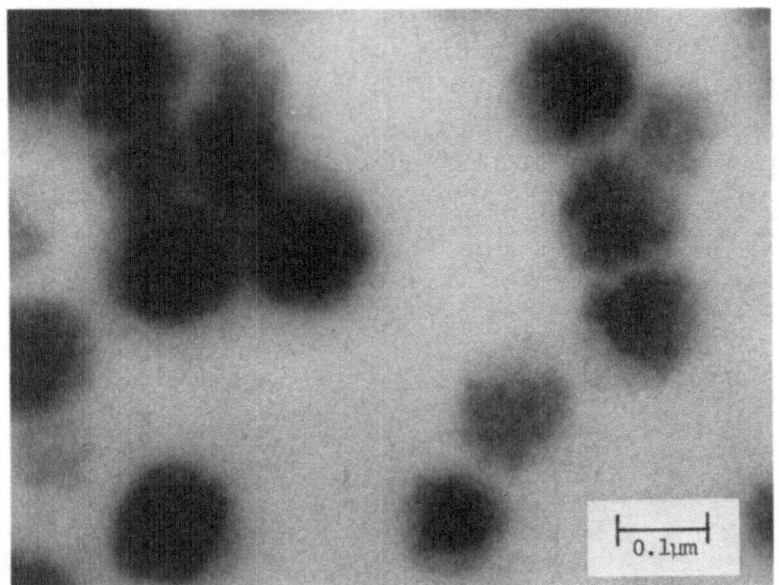

Figure 5: *OsO₄-Stained Cross-Sections of Reverse IPN Latex Particles: 50 Polymer II (S/B: 52.5/47.5)//50 Polymer I (nBA/S/AA/X-Linker: 65/31/4/0.4), Polymer I - Rich (light) and Polymer II - Rich (dark).*

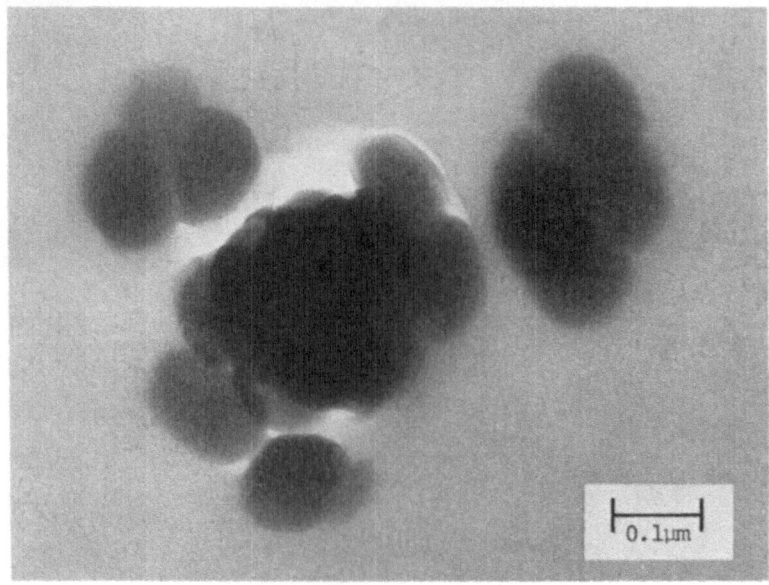

Figure 6: *OsO₄-Stained Cross-Sections of Homogeneous Latex Particles: (nBA/S/B/AA/X-Linker: 32.5/41.75/23/75/2/0.2).*

Figures 7 and 8 show TEM's the OsO_4-stained cross-sections of full IPN latex particles [50(nBA/S/AA: 65/31/4)//50(S/B: 60/40)] synthesized with two different amounts of crosslinker, 0.2 and 0.6 part per 100 parts polymer I, respectively. These micrographs show the effect of cross-linking of polymer I on the size of polymer II (S/B copolymer) microdomains. It can be seen that the size of microdomains decreases and their population within the polymer I particles increases with increasing crosslinking of polymer I.

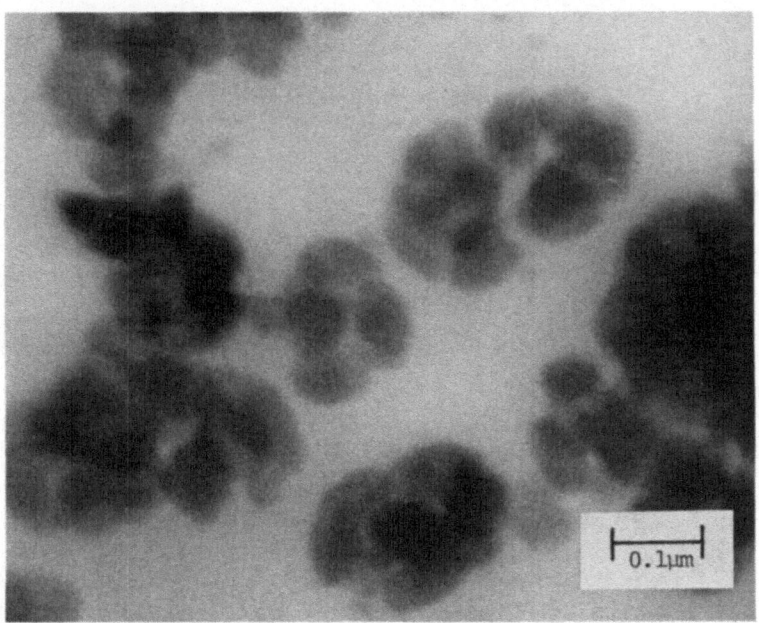

Figure 7: OsO_4-Stained Cross-Sections of Full IPN Latex Particles: 50 Polymer I (nBA/S/AA/X-Linker: 65/31/4/0.2)//50 Polymer II (S/B: 60/40), Polymer I - Rich (light) and Polymer II - Rich (dark).

54

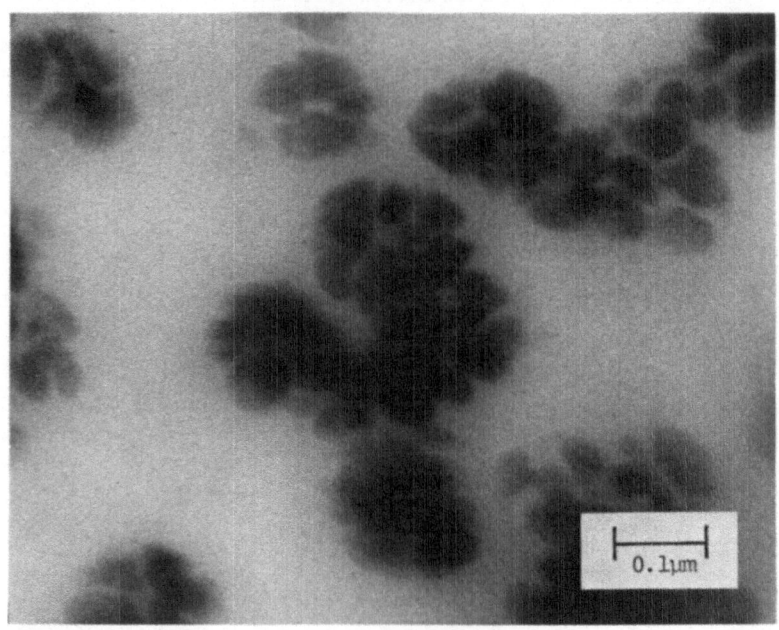

Figure 8: OsO₄-Stained Cross-Sections of Full IPN Latex Particles: 50
Polymer I (nBA/S/AA/X-Linker: 65/31/4/0.6)//50 Polymer II (S/B: 60/40),
Polymer I - Rich (light) and Polymer II - Rich (dark).

5.2. Differential Scanning Calorimetry (DSC) for Semi-IPN Latex Polymers

The DSC curves in Figure 9 show the effect of the crosslinking of
polymer I (nBA/S/AA: 65/31/4) on their glass transition temperatures,
Tg's. The Tg increases slightly with increasing crosslinking of polymer
I. The DSC curves in Figure 10 show the effect of the crosslinking of
polymer I's on the thermal properties of semi-IPN latex films, [50
polymer I//50 polymer II (polystyrene)]. As can be seen from Figure 10,
two distinctive Tg's of polymer I's and polystyrene gradually disappear
and converge into a broad glass transition range of temperatures, as the
crosslinking of polymer I's increases. This change in their thermal
behaviors suggests that the size of microdomains decreases with
increasing crosslinking of polymer I, thus increasing interactions
between polymers I and II. This observation is in good agreements with
the findings obtained from the electron microscopy study.

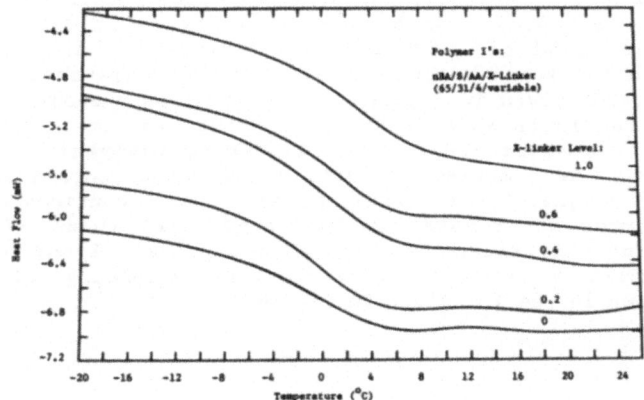

Figure 9: The Effect of the Crosslinking of Polymer I's on Their Glass Transition Temperatures, Tg's.

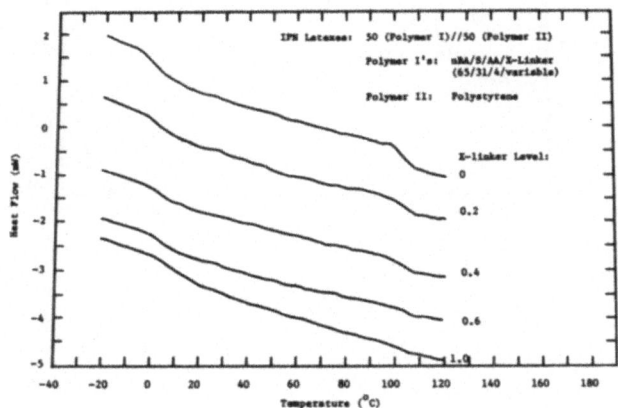

Figure 10: The Effect of the Crosslinking of Polymer I's on the Thermal Properties of IPN Latex Films.

5.3. Dynamic Mechanical Spectroscopy (DMS) for Semi-IPN Latex Polymers

Figures 11 and 12 show the dynamic mechanical properties of semi-IPN latex polymers synthesized without and with crosslinker in polymer I, respectively. Samples were prepared in the form of bars at high temperatures (150°C). Therefore, the original latex particle morphologies were somewhat altered. It can be seen from Figure 11 that the sample behaved like a blend of soft and hard polymers. This behavior is consistent with its latex morphology: an inverted core-shell structure, as shown in the TEM given in Figure 2. Also, the shear modulus curve (G') suggests that the inverted core-shell structure was re-inverted during molding at high temperatures, and polystyrene (polymer II) became a continuous phase. As a matter of fact, a low temperature-molded sample bar showed that polymer I (soft shell polymer) was a continuous phase. G' and tan δ curves of Figure 12 indicated that there were strong interactions and significant mixing between polymer I and polymer II. Again, this result is consistent with the latex morphology of the sample polymer, as shown in the TEM given in Figure 3.

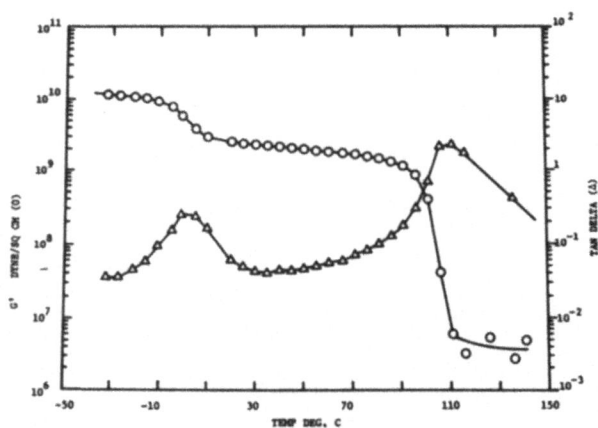

Figure 11: Dynamic Mechanical Properties of Semi-IPN Latex Polymer Synthesized Without Crosslinker In Polymer I: 50 Polymer I (nBA/S/AA/X-Linker: 65/31/4/0)//50 Polymer II (PS). The Measurement was made at a frequency of 1 Hz.

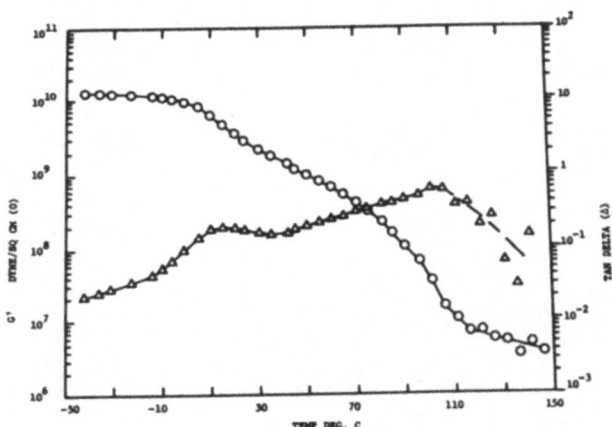

Figure 12: Dynamic Mechanical Properties of Semi-IPN Latex Polymer Synthesized With Crosslinker In Polymer I: 50 Polymer I (nBA/S/AA/X-Linker: 65/31/4/1)//50 Polymer II (PS). The Measurement was made at a frequency of 1 Hz.

5.4. Grafting Aspects of IPN Latexes

Hughes and Brown[26], Lee[27], Min et al[28], Dimonie et al[29], Vanderhoff et al[30], and Stutman et al[31] have studied grafting in various structured latex systems and found significant amounts of grafted copolymers. The fact that our present latex systems were very similar to those studied by Min et al and Stutman et al suggests the possibility of grafting reactions in our IPN latex systems.

5.5. Mechanical Properties of Full IPN Latex Films

Table 1 shows the mechanical properties of full IPN, reverse IPN (core-shell), blend, and homogeneous latex films, along with those of polymer I and polymer II latex films. It can be seen from Table 1 that although they have the same overall composition, a variety of mechanical properties can be obtained by different combinations of polymer I and polymer II. Figure 13 shows the tensile strength and % elongation at break of polymer I latex films as a function of crosslinking. Figures 14 and 15 show the tensile strength and % elongation at break of full IPN latex films as a function of both polymer II composition and polymer I crosslinking, respectively.

58

TABLE 1. Mechanical Properties of Full IPN, Reverse IPN (Core-Shell),
Blend, and Homogeneous Latex Films Along with Those of
Individual Component Latex Films

Latex Type	Tensile Strength lbs/in^2	% Elongation at Break
Polymer I Latex (nBA/S/AA/X-Linker: 65/31/4/0.4)	1291	604
Polymer II Latex (S/B: 52.5/47.5)	120	No Break
Full IPN Latex (50 Polymer I// 50 Polymer II)	903	464
Reverse IPN Latex (50 Polymer II// 50 Polymer I)	1066	344
Blend Latex (50 Polymer I Latex/ 50 Polymer II Latex)	1035	192
Homogeneous Latex (nBA/S/B/AA/X-Linker: 32.5/41.75/23.75/2/0.2)	497	612

1. Latexes were cast on Teflon-coated plates and dried at room
   temperature.

2. Stress-strain curves were obtained using an Instron
   instrument.

3. Cross-head speed was 20 inches per minute.

As can be seen from Figures 13, 14, and 15, the mechanical properties of
full IPN latex films are almost independent of the crosslinking of
polymer I unlike those of polymer I latex films, but are strongly
dependent on the composition of polymer II. It is interesting to note
that although the crosslinking of polymer I affects the size of polymer
II microdomains, as shown in the TEM's of Figures 7 and 8, it does not
affect the mechanical properties of full IPN latex films. This suggests
that the mechanical properties of IPN latex films may be insensitive to
the size of polymer II microdomains, as long as their size is small
enough to form IPN latexes. Figures 16 and 17 show the tensile strength
and % elongation at break of full IPN latex films as a function of both
stage ratio and the extent of polymer I crosslinking, respectively.
Again, the mechanical properties of these IPN latex films are found to be
independent of the crosslinking of polymer I, except at low contents of
polymer II. It is also interesting to note that the mechanical properties

of full IPN latex films are quite different from those of latex blend films, i.e., for tensile strength, IPN latexes exhibit average values of their respective component polymers over a wide range of stage ratios, and for elongation, they behave like more crosslinked polymers, compared to their latex blend counterparts. These findings are not surprising from the fact that full IPN latexes are composed of two polymers in network form despite phase separation.

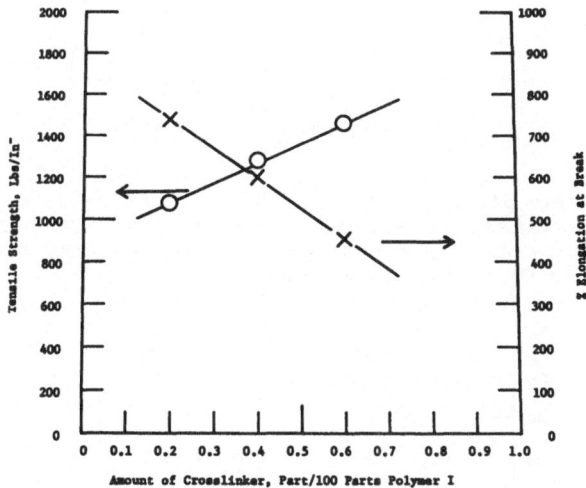

Figure 13: Tensile Strength and % Elongation of Polymer I (nBA/S/AA/X-Linker: 65/31/4/Variable) Latex Films as a Function of the Amount of Crosslinker. Drying Condition: Room Temperature.

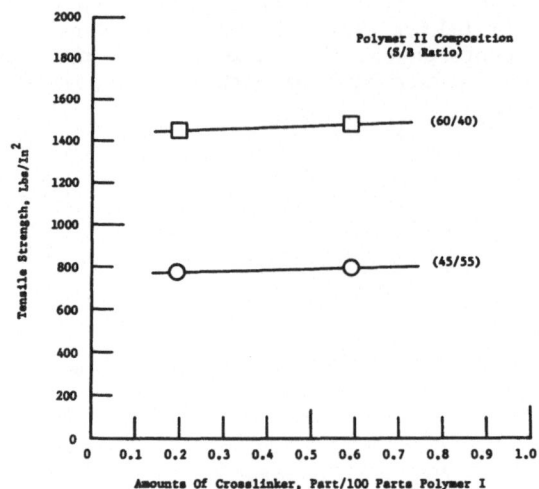

Figure 14: *Tensile Strength of Full IPN Latex Films as a Function of the Amount of Crosslinker In Polymer I and Polymer II. Composition: 50 Polymer I (nBA/S/AA/X-Linker: 65/31/4/Variable)//50 Polymer II (S/B: Variable). Drying Condition: Room Temperature.*

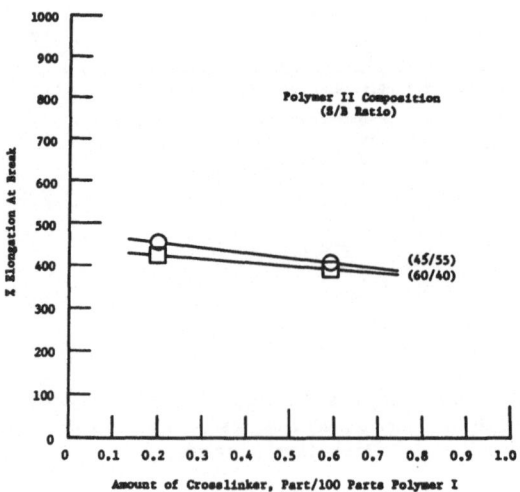

Figure 15: *% Elongation at Break of Full IPN Latex Films as a Function of the Amount of Crosslinker in Polymer I and Polymer II. Composition: Same as Figure 2.*

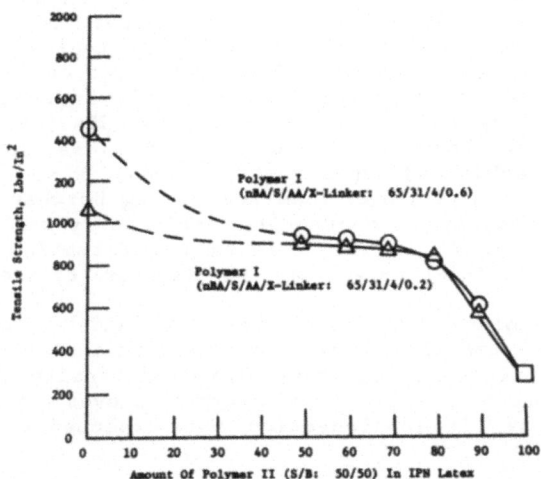

Figure 16: *Tensile Strength of Full IPN Latex Films as a Function of Polymer II Content and the Crosslinking of Polymer I.*

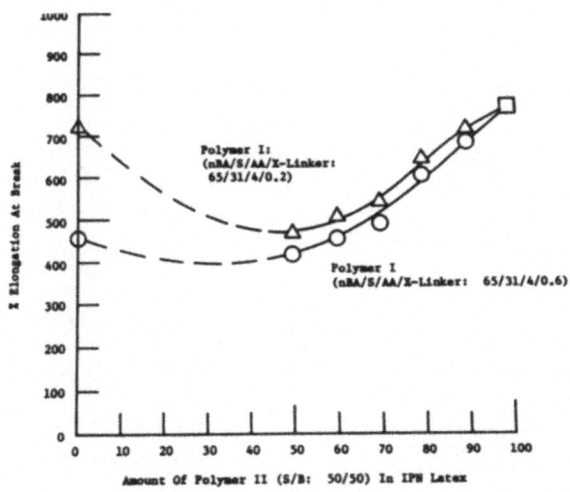

Figure 17: *% Elongation at Break of Full IPN Latex Films as a Function of Polymer II Content and the Crosslinking of Polymer I.*

6. SUMMARY

Semi- and full IPN latexes were synthesized by emulsion polymerizing styrene and mixtures of styrene and butadiene, respectively, in the presence of crosslinked, carboxylated polymer latexes as seeds. Their particle morphology was studied by electron microscopy in conjunction with both hydrazine/OsO_4- and OsO_4- staining techniques. It was found that their particle morphology was a cellular structure typical of IPN's composed of two immiscible polymers, and the domain size of polymer II decreased with increasing crosslinking of polymer I. Indeed, as the crosslinking of polymer I seed latexes increased, two distinctive Tg's gradually disappeared and converged into a broad glass transition temperature. Dynamic mechanical spectroscopy (DMS) also provided information on the extent of interactions and mixing between the two polymers I and II. Differential scattering calorimetry and dynamic mechanical spectroscopy provided very valuable information on the morphology of IPN latex particles as well as the structure of sample bars made from IPN latexes.

A variety of mechanical properties was obtained by different combinations of polymers I and II. The mechanical properties of IPN latex films were found to be unique in that they exhibited tensile strengths intermediate between those of polymer I and polymer II over a wide range of stage ratios and showed lower elongation than expected from their blend counterparts.

7. ACKNOWLEDGMENTS

The authors would like to thank R. Czerepinski for DSC data, L. Morford and J. Marshall for electron microscopy, and C. Berglund for DMS data. They also would like to extend their appreciation to C. McDonald for his delineation of latex particle morphology.

REFERENCE

1. Sperling LH: *Interpenetrating Polymer Networks and Related Materials*, Plenum Press, (1981).
2. Sperling LH; Chiu Tai-Woo; Hartman CP and Thomas DA: *Intern. J. Polymeric Mater.*, 1, 331 (1972).
3. Sperling LH; Chiu Tai-Woo and Thomas DA: *J. Appl. Polym. Sci.*, 17, 2443 (1973).
4. Sperling LH; Thomas DA; Lorenz JE and Nagel EJ: *J. Appl. Polym. Sci.*, 19, 2225 (1975).
5. Sperling LH: *J. Polym. Sci.*, Macromol. Rev., 12, 141 (1979).
6. Sionakids J; Sperling LH and Thomas DA: *J. Appl. Polym. Sci.*, 24, 1179 (1979).
7. Grancio MR and Williams DJ: *J. Polym. Sci.*, Part A-1, 8, 2617 (1970).
8. Williams DJ: *J. Elastoplastics*, 3, 187 (1971).
9. Keusch P and Williams DJ: *J. Polym. Sci.*, Polym. Chem. Ed., 11, 143 (1973).
10. Keusch P; Prince J and Williams DJ: *J. Macromol. Sci. Chem.*, 7, 623 (1973).
11. Keusch P; Graff RA and Williams DJ: *Macromolecules*, 7, 304 (1974).
12. Vanderhoff JW: Proc. Water-Borne and High Solids Coatings Symp. 3, Paper 6, New Orleans, 1976.
13. Narkis M; Talmon Y and Silverstein M: *Polymer*, 26, 1359 (1985).
14. Lee DI; Takamura T and Stevens EF: U.S. Patent No. 4,569,964 (1986).
15. Matsumoto T; Okubo M and Imai T: Kobunshi Ronbunshu, *Eng. Ed.*, 3, 1814 (1974).
16. Matsumoto T; Okubo M and Shibao S: Kobunshi Ronbunshu, *Eng. Ed.*, 5, 784 (1976).
17. Eliseeva VL; Titova NV; Gevorkyan AV; Nazaryan LG; Chalykh AY and Matveev VV: *Dokl. Akad. Nauk. SSSR*, 261, 402 (1981).
18. Ishikawa T and Lee DI: U.S. Patent No. 4,325,856 (1982).
19. Lee DI and Ishikawa T: *J. Polym. Sci.*, Polym. Chem. Ed., 21, 147 (1983).
20. Muroi S; Hashimoto H and Hosoi K: *J. Polym. Sci.*, Polym. Chem. Ed., 22, 1365 (1984).
21. Muroi S: *J. Appl. Polym. Sci.*, 10, 713 (1966).
22. Cho I and Lee KW: *J. Appl. Polym. Sci.*, 30, 1903 (1985).
23. Kato K: *Polym. Lett.*, 4, 35 (1966).
24. Kanig G and Neff H: *Colloid Polym. Sci.*, 253, 29 (1975).
25. Misra SC; Pichot C; El-Aasser MS and Vanderhoff JW: *J. Polym. Sci.*, Polym. Lett. Ed., 17, 567, (1979).
26. Hughes LJ and Brown GL: *J. Appl. Polym., Sci.*, 5, 580 (1961).
27. Lee DI: Unpublished data on two-stage (PS//S-B) latexes, (1976).
28. Min TI; Klein A; El-Aasser MS and Vanderhoff JW: *J. Polym. Sci.*, Polym. Chem. Ed., 21, 2845 (1983).
29. Dimonie VL; El-Aasser MS; Klein A and Vanderhoff JW: *J. Polym. Sci.*, Polym. Chem. Ed., 22, 2197 (1983).
30. Vanderhoff JW; Dimonie VL; El-Aasser MS and Klein A: *Makromol. Chem. Suppl.*, 10/11, 391 (1985).
31. Stutman D; Klein A; El-Aasser MS and Vanderhoff JW: *Ind. Eng. Chem. Prod. Res. Dev.*, 24, 404 (1985).

EMULSION COPOLYMERIZATION: SIMULATION OF PARTICLE MORPHOLOGY

J. GUILLOT

Laboratoire des Materiaux Organique
CNRS
69390 Vernaison
FRANCE

1. INTRODUCTION

Despite the fact that emulsion polymers are generally copolymers, only few fundamental works have been devoted to the kinetics of emulsion copolymerization. One reason is, certainly, that copolymerization is thought to be even more complex. However, in many cases copolymerization proved to be a powerful tool for mechanistic investigations; a good choice of the comonomer making it something like a molecular "sensor". In the past few years an increasing number of publications dealt with emulsion copolymerization processes: batch, semicontinuous with a simple or complex monomer addition.

Another noteworthy tendency is the development of computer simulations to describe, at least, the macromolecular behaviour of an emulsion copolymerization, taking into account the many aspects of such an heterogeneous polymerization process, in which monomer partition, monomer-monomer and monomer-polymer interactions must be considered on the basis of partition coefficients or thermodynamics; although both approaches, of course, are closely related. Another interesting and quite new aspect, allowed by the simulation, is the possibility to know quantitatively the copolymerization in the aqueous phase; which, in turn, could be of a major interest to understand the nucleation or flocculation mechanisms. We have also investigated in our laboratory whether emulsion copolymerization allows the determination of reliable propagation rate constants, necessary to describe the process versus time, or to control it.

On the other hand, a lot of work have been devoted to the synthesis of latex with structured particles with composition gradients, core-shell, multilayers and many other complex morphologies, and the properties of their corresponding films. A good review was recently given by J. C. Daniel in 1984[1], which showed that the particle morphology could be related to the polymerization process, the nature of the monomers, their relative polarities, and thermodynamic compatibility of the components. But in many cases, the structured latexes were prepared by multiple stage polymerization, i.e., schematically, by homo-polymerization onto more or less structured seed particles. It also appeared from these works that the morphology which is developed during the polymerization, very often is not a stable one, except if some actual copolymer or grafting do exist. That could be regarded as an evidence that thermodynamics plays a definite role with respect to the particle morphology.

This article mainly deals with actual copolymerization and with the possibility to model and simulate the occurence of structured latex particles, as the copolymerization proceeds. As compatibility and

diffusion certainly have an important influence, a quite new
thermodynamic approach is proposed. It is believed, indeed, that useful
information could be easily derived from this kind of simulation for
developing other processes for synthesizing structured latexes.

The particle's structure and its control presuppose that the
synthesis results in heterogeneous macromolecules, i.e. with a large
composition drift, depending on reactivity ratios, monomer feed, monomer
partition, and polymerization process. In radical polymerization, indeed,
the lifetime for a growing macroradical is very short, so the
microstructure of the chains obtained at a given moment and locus is the
image of the monomer feed at that time and location. Consequently, the
final particle contains a more or less complex mixture of individual
macromolecules which will tend to be organized according to their
thermodynamic compatibilities and the energy allowed. It is also
reasonable to think that a copolymer should improve the overall
compatibility, as should do some grafting of very different chains, by an
emulsifier-like mechanism.

It is clear, also, that the nucleation and flocculation processes
play a determining role on the final particle morphology.

Since polymer phase separation seems to be closely related to the
polarity, it is necessary to develop accurate kinetic studies and
simulations for both the understanding of the behaviour of such an
heterogeneous polymerization process and the actual control of the
synthesis.

2. SIMULATIONS

Most of the recent work is mainly focused on the prediction of
copolymer compositional changes under the prevailing experimental
conditions for various monomer pairs such as methyl methacrylate(MMA)-
styrene(S)[2], acrylonitrile(AN)-S[3], S-butadiene[4]. For example if molecular
weights or particle size distribution are included, it is usually done on
the basis of somewhat oversimplified assumptions concerning nucleation
and flocculation. However, since quite elaborate and satisfying theories
have been developed recently for the basic latex characteristics[5,6,7,8,9]
in homopolymerization, some progress is expected for copolymerization.

3. SIMULATION OF COPOLYMER MACROMOLECULES

A general difficulty encountered in an ambitious simulations is the
huge number of parameters whose accurate determination is not always
obvious. When only estimations or, worst yet, adjustable values are
assigned to these parameters, the lower the number of such parameters,
the more reliable the simulation will be. Two approaches have been used,
based on: (a) partition coefficients, when the main objective is the
copolymer macromolecules; and (b) thermodynamics, if colloidal aspects
are taken into account.

3.1. Partition Coefficients[2,3,4,10]

The partition coefficients can be accurately determined from
experiments. They can be a function of experimental conditions such as
temperature, monomer feed composition, and free emulsifier. This approach
has been extensively used in our laboratory. It is possible to derive a
simulation for copolymerization of several monomers[11]. The developed
computer programs are quite simple and give a good description of the
copolymer macromolecules. They also lead to some quite new and
interesting results related to the present purpose.

3.1.1. <u>Emulsion reactivity ratios (r_{ij}) are "apparent" kinetic parameters.</u>

The simulation shows that reactivity ratios in emulsion copolymerization of monomers of different water solubility must be regarded as apparent kinetic parameters, the values of which are dependent upon monomer polarity and experimental conditions, above all the monomer/water ratio (M/W)[12]. Any factor which influences the monomer water solubility, such as temperature, free emulsifier (solubilization) will also contribute to their changes. It appears that, at a given M/W, the usual methods to determine reactivity ratios are apparently valid which explains the large range of emulsion r_{ij} values reported in the literature. The general trends of the variation in r_{ij} from Bulk to Solution are to lower and lower values for the more water soluble monomer when M/W is decreased, and to higher and higher values for the less water soluble monomer. It is only by extrapolating to M/W → ∞ (W/M → 0) that one would get the actual constant reactivity ratios.

The knowledge of the variation of r_{ij}'s (See Figure 1) can be useful in predicting the copolymer composition at low conversion, for a given solid content. Such curves also allows one to predict if azeotropic copolymers can exist; and what happens when the apparent r_{ij} are both greater or lower than unity. A relationship is found between the overall monomer feed and the optimum M/W in the batch, which results in continuous quasi-azeotropic copolymers from the unique azeotropic composition for Bulk/Solution to the homopolymerization of the more water-soluble monomers[12]. The discrepancy in the monomer polarities increases the effect. Experimental works have confirmed this theoretical results[13,14].

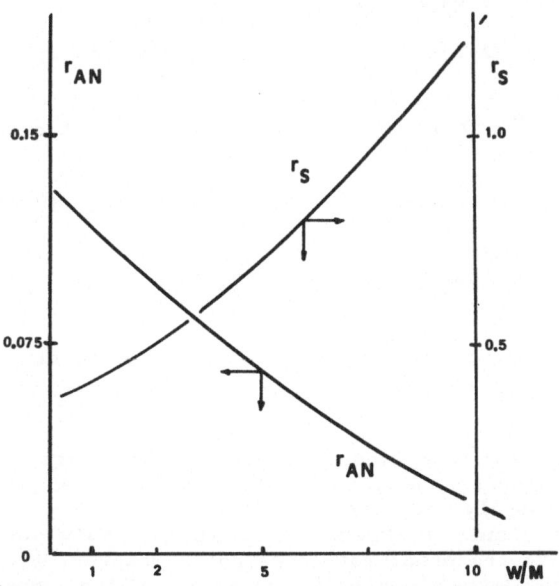

Figure 1:   Acrylonitrile(AN) - Styrene(S) Batch Emulsion Copoly-merization. Variation of r_{ij} vs. water/monomer ratio(W/M).

3.1.2. <u>Determination of propagation rate constants</u>.[16]

For many applications, it is convenient to plot a function versus the overall conversion which allows the introduction in the simulation of only reactivity ratios, which are quite accurate parameters. However, when functions must be plotted versus time or when monomers are added in semi-batch, power feed process[15] or more complex process, it is essential to use in the simulation the absolute rate constants, such as propagation rate constants, kp_{ij}. However, it is generally observed that the literature values of kp_{ij}, if given, are usually inaccurate. A series of Styrene-alkyl acrylate copolymerizations were studied under "mild" conditions (50°C, low persulfate concentrations). A comparison between the overall polymerization rate, R_p, and particle number, N_p, at various times, has shown that the kinetic behaviour was very close to a classical one, particularly the average number of radicals per particle, $\bar{n}$, which was close to 0.5, at intermediate conversion. Indeed, the overall polymerization rate, R_p, can be written as:

$$R_p = k_p[M]\bar{n}N_p \qquad (1)$$

where [M] is the monomer concentration within polymer particle.

From the theory of radical polymerization and the "steady state assumption" the overall rate constant in a copolymerization, k_p, is derived as a function of reactivity ratios, monomer feed composition and absolute propagation rate constants for both homopolymerizations, k_{ss} and k_{aa} (s is referred to styrene and a to the comonomer). If copolymerization proceeds mainly in the particle phase and if all the particles are considered as a whole, the propagation rate constant of copolymerization can be expressed as follows:

$$k_p = k_{sa}/[1+k_{sa}/k_{as}(A/S)] \times \{[r_A(A/S)+1] \ f_A + [r_S+(A/S)]f_S\} \qquad (2)$$

$k_{sa} = k_{ss}/r_S$ (ℓ mole^{-1} s^{-1}) where r_S is the reactivity ratio for S radicals;

$k_{as} = k_{aa}/r_A$ (ℓ mole^{-1} s^{-1}) where r_A is the reactivity ratio for A radicals;

[M] is the overall monomer concentration, mole ℓ^{-1}; and

(A/S) is the ratio of monomer concentrations within particle phase.

It is better to measure R_p when the nucleation stage is achieved and before the droplet phase has disappeared, i.e., the range of 20 - 30 mole % conversion is probably the best.

If k_{sa} (for styrene) is known, in principle only one experiment is necessary to derive the other rate constant. However, it seems better to carry out several experiments at various initial monomer feeds. A linearized form of the latter equation can be written if one sets,

$$F = [r_A(A/S) + 1] \ f_A + [r_S + (A/S)] \ f_S \qquad (3)$$

and

$$K - R_p N_a/(F [M] N_p) \qquad (4)$$

i.e.

$$\frac{1}{K} - \frac{1}{\bar{n} k_{sa}} + \frac{1}{\bar{n} k_{as}} \frac{A}{S} \qquad (5)$$

A graphic or regression procedure can give $\alpha - 1/(\bar{n}k_{sa})$ as the intercept and $\beta - 1/(\bar{n}k_{as})$ as the slope, in a linear diagram of $1/K$ versus $f(A/S)$ such as in Figure 2.

Since k_{sa} is known for styrene, α is a good estimate of the deviation (i.e. $\bar{n} \neq 0.5$) from the theory of emulsion polymerization. The unknown propagation rate constant $k_{as}(k_{aa})$ is readily derived either from β and $\bar{n}$ or from the ratio of α/β.

The same method was used to derive the propagation rate constants for methyl acrylate, ethyl acrylate, acrylonitrile, and vinyl acetate, which are given in Table 1. On the other hand, when conditions are changed, comparison between R_p and N_p gives $\bar{n}$. But if $\bar{n}$ should be assumed to remain close to 0.5, the simulation gives an estimation of the actual active number of particles. It is necessary to know the absolute rate constants to consider the occurrence of diffusion controlled mechanism.

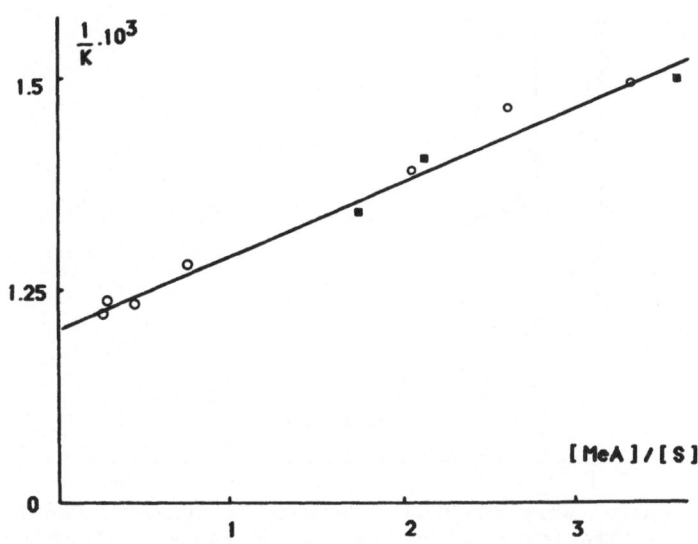

Figure 2: Methyl acrylate (MeA) - Styrene(S) Emulsion Copolymerization. Determination of propagation rate constants k_{pi} $1/K$ vs. ratio of the monomer concentrations within the swollen polymer particles, [MeA]/[S]

TABLE 1. Propagation Rate Constants from Emulsion Copolymerizations

Monomer	Styrene		BuA	EA	MeA	VAc	AN
T(^{o}C)	50	70	70	60	50	60	55
k_p (ℓ mol^{-1} s^{-1})	130	405	305	250	335	~4,000	~4,000

BuA: butyl acrylate; EA: ethyl acrylate; MeA: methyl acrylate; VAc: vinyl acetate; AN: acrylonitrile

3.1.3. Polymerization in the Water Phase.

Accurate measurements of copolymer composition at the very beginning of copolymerization[11] have shown that, with very reactive radicals (e.g., VAc, AN, MMA, MeA), the resulting copolymer is enriched in the corresponding monomer, which is also the more water-soluble. Above 10-15% conversion, the copolymer composition is very close to the one computed when polymerization in water phase is assumed to be negligible (Figure 3). This deviation can be reasonably attributed to copolymerization in the aqueous phase and can be easily introduced in the simulations. The comparison between experimental and simulation curves gives quantitative informations which are useful for mechanistic investigations.

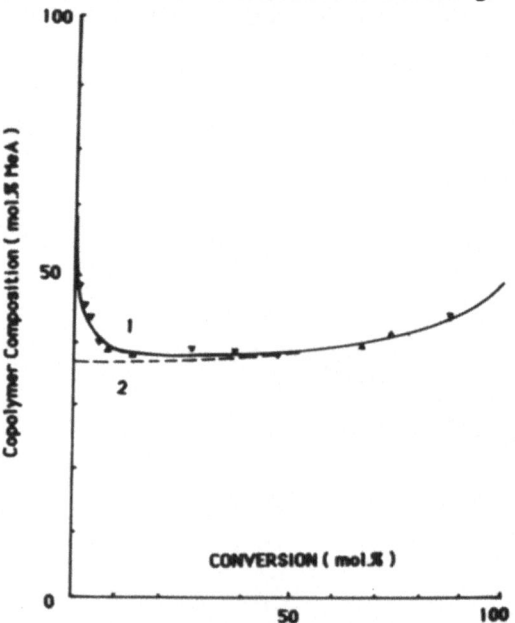

Figure 3: Methyl acrylate(MeA) - Styrene(S) Emulsion Copolymerization. Effect of polymerization in the water phase. Overall copolymer Composition (MeA mol.%) vs. conversion:
  1. *Simulation of Polymerization taking into account the Polymerization in the water phase*
  2. *Simulation assuming only Polymerization in the particle phase*

If the amount of hydrosoluble low molecular weight chains is determined to be small, this means that the oligoradical capture rate by particles is large or that nucleation proceeds by a "coagulative nucleation" mechanism as recently suggested by Gilbert, et al[8]. Consequently it is expected that this should have an impact on the particle size distribution and even on the particle morphology.

3.2. Thermodynamic Approach[10,17]

The fact that partition coefficients are functions of many experimental parameters is an evidence that thermodynamics governs, to a large extent, the process. Consequently, when it is possible to get the right interaction parameters, thermodynamic simulation gives the same previous results, and in addition, a description of the colloidal aspects involved in the polymerization.

Swelling[18] and monomer transfer[19] have been shown to be controlled by the thermodynamics. More recently, it was suggested that this approach could be used for copolymerization[10] and, even, that a simulation based on thermodynamics could give an accurate description of copolymerization of monomers of various polarities by different processes. It is assumed that the chemical potential of the monomers, μ_i, is the same in every phase (droplets, water, polymer), i.e., the process is not a diffusion controlled one[20]. It is also assumed that particle swelling by monomers follows the Flory-Huggins theory. However, monomer molecules have to overcome the interfacial forces (γ) since the curvature of colloid particle is large (1/r); an additional "kelvin" term is, then, necessary.

$$2 \ V_m \ \gamma/r \tag{6}$$

where V_m is the molar volume, γ is the surface tension, and r is particle radius. The polymerization within particles makes the μ_i^P to change continuously. The result is a difference in the μ of the component in water and that in the droplet phase, and in the transfer of a quantity of monomer towards particles from water or droplets, in order to reestablish the equality of the potentials in all phases.

Furthermore, the thermodynamic approach also allows one to consider the surface charge density, whose influence can become effective when the macroradicals generated in the aqueous phase are charged (persulfate residue), since any charge carrier must possess an extra energy to cross the charged interphase:

$$Z_i \ F \ (-d\epsilon/dx)$$

where Z_i is the electric charge, F is the Faraday, $-d\epsilon/dx$ is the surface electric field.

Crosslinking can also be taken into account since, according to Flory's theory of elasticity, the following term has to be added to the polymer solution:

$$(1/Ln) \ (\Phi_p^{1/3}-0.5\Phi_p) . \tag{7}$$

where Ln is the average distance between two crosslinks, which can be derived from the copolymerization theory[21], Φ_p is the volume fraction of polymer within the particle. It is worthy to note that crosslinking, to some extent, results in changing the particle morphology (styrene/ divinylbenzene).

A general expression for the total chemical potential of a species i, within the particle is given by the following equation.

$$\Delta\mu_i = RT \{ \ln\Phi_i + [1 - (1/X_n)] \ \Phi_p + \chi_i \ \Phi_p^2 \} \quad \text{Polymer solution}$$

$$+ 2V_m\gamma/r \qquad\qquad \text{Surface tension}$$

$$\text{(8)}$$

$$+ Z_i F \ (-d\epsilon/dx) \qquad\qquad \text{Surface electric charge}$$

$$+ (1/Ln)(\Phi_p^{1/3} - 0.5 \ \Phi_p) \qquad \text{Crosslinking}$$

where Φ_i is the monomer volume fraction within the particle, χ is the Flory-Huggins interaction parameter, and X_n, is the mean degree of polymerization.

The main problem in this approach is the knowledge of interaction parameters, which are only scarcely reported in the literature. However, for non-polar monomers, an estimation can be derived from the solubility parameters. Otherwise, experiments are necessary: vapor pressure, swelling, or partition coefficient measurements.

4. SIMULATION OF PARTICLE MORPHOLOGY

On the basis of the previous approach, a simulation of the particle morphology was tentatively developed for vinyl acetate (VAc) - butyl acrylate (BuA) batch copolymerization, which can be considered as a model due to the large discrepancy in reactivity ratios, poor polymer compatibility, and very different Tg values[22].

A homogeneous monomer distribution within the particle is not assumed. On the contrary it is proposed that owing to discrepancies in thermodynamic interactions between each monomer and copolymer and/or emulsifier, a heterogeneity could occur in monomer location. At the very beginning, the more hydrophobic monomer is assumed to be preferentially located in the inner part, and the more hydrophilic monomer in the outer layers of the initial micelles or primary particles. A composition gradient would occur within the swollen polymer particles which originates the formation of a "structured" morphology in the final latex particles.

As the copolymerization proceeds, it is possible to simulate the change in monomer composition gradient, by computing at any polymerization increment, the new local value of the chemical potentials (μ_i) of both monomers, and the amounts of monomers which must diffuse to reestablish the equality in the μ_i. Figure 4 gives the principle of the computations. Any polymerization increment results in a new spherical outer polymer-monomer layer, on which emulsifier will adsorb. For a given layer, the monomer polymerizes at a rate which is proportional to its local concentration, and the instantaneous copolymer reflects the monomer feed composition in that phase. At any step, the changing kinetic parameters are utilitzed to recalculate the local μ_i and $\Delta\mu_i$. It is, then, possible to plot, as a function of the overall conversion, the mean and instantaneous copolymer compositions in any such layer of a thickness, h, at a distance R from the center of the latex particle.

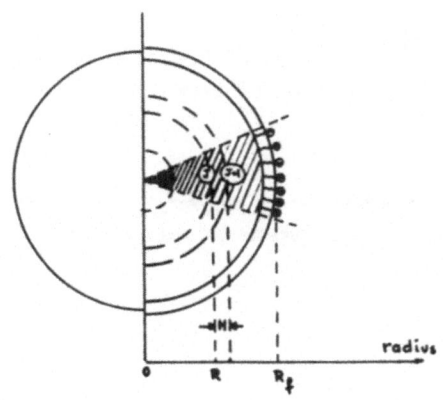

$$\left\{ \begin{array}{l} -(\Delta\mu_1)^j_{j+1} = \dfrac{\partial\Delta\mu_1}{\partial\Phi_1} \cdot d\Phi_1 \; + \; \dfrac{\partial\Delta\mu_1}{\partial\Phi_2} \cdot d\Phi_2 \\[3mm] -(\Delta\mu_2)^j_{j+1} = \dfrac{\partial\Delta\mu_2}{\partial\Phi_1} \cdot d\Phi_1 \; + \; \dfrac{\partial\Delta\mu_2}{\partial\Phi_2} \cdot d\Phi_2 \end{array} \right.$$

Figure 4:   Principle of "CORE-SHELL" Morphology Simulation

A simulation of a 50/50 mole % VAc/BuA batch copolymerization is reported in Figure 5.   The copolymer gradient appears to be quite steep but is in rather good agreement with data from "soap titration" technique proposed by Maron[23] and other characterization methods.   The shape of the gradient changes with the initial monomer feed.   However, in every case the simulation shows that right in the center of the particle the copolymer is richer in VAc than the rest of the "core", and that at high conversion a small amount of BuA, probably trapped deep in the core, is polymerized at the surface.

74

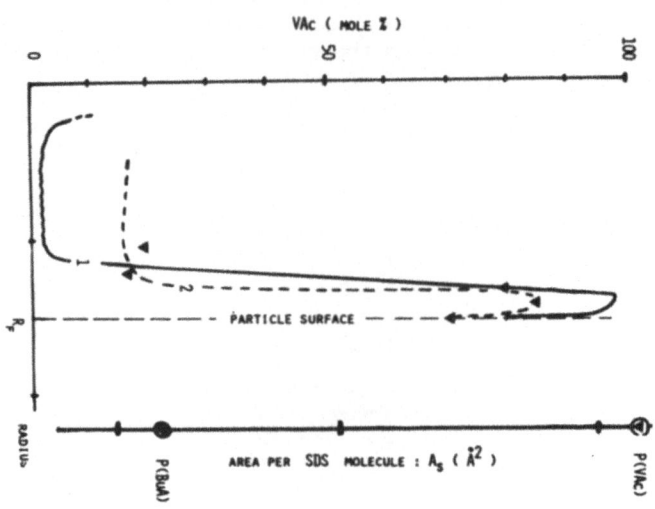

*Figure 5: Butyl acrylate - Vinyl acetate Batch Emulsion Copolymers
(50/50 mol.%).*
*Curve 1: Theoretical composition gradient vs. particle radius (R_f:final
size)*
*Curve 2: Surface area occupied per SDS molecule, measured on samples
withdrawn at various conversions*

The simulation which is based on the simple assumption of an original
monomer gradient is certainly somewhat unrealistic as is the simulation
which is based on the hypothesis of a random walk for the growing
macroradical within the whole particle. There is thermodynamic arguments
against this assumption which would predict something like an anchoring
of the growing macromolecule near the surface and diffusion of monomer
molecules towards the active center. Instead, recent measurements of the
radius of gyration of deuterated polystyrene polymerized onto a normal
polystyrene seed latex show that the deuterated polymer chain occupies
the whole particle[24] which is in agreement, to a large extent, with our
assumption.

In any case, this work must be regarded as a rough approach to be
improved, since many other interactions can play a definite role in the
formation of structured copolymer particles, even during the
copolymerization process. Among them, monomer-emulsifier, monomer-
monomer, monomer-polymer, polymer-polymer compatibilities and diffusion
phenomena, above all, when starved conditions are used to enhance the
designed particle structure (e.g., core-shell, multilayer). It is also
obvious that the nucleation and/or flocculation mechanism must be
considered in a complete simulation.

5. DIFFUSION

When diffusion occurs, the chemical potentials are no longer the same from one point to another. The gradient in μ generates a driving force which makes a given species to move in a determined direction and at a definite rate. The problem then is to know the μ field and the diffusion coefficients at any location. Indeed, the driving force is related to the rate of entropy change, $^\circ S$, and to the transport rate of the species k under consideration, J_k:

$$\text{Driving force: } \Delta X_k = {}^\circ S/J_k \tag{9}$$

The driving force is defined depending on the operating phenomenon, according to following equations,

heat transfer: $\quad \Delta X_k = (-1/T) \cdot (dT/dz)$ (10)

charge transfer: $\quad \Delta X_k = (-1/T) \cdot (d\epsilon/dz)$ (11)

fluid transfer: $\quad \Delta X_k = (-1/T) \cdot (dP/dz)$ (12)

diffusion: $\quad \Delta X_k = (-1/T) \cdot (d\mu^c/dz)$ (13)

From non-equilibrium thermodynamics, the flow of a species k through a surface is proportional to its rate of diffusion and to its local concentration c_k. On the other hand, this rate is proportional to the driving force and to the mobility m_k of the species k

$$J_k \propto m_k c_k \text{ (driving force)} \tag{14}$$

hence, $\quad J_k \propto m_k c_k (-\Delta\mu_k/\Delta z)$

For a surface with an area S the flow of k through the surface is given by:

$$J_k = SD_k c_k (1/\Delta z)(-\Delta\mu_k/RT)$$

where D is the diffusion coefficient ($cm^2 \ s^{-1}$) = $m_k RT$ (15)

At any time and locus, the program can compute for an increment of polymerization dP, the driving force $\Delta\mu/RT = f(\Phi_k, c_{ij}, \gamma, r, \text{charge density})$. The practical difficulty is to determine the diffusion coefficients and their changes with temperature and polymer volume fraction.

It becomes possible to calculate, for example, the maximum monomer diffusion relative to the polymerization of a given amount of monomer:

$$(dM)_{diff}/(dM)_{polym} = SDc(1/\Delta z)(-\Delta\mu/RT)(dt/dP) \tag{16}$$

If several driving forces are simultaneously operating temperature, monomer and copolymer distribution, electrical field), the actual driving force which controls the motion of any species, say a monomer molecule, is the vector sum of all the component forces. Furthermore, some coupling could exist between the various flows, and in this case their coefficients are even more difficult to know.

In emulsion polymerization, the temperature is usually kept constant. The electrical component is operative only at the particle interface and could be described according DLVO theory. Diffusion coefficients can be found in the literature or measured from experiments (permeability studies of copolymer films).

The same simple mathematical treatment also applies to the diffusion of macromolecules within the particle. Without doubt, this treatment presents a useful tool to understand at least the phenomenological changes versus time that lead to the formation of structured latex particles, as well as during the film formation. Even if the macromolecular diffusion coefficients are small, the overall driving force (so called "compatibility") can be high enough to make a chain gather in a separate domain.

6. CONCLUSION

Thus, a theoretical simulation with the primary objective of predicting the particle morphology must start with a sound polymerization mechanism. Structured particles very often result from polymerization of comonomers with different polarities; hence different water solubility, reactivity, and polymer compatibility. Consequently, the reactivity ratios, which determine the microstructure of individual macromolecules, must be known with great accuracy, even though they may appear as "apparent" parameters in the simulation of emulsions. Similarly, with polar monomers the polymerization within the water phase is no longer negligible. Thus, water-phase polymerization must be taken into account, since it generates a fraction of copolymer molecules with microstructure which is different from molecules originated by polymerization within the particles. A further step in the simulation, which is worthy of further development, is the organization of all these polymer molecules within the latex particle. The thermodynamics seems to be a powerful tool for this purpose. Moreover, non-equilibrium thermodynamics is certainly the right approach to introduce the diffusion effect into the simulation. This should lead to predictions of the particle structure as well as the stability of the particle morphology during the polymerization process and aging. The power of this approach resides in the fact that it is possible to determine the driving forces which are acting on molecules, or segments of molecules, at various times or locations.

REFERENCES

1. Daniel JC: *Makromol. Chem. Suppl.* **10/11** 359 (1985).
2. Nomura M; Fujita K: *Makromol. Chem. Suppl.* **10/11** 25 (1985).
3. Omi S; Negishi M; Kushibiki K; Iso M: *Makromol. Chem. Suppl* **10/11** 149 (1985).
4. Brodhead TO; Hamielec AE; MacGregor JF: *Makromol. Chem. Suppl.* **10/11** 105 (1985).
5. Fitch RM; Tsai CH: In *"Polymer Colloids"*, Fitch RM Ed. Plenum Press, New York, 1971.
6. Elisseva VI: *Acta Polymerica* **32** 355 (1981).
7. Hansen FK; Ugelstad J: In *"Emulsion Polymerization"*, Piirma I.(ed) Academic Press, New York, 1982.
8. Feeney PJ; Napper DH; Gilbert RG: *Macromolecules* **17** 2520 (1984).
9. Bovens JL Thesis, University of Liege, Belgium (1986).
10. Guillot J: *Makromol. Chem. Suppl.* **10/11** 235 (1985).
11. Guillot J: To be published in *Makromol. Chem.*
12. Guillot J: *Nouv. J. de Chimie* (in press)
13. Dimonie V; El-Aasser MS; Klein A; Vanderhoff JW: *J. Polymer Sci. Chem. Ed.* **22** 2197 (1984).
14. Djekhaba S; Graillat C; Guillot J: *European Polymer J.* **22(9)** 729 (1986).
15. Bassett DR; Hoy KL: *ACS Symposium Series* **165** 371 (1981).
16. Guillot J; Bonardi C: To be published.
17. Delgado J; El-Aasser MS; Silebi CA; Vanderhoff JW and Guillot J: Submitted for publication, *J. Polymer Sci.*
18. Krigbaum WR; Charpentier DK: *J. Polymer Sci.* **14** 241 (1954).
19. Ugelstad J; Mork PC; Nordhuus I; Mfutakamba H: *Makromol. Chem. Suppl.* **10/11** 215 (1985).
20. Gardon JL: *J. Polymer Sci.* **3** 241 (1973).
21. Guillot J: *Makromol. Chem.* **183** 625 (1982).
22. Guillot J; Pichot C: IUPAC Meeting, Amherst 1982. Preprints p.675.
23. Maron SH; Elder ME; Ulevitch W: *J. Colloid Sci.* **9** 89 (1954).
24. Line MA; Klein A; Sperling LH: *ACS Symposium, New York*, **54** 593 (1986).

MONOMER DISTRIBUTION AND TRANSPORT IN MINIEMULSION COPOLYMERIZATION

Joaquin Delgado, Mohamed S. El-Aasser, Cesar A. Silebi, John W. Vanderhoff and Jean Guillot*

Emulsion Polymers Institute
Departments of Chemical Engineering and Chemistry
Lehigh University
Bethlehem, PA 18015

1. INTRODUCTION

Miniemulsions are oil-in water emulsions prepared using a mixed emulsifier system comprised by an ionic surfactant and a cosurfactant, such as a fatty alcohol or a long chain alkane[1]. The two main characteristics of the miniemulsions are their good stability and droplet size, ranging from 50 to 400 nm in diameter. From the latter characteristic arises the term miniemulsion, to distinguish them from the conventional emulsions or macroemulsions with droplets larger than 1 μm in diameter and from the microemulsions with droplets less than 0.1 μm in diameter.

Ugelstad, El-Aasser and Vanderhoff[2] pioneered the research in minemulsion polymerization when they reported some results on the emulsion polymerization of styrene with a mixed emulsifier system consisting of sodium lauryl sulfate and hexadecanol. They showed that the addition of a small amount of fatty alcohol led to a more efficient emulsification of the monomer, resulting in a more stable emulsion. It was also indicated that the principal locus of polymerization was the monomer droplets.

Several works followed that of Ugelstad, El-Aasser and Vanderhoff, centering on two aspects of the miniemulsions, one on the mechanism of formation and stabilization of the miniemulsions[3-6], the other on their polymerization kinetics [7-9]. First Chamberlain et al.[10] and later Choi et al.[11] presented experimental evidence for the polymerization of styrene miniemulsions with water soluble initiator, indicating that radical entry to styrene droplets, and the subsequent particle formation, was a process of low efficiency and that only a fraction of the original number of droplets was initiated. Moreover, Choi et al. also showed that this was still the case with oil soluble initiators although the fraction of monomer droplets initiated increased with increasing initiator concentration.

Delgado et al.[12] showed that in the case of the miniemulsion copolymerization of vinyl acetate and butyl acrylate, using the mixed emulsifier system of sodium hexadecyl sulfate and hexadecane, the presence of the cosurfactant made the miniemulsion copolymerization process different from the conventional emulsion copolymerization process in many aspects, such as rate of polymerization, final particle size, locus of polymerization, evolution of the copolymer composition and mechanical and physical properties of the copolymers formed.

*Current address: CNRS Laboratoire des Materiaux Organiques, B.P. 24, 69390 Vernaison, France

In an emulsion copolymerization process, particle morphology is key to the application performance of the latex. Particle morphology is known to be determined by the comonomer composition and concentration at the locus of polymerization as well as the reactivity ratios and water solubilities of the monomers. Obviously the distribution of the monomers between the different phases during the course of the polymerization is key to the particle morphology. Thus, knowledge of the transport mechanism and the rate of monomer transport between the various phases are essential for any study in emulsion copolymerization which is aimed at controlling particle morphology. In the miniemulsion copolymerization process, where the presence of the cosurfactant affects so many aspects of the copolymerization, the development of a monomer transport model is very helpful to understand the role of the cosurfactant in the polymerization kinetics.

Monomer transfer during an emulsion polymerization process was first treated by Brooks[13,14] from a diffusional point of view. In his treatment, the diffusion of monomer from the monomer droplets to the aqueous phase was neglected as a rate determining step on the basis that in conventional emulsion systems only a very small proportion of the surfactant is adsorbed onto the droplets to hinder diffusion of the monomer out of the droplets. It was concluded that the only possible resistance to mass transfer, and consequently the only possible rate-determining step, was the resistance to diffusion at the interface between the monomer-swollen polymer particles and the aqueous phase.

This may not be the case for the miniemulsions, where most of the surfactant is adsorbed onto the droplets[12], and sometimes an interfacial complex with viscoelastic properties may be present[5]. Therefore, the resistance to the mass transfer from the monomer droplets to the aqueous phase cannot be neglected without further analysis.

Ugelstad et al.[15] treated the transfer of monomer from the monomer phase to the surface of the latex particles during swelling processes. Smoluchowski's equation[16] was used to relate the monomer flow to the size of droplets and particles. It was assumed that Henry's law could be applied to calculate the concentration of the monomer at the interfaces around monomer droplets and latex particles. Henry's law may be applicable for cases where the monomer has a low water solubility e.g. styrene, but certainly cannot be applied for monomers of high water solubility, e.g., vinyl acetate, acrylonitrile, methyl acrylate or acrylic acid.

Guillot[17] was the first to apply successfully a thermodynamic treatment to obtain the equilibrium monomer partitioning and copolymer composition as a function of conversion during the course of a conventional emulsion copolymerization process. The limitation in applying a basically thermodynamic treatment is that, owing to the nature of the thermodynamics, it is not possible to have a dimensional time frame for the swelling process. This can be overcome by coupling the thermodynamics, which gives the magnitude of the driving force promoting the process, with a dynamic treatment which can supply the time frame.

In this paper a mathematical model is presented to account for the interparticle or interphase monomer transport which occurs during the course of a miniemulsion copolymerization process. The miniemulsion copolymerization of vinyl acetate (VAc) and n-butyl acrylate (BuA) is used as an example. Equilibrium swelling thermodynamics are used to estimate the magnitude of the driving force for the mass transfer process.

2. MATHEMATICAL MODEL FOR THE MONOMER TRANSPORT

In a monomer transport process where simultaneous reaction occurs, if the time required to transport a given amount of monomer to the locus of reaction is greater than the time necessary to consume it, the process becomes mass transport controlled. If the transport and consumption rates are equal, the process will be at steady state conditions and the concentration at the locus of polymerization will be the initial one, as long as monomer continues to be supplied. On the other hand, if the transport rate is faster than the reaction rate, the concentration of monomer in the phase where the polymerization takes place will increase up to a maximum, given by the equilibrium swelling concentration. The faster the transport rate with respect to the reaction rate, the sooner the equilibrium concentration will be reached.

It is postulated that in an agitated system, in the absence of a chemical reaction, the amount of mass of a compound i transported per unit time from a given phase q to an adjacent one r is proportional to the transfer area, A_q, and to the difference between the equilibrium concentration of the material in the phase under consideration, $C_{ei,q}$, and its actual concentration, $C_{i,q}$. This can be expressed as

$$dN_{i,q}/dt = K_{i,q-r}A_q(C_{ei,q}-C_{i,q}) \tag{1}$$

where $dN_{i,q}/dt$ is the molar flow of component i in or out of phase q in mole/sec, $K_{i,q-r}$ is the mass transfer coefficient of i between phase q and r in cm/sec; and the concentrations are given by the number of moles of the component in the phase under consideration divided by the volume of the phase. If the surfactant forms an interfacial complex around the droplets, the mass transfer coefficient for the monomer between the monomer droplets and the continuous phase may be expressed as D_i/δ, where D_i is the diffusion coefficient of i in the interfacial complex layer and δ is the thickness of the mass transfer resistance layer around the monomer droplets.

The volume of the phase under consideration will be given by

$$V_q = \sum_{i=1}^{n} \bar{V}_i N_{i,q} \tag{2}$$

where V_q is the total volume of phase q, $\bar{V}_i$ is the molar volume of i and $N_{i,q}$ is the number of moles of component i in phase q.

For the case where the system is undergoing a polymerization reaction, the material balance for the monomer in any given phase is given by the following equation

$$dN_{i,q}/dt = K_{i,q-r}A_q(C_{ei,q}-C_{i,q})-R_{p(i,q)}V_q \tag{3}$$

where the first term on the right hand side of the equation expresses the convective molar flow of i and the second term expresses the rate of polymerization of monomer i in the phase under consideration in moles/sec.

The polymerization reaction may take place in the monomer-swollen polymer particles and/or in the aqueous phase. When the copolymerization occurs in the monomer-swollen polymer particles the material balance for the monomer present in them will be given by

$$dN_{i,p}/dt = K_{i,a-p}A_p(C_{ei,p}-C_{i,p}) - \sum_{j=1}^{n} k_{p(i,j)}C_{i,p}\bar{n}_jN_p/N_A \qquad (4)$$

where $K_{i,a-p}$ is the mass transfer coefficient for i between the aqueous phase (a) and the monomer-swollen polymer particles (p); $k_{p(i,j)}$ is the propagation rate constant between monomer i and radical j, N_p is the total number of particles, $\bar{n}_j$ is the average number of radicals type j per particle and N_A is Avogadro's number.

If particle nucleation occurs in the monomer droplets, the material balance for the monomer in the polymer particles during this interval will contain a term expressing the change in the number of moles of monomer i in the polymer particles due to the change in status of the monomer in the monomer droplets when the monomer droplets become polymer particles upon entry of a radical. On the other hand, when droplet-particle coalescence occurs, the monomer contained in the monomer droplets becomes part of the polymer particles upon coalescence and this effect has to be also included. The resulting expression will be Equation (4) plus the term $(N_{i,d}/N_d)R_{dp}$, where N_d is the total number of monomer droplets present in the system, $N_{i,d}$ is the number of moles of component i in the monomer droplets and R_{dp} is the rate of change of monomer droplets to polymer particles. R_{dp} is therefore given by

$$R_{dp} = \rho_D N_d + k_1 N_p N_d \qquad (5)$$

where ρ_D is the rate of radical entry into the monomer droplets and k_1 is a coagulation rate constant.

For the case that polymerization occurs in the aqueous phase the material balance for the monomers in the aqueous phase is given by

$$dN_{i,a}/dt = K_{i,d-a}A_d(C_{i,d}-C_{ei,d}) - K_{i,a-p}A_p(C_{ei,p}-C_{i,p}) - \sum_{j=1}^{n} k_{p(i,j)}C_{i,a}[R\cdot]_{j,a}V_a \qquad (6)$$

where the subscript a represents the aqueous phase and d the monomer droplets; $[R\cdot]_{j,a}$ is the concentration of radicals j in the aqueous phase and V_a is the volume of the aqueous phase.

The term $K_{i,d-a}A_d(C_{i,d}-C_{ei,d})$ which expresses the monomer flow out of the droplets is equivalent to the expression $-K_{i,d-a}^*A_d(C_{ei,a}-C_{i,a})$ and either one may be used.

When writing the material balance of the monomers in the monomer droplets for the case where particle generation results from oligoradical entry or capture by the monomer droplets, two processes have to be taken into consideration. The first is the monomer transported due to the convective molar flow which can be expressed by Equation (1). The second is the transfer of monomer when (i) a monomer droplet becomes a polymer particle due to the entry of a radical, and therefore the monomer originally belonging to a monomer droplet belongs now to a polymer particle and when (ii) droplet-particle coalescence occurs. These processes are expressed mathematically by

$$dN_{i,d}/dt = K_{i,d-a}A_d(C_{ei,d}-C_{i,d}) - \frac{N_{i,d}}{N_d} R_{dp} \qquad (7)$$

The change in the number of droplets as the polymerization proceeds is given by the disappearance of droplets due to the particle nucleation and the disappearance of droplets due to coagulation with themselves and the already formed polymer particles:

$$-dN_d/dt = \rho_D N_d + k_2 N_d^2 + k_1 N_p N_d \qquad (8)$$

All these material balances have to be coupled with expressions for the volume of the phases. In general, these volumes can be calculated from Equation (2) if no reaction occurs in the phase. In the case that polymerization occurs the following expression applies

$$V_q = \sum_{i=1}^{n} \overline{V}_i N_{i,q} + \sum_{i=1}^{m} N_{io} X_{i,q} \overline{V}_{pi} \qquad (9)$$

where N_{io} is the initial number of moles of monomer i in the system, $X_{i,q}$ is the fractional conversion of monomer i in phase q and $\overline{V}_{pi}$ is the molar volume of the monomeric units of i in the polymer formed.

2.1 The Equilibrium Concentrations

At this point there is one parameter which is unknown: the equilibrium concentrations of the different species in each one of the phases. This parameter can be determined from the equilibrium swelling thermodynamics.

The equilibrium swelling thermodynamics of polymer particles was initially treated by Morton[18] and subsequently by several authors. Ugelstad et al.[15] studied and treated thermodynamically the effect of adding a low molecular weight water-insoluble compound into the latex particles on the swelling capacity of the latex. Tseng et al.[19] treated thermodynamically the effect of the presence of water in the latex particles on their swelling capacity. Guillot[17] applied a thermodynamic treatment to obtain the comonomer distribution as a function of conversion during the course of a conventional emulsion polymerization process. In this section the thermodynamic equations describing the equilibrium swelling for the miniemulsion copolymerization and the conventional emulsion copolymerization processes are presented. The system is considered to be comprised of three phases: the monomer-swollen polymer particles, also referred as polymer particles, the monomer droplets and the continuous aqueous phase.

The partial molar free energy of mixing of a component i in a given phase q is given by the Flory-Huggins lattice theory of polymer solutions[20], with the addition of an interfacial energy term when the phases are in the form of spheres[18], as expressed by Ugelstad[21]:

$$(\overline{\Delta G}/RT)_{i,q} = \ln\phi_{i,q} + \sum_{j=1}^{n} (1-m_{ij})\phi_{j,q} + \sum_{j=1,j\neq i}^{n} \chi_{i,j}\phi_{j,q}^2 \qquad (10)$$

$$+ \sum_{j=1,j\neq i}^{n-1} \sum_{k=j+1,k\neq i}^{n} \phi_{j,q}\phi_{k,q}(\chi_{ij}+\chi_{ik}+\chi_{jk}m_{ij}) + 2\gamma\overline{V}_i/(r\ R\ T)$$

where χ_{ij} is the Flory-Huggins interaction parameter; $\phi_{i,q}$ is the volume fraction of component i in phase q; m_{ij} is the ratio of the equivalent number of molecular segments between i and j, usually expressed by the ratio of the molar volumes $(\overline{V}_i/\overline{V}_j)$, R is the gas constant, T the temperature, r the radius of the phase and γ the interfacial tension.

The composition of every phase is as follows for the miniemulsion copolymerization process: the monomer-polymer particles are comprised of VAc, BuA, hexadecane (HD) and the copolymer formed; in the monomer droplets, only VAc, BuA and hexadecane are considered to be present; and due to the water solubility of VAc and BuA both of them are considered to be present in the continuous aqueous phase along with water.

The use of the Flory-Huggins lattice theory to express the partial molar free energy of mixing in mixtures involving small molecules as in the case of the monomer droplets and the aqueous phase is open to doubt. Ugelstad[21] proposed the use of the Flory-Huggins equation provided the values used for the interaction parameters (χ_{ij}) and the ratio of the equivalent number of molecular segments (m_{ij}) are those obtained experimentally (χ^*_{ij}, m^*_{ij}). It should be pointed out that under these circumstances m^*_{ij} is not necessarily equal to the ratio of the molar volumes of the two components. Keeping this in mind, the expressions accounting for the partial molar free energy of mixing of the monomers in every phase are:

i) <u>monomer-polymer particles:</u>

$$(\overline{\Delta G}/RT)_{1,p} = \ln\phi_{1,p}+(1-m^*_{12})\phi_{2,p}+(1-m^*_{1H})\phi_{H,p}+(1-m_{1P})\phi_{P,p}$$

$$+\chi^*_{12}\phi^2_{2,p}+\chi^*_{1H}\phi^2_{H,p}+\chi_{1P}\phi^2_{P,p}$$

$$+\phi_{2,p}\phi_{H,p}(\chi^*_{12}+\chi^*_{1H}-\chi^*_{2H}m^*_{12})+\phi_{2,p}\phi_{P,p}(\chi^*_{12}+\chi_{1P}-\chi_{2P}m^*_{12})$$

$$+\phi_{H,p}\phi_{P,p}(\chi^*_{1H}+\chi_{1P}-\chi_{HP}m^*_{12})+2\gamma_p\overline{V}_1/(r_pR\ T) \qquad (11)$$

where 1 and 2 refer to VAc and BuA respectively; H stands for hexadecane, P for the copolymer and p for the monomer-polymer particles. Equation (11) is for monomer 1 (VAc), and a similar expression is used for monomer 2 (BuA).

ii) <u>monomer droplets:</u>

$$(\overline{\Delta G}/RT)_{1,d} = \ln\phi_{1,d}+(1-m^*_{12})\phi_{2,d}+(1-m^*_{1H})\phi_{H,d} \qquad (12)$$

$$+\chi^*_{12}\phi^2_{2,d}+\chi^*_{1H}\phi^2_{H,d}$$

$$+\phi_{2,d}\phi_{H,d}(\chi^*_{12}+\chi^*_{1H}-\chi^*_{2H}m^*_{12})+2\gamma_d\overline{V}_1/(r_dR\ T)$$

where d stands for monomer droplets and γ_d is the interfacial tension between the monomer droplets and the aqueous phase. Equation (12) is for monomer 1 (VAc) and a similar expression is used for monomer 2 (BuA).

iii) <u>continuous aqueous phase</u>:

$$(\overline{\Delta G}/RT)_{1,a} = \ln\phi_{1,a}+(1-m_{12}^*)\phi_{2,a}+(1-m_{1w}^*)\phi_{w,a}$$
$$+\chi_{12}^*\phi_{2,a}^2+\chi_{1w}^*\phi_{w,a}^2+\phi_{2,a}\phi_{w,a}(\chi_{12}^*+\chi_{1w}^*-\chi_{2w}^*m_{12}^*)$$

(13)

where a stands for the continuous aqueous phase and w for water.
Equation (13) is for monomer 1 (VAc) and a similar expression is used for
monomer 2 (BuA).

Similar to the approach used by Guillot[17], a mean of the homopolymer
interaction parameters was used as the copolymer interaction parameter.
The interaction parameters between the monomers, hexadecane and the
copolymers, χ_{iPi} were either obtained from the solubility parameters or
calculated from monomer partitioning studies. The reader is referred to
reference (22) for the details of the calculations. Table 1 shows the
values of the different parameters used in the simulations.

TABLE 1. Values for the interaction parameters and segment volume ratios
used in the simulations

χ_{1PA}	0.37	χ_{2H}^*	0.19
χ_{1PB}	0.36	χ_{1H}^*	0.53
χ_{2PA}	0.46	χ_{1w}^*	3.20
χ_{2PB}	0.35	χ_{2w}^*	6.43
χ_{HPA}	1.22	χ_{12}^*	0.15
χ_{HPB}	0.63	m_{12}^*	0.33
χ_{21}^*	χ_{12}^*/m_{21}^*	m_{1H}^*	$\overline{V}_1/\overline{V}_H$
m_{21}^*	$1/m_{12}^*$	m_{2H}^*	$\overline{V}_2/\overline{V}_H$

The formation of the polymer particles from monomer droplets and the
subsequent polymerization drives the system to a non-equilibrium
condition, causing a monomer flow between the phases to reestablish
equilibrium. The equilibrium condition is attained when the partial
molar free energies of mixing of the monomers in the different phases are
the same. Therefore, the equilibrium condition can be expressed by a
system of non-linear equations where thermodynamic and material balances
are included. The following expressions describe the system of non-
linear equations:

Equilibrium between monomer-polymer particles and the aqueous phase

$$(\overline{\Delta G}/RT)_{1,p} - (\overline{\Delta G}/RT)_{1,a} = 0 \tag{14}$$

$$(\overline{\Delta G}/RT)_{2,p} - (\overline{\Delta G}/RT)_{2,a} = 0 \tag{15}$$

Equilibrium between the aqueous phase and the monomer droplets

$$(\overline{\Delta G}/RT)_{1,a} - (\overline{\Delta G}/RT)_{1,d} = 0 \tag{16}$$

$$(\overline{\Delta G}/RT)_{2,a} - (\overline{\Delta G}/RT)_{2,d} = 0 \tag{17}$$

Material balances for the phases:

Material balance for the monomer-polymer particles

$$\phi_{1,p} + \phi_{2,p} + \phi_{H,p} + \phi_{P,p} = 1 \tag{18}$$

Material balance for the monomer droplets

$$\phi_{1,d} + \phi_{2,d} + \phi_{H,d} = 1 \tag{19}$$

Material balance for the aqueous phase

$$\phi_{1,a} + \phi_{2,a} + \phi_{w,a} = 1 \tag{20}$$

Material balances for the components:

Material balance for VAc

$$A_o \overline{V}_1 (1-X_1) = \phi_{1,p} V_p + \phi_{1,d} V_d + \phi_{1,a} V_a \tag{21}$$

Material balance for BuA

$$B_o \overline{V}_2 (1-X_2) = \phi_{2,p} V_p + \phi_{2,d} V_d + \phi_{2,a} V_a \tag{22}$$

Material balance for the copolymer

$$A_o \overline{V}_{PA} X_1 + B_o \overline{V}_{PB} X_2 = \phi_{P,p} V_p \tag{23}$$

Material balance for water

$$W_o \overline{V}_w = \phi_{w,a} V_a \tag{24}$$

Material balance for hexadecane

$$z \quad H_o \overline{V}_H = \phi_{H,p} V_p \tag{25}$$

$$(1-z) \quad H_o \overline{V}_H = \phi_{H,d} V_d \tag{26}$$

where A_o, B_o, H_o, and W_o are the initial number of moles of VAc, BuA, hexadecane and water, respectively. X_1 and X_2 are the molar conversions of VAc and BuA respectively and z is the fraction of droplets converted to polymer particles (N_p/N_{do}).

The solution to this set of equations (Equations (14) to (26)) will provide the volume fraction of every component in the three phases and the volume of the phases at equilibrium as a function of conversion, the number of particles and the number of monomer droplets. The equilibrium concentrations will be expressed as:

$$C_{ei,q} = \phi_{i,q}/\overline{V}_i \tag{27}$$

The expressions accounting for the partial molar free energy of mixing of the monomers in the different phases for the conventional emulsion copolymerization of VAc and BuA are similar to those for the miniemulsion process, with the exception that hexadecane is not considered and therefore, the volume fractions of hexadecane in the monomer-polymer particles and in the monomer droplets are zero. For the special case where hexadecane is present in the monomer droplets but not in the polymer particles, the expressions for the partial molar free energy of mixing of the monomers in the three phases are similar to those for the miniemulsion process, only now hexadecane is not in the polymer particles and therefore all the terms containing hexadecane are omitted from the expression.

2.2 Integration Algorithm

In order to solve in an efficient way the system of mathematical equations composed by the differential equations expressing the particle generation, monomer transfer and polymerization and the system of non-linear equations expressing the equilibrium states, the physical events occurring during the copolymerization have been split into a succession of elementary events which can be treated as mathematically independent. Figure 1 shows the schematic representation of the physical and mathematical events considered in the development of the integration algorithm.

INTEGRATION ALGORITHM

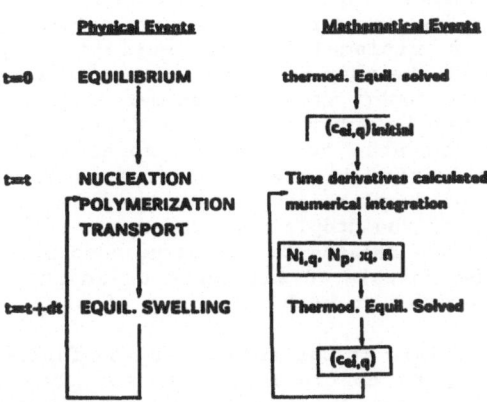

Figure 1: Integration algorithm.

Initially it is considered that before the initiator is added, the system, comprised of monomer droplets and the continuous aqueous phase, is at equilibrium. The concentration of the components in each of the phases is calculated by solving a set of non-linear equations describing the thermodynamic equilibrium between monomer droplets and the aqueous phase. Once the initiator is added and the polymerization starts, the process is divided in a sequence of small time intervals (dt) during which particle nucleation, polymerization and monomer transfer occur simultaneously. At the beginning of each of the time intervals the equilibrium swelling concentrations of the monomers in the three phases (monomer droplets, polymer particles and aqueous phase) are calculated by solving the thermodynamic system of non-linear equations (Equations (14) to (26)) using the conditions of the system at this point (N_p, N_d, X_1 and X_2). The equilibrium concentrations are then used as input to estimate the values of the time derivatives of the equations describing the monomer transport and polymerization. These values are then used in the numerical integration[23]. All simulations were performed using a Cyber 850 mainframe computer.

3. **EFFECT OF THE MASS TRANSFER COEFFICIENTS AND TRANSFER AREA ON THE RATE OF POLYMERIZATION**

When the mass transfer process is multiphase and can be segregated into a series of steps, the slowest step will control the overall mass transfer process. In the case of an emulsion polymerization where the polymerization takes place in polymer particles, the monomer consumed in the polymer particles is replaced with monomer present in the aqueous phase and this monomer is replaced by monomer transported from the monomer droplets. The transport process may thus be divided into three steps: (i) transport from the aqueous phase to the polymer particles, (ii) transport through the aqueous phase and (iii) transport from the monomer droplets to the aqueous phase. The second step usually can be neglected as a rate determining step because of the existence of agitation, which results in a convective mass transfer process through the aqueous phase instead of a diffusive one and the relatively high water solubility of the monomers (VAc and BuA). Thus, either monomer transport out of the monomer droplets or entrance of the monomers to the polymer particles has to be considered as the rate determining step.

In the case of a miniemulsion polymerization process it is reasonable to assume that the most probable monomer transport controlling step is the exit of the monomers from the monomer droplets to the aqueous phase. Two reasons may be advanced. The first is the continuous decrease in the transfer area between the monomer droplets and the aqueous phase. The reduction in the transfer area is the result of a decrease in the number of droplets as they become polymer particles and a reduction in the radius of the droplets as the monomer is transported out of them. The second reason relates to the large adsorption of surfactant on the droplets and the postulated existence of an interfacial complex which might act as a barrier to both the entry of oligoradicals and the exit of monomers.

Based on this reasoning, it was assumed in the following simulations that the monomer transport controlling step was the transfer of the monomers from the monomer droplets to the aqueous phase. Under these circumstances, the transfer of the monomer from the aqueous phase to the polymer particles and vice versa is fast enough such that the aqueous phase and the polymer particles establish a pseudo-equilibrium between them. Thus, the monomer transported out of the monomer droplets partitions between the aqueous phase and the polymer particles with a partition coefficient given by $C_{ei,a}/C_{ei,p}$ which is calculated from the equilibrium swelling thermodynamics.

The simulations were carried out for the miniemulsion copolymerization of 50:50 molar ratio VAc-BuA monomer mixture with an initiator concentration of 2.2 mM. The monomer droplets were assumed to be 100 nm in radius and a coefficient for radical entry into the monomer droplets of 3.3×10^{-4} sec^{-1} was used. The value of the mass transfer coefficient from the droplets to the aqueous phase was varied and its effect on the kinetics was analyzed.

Figure 2 shows the conversion-time curves obtained varying the mass transfer coefficient assuming that it was the same for the two monomers. For values of the mass transfer coefficients greater than 10^{-4} cm/sec the time conversion curves are similar to those obtained assuming that the polymer particles are at any moment at equilibrium swelling condition. This means that under these conditions the rate at which the monomer is transported out of the monomer droplets is much faster than the rate at which the monomer is consumed, and therefore all the phases are at equilibrium swelling condition.

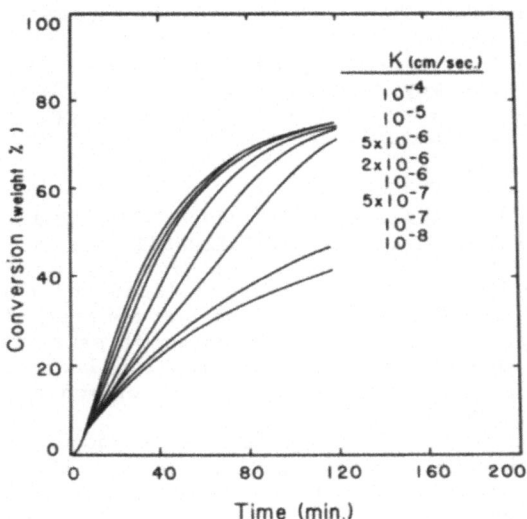

Figure 2: Effect of the value of the mass transfer coefficient on the rate of polymerization.

For values of the mass transfer coefficient between 10^{-4} to 10^{-8} cm/sec the polymerization becomes gradually more monomer transport-controlled and as the value of the mass transfer coefficient is decreased the rate of polymerization slows down. Finally, for values smaller than 10^{-8} cm/sec, the monomer is transported so slowly to the polymer particles that the copolymerization becomes controlled by the rate of radical entry to the monomer droplets and the only polymerization process taking place is the consumption of the monomer contained in the droplets once initiated.

When the mass transfer coefficient is different for each monomer, the monomer with lower mass transfer coefficient controls the overall mass transfer process to some extent and the rate of polymerization lies between those obtained with the lower and higher mass transfer coefficient.

It is important to realize that, even though in this discussion all the effects have been related to the mass transfer coefficient, the controlling variable is really the product of the overall mass transfer coefficient and the transfer area. Therefore, the stated values for the mass transfer coefficient will have to be corrected if different numbers and sizes of monomer droplets are assumed or determined. Also, a homogeneous monomer composition in the droplets has been assumed in the simulation because that was the starting point in the copolymerizations carried out. This presupposes that the system is at equilibrium condition before the polymerization is started. An interesting effect to consider theoretically and experimentally would be the case where miniemulsions of the individual monomers are mixed together and the polymerization started after different mixing periods. In this case the polymerization starts from a non-equilibrium situation and the monomer transport would play a more important role in the kinetics of the process and copolymer properties. Preliminary experiments carried out on this subject[24] with butyl acrylate, styrene and 2-ethyl-hexyl acrylate have shown differences in the appearance and properties of films formed from latexes prepared in the way described previously. The importance in obtaining experimental values for the mass transfer coefficients is obvious, and efforts are currently being made for their determination[25].

4. THE DISTRIBUTION OF THE MONOMERS
4.1 Differences Between the Miniemulsion Copolymerization Process and the Conventional Emulsion Copolymerization Process

The distribution of the monomers between the monomer droplets, aqueous phase and polymer particles was simulated for the miniemulsion copolymerization of 50:50 molar ratio VAc-BuA monomer mixture and the conventional emulsion copolymerization process of same monomer mixture. In the simulations instantaneous equilibrium swelling conditions were assumed and the simulations show an upper limit of monomer concentration in the polymer particles.

Monodisperse monomer droplets with 100 nm in radius were used in the simulation of the miniemulsion copolymerization process. Polymerization was assumed to be initiated in 20% of the droplets, thereby becoming polymer particles. These droplets are referred to as "active droplets". The remaining monomer droplets act as monomer reservoirs supplying monomer to the polymer particles. The concentration of hexadecane used in the simulations is 8 mM corresponding to that used in most of the experimental kinetic runs. For the simulation of the conventional emulsion copolymerization process it was supposed that the number of particles present during the copolymerization process was constant and equal to the final number of particles determined experimentally.

Figures 3 to 5 show the distribution of the monomers between phases expressed as the percentage of the initial number of moles of the monomers in any given phase ($100\ N_{i,q}/N_{io}$) versus the overall conversion. In Figures 3 and 4 the distribution of the monomers between monomer droplets and polymer particles for the miniemulsion process and the conventional emulsion polymerization process is respectively shown.

Figure 5 shows the distribution of the monomer in the aqueous phase for the two processes. It is observed in Figures 3 to 5 that the distributions profiles of VAc and BuA in each phase is different from each other. The main reason for this different behavior is the higher reactivity of BuA which results in a faster consumption rate for this monomer. Also differences in the distribution of the monomers between the miniemulsion copolymerization process and the conventional one are noted.

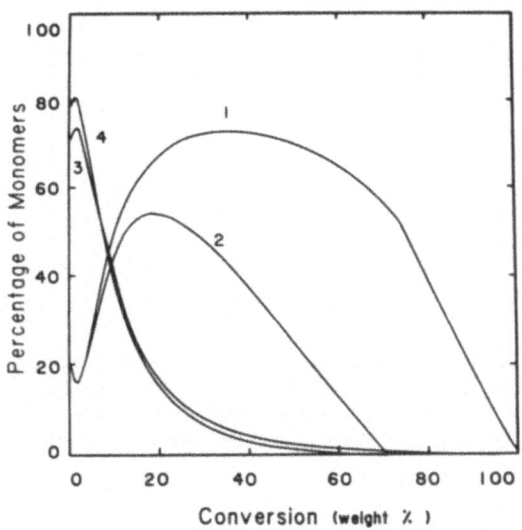

Figure 3: Distribution of the monomers between monomer droplets and polymer particles for the miniemulsion copolymerization of 50:50 molar ratio VAc-BuA. (1) VAc in polymer particles; (2) BuA in polymer particles; (3) VAc in monomer droplets; (4) BuA in monomer droplets.

The first difference observed is the disappearance of monomer droplets in the conventional process at around 25% conversion, while in the miniemulsion process, monomer droplets remain until the end of the reaction. The disappearance of the monomer droplets in the conventional process results from the transfer of the monomers from the monomer droplets to the locus of polymerization, the polymer particles. The presence of hexadecane in the monomer droplets in the miniemulsion process reduces the free energy of mixing of the monomers in the monomer droplets. Therefore, the difference in free energy of mixing between monomer droplets and polymer particles in the miniemulsion copolymerization case is less than in the conventional case. The result is a smaller flow of monomer from the monomer droplets to the polymer particles during the miniemulsion copolymerization process. Physically, the effect is that hexadecane cannot be transported from the monomer droplets to the polymer particles because of its extremely low water solubility. As a result, the concentration of hexadecane in the monomer droplets is greater than in the polymer particles and monomer is retained in the monomer droplets to minimize the hexadecane concentration gradient. It should be pointed out that this does not mean that the

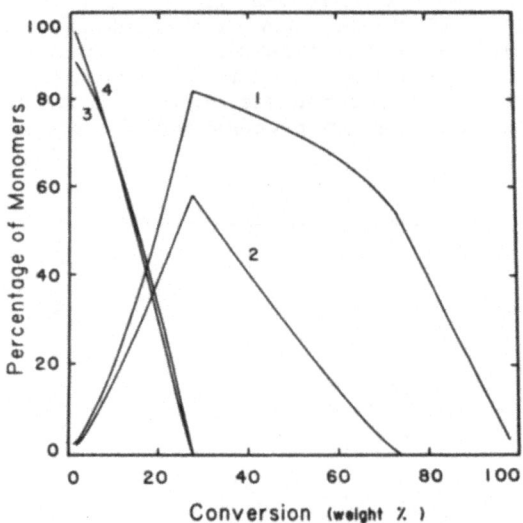

Figure 4: Distribution of the monomers between monomer droplets and polymer particles for the conventional emulsion copolymerization of 50:50 molar ratio Vac-BuA. (1) VAc in polymer particles; (2) BuA in polymer particles; (3) VAc in monomer droplets; (4) BuA in monomer droplets.

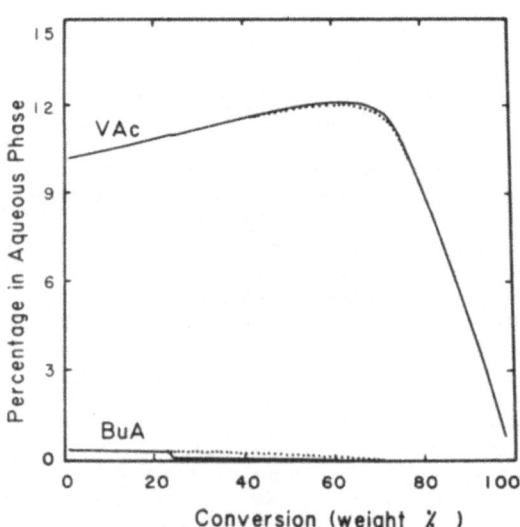

Figure 5: Distribution of VAc and BuA in the aqueous phase. Continuous line: conventional process; dotted line: miniemulsion process.

polymerization is monomer-transport controlled, as the transport of the monomer is still faster than the polymerization reaction, but the presence of hexadecane in the monomer droplets lowers the magnitude of the monomer flow between monomer droplets and polymers particles. A corollary of this conclusion is that, theoretically the monomer droplets will not disappear as a result of a diffusional mass transfer process when hexadecane is present in them.

The disappearance of monomer droplets in the conventional emulsion copolymerization process means that from this point on (25% conversion), the volume fraction of monomer in the conventional process polymer particles is higher than that for the particles of the miniemulsion process. Conversely, the volume fraction of the copolymer in the polymer particles is higher for the miniemulsion process. Hence, any type of transfer reaction to the polymer chain (causing branching and crosslinking) will be favored in the miniemulsion copolymerization process. This fact could explain the insolubility of some of the copolymers prepared in the miniemulsion process [12]

The prediction of the model for the total percentage of monomer (VAc + BuA) in the form of monomer droplets during the conventional emulsion copolymerization of 50:50 molar ratio of VAc and BuA is compared in Figure 6 (solid line) with experimental results obtained from three different kinetic runs. The experimental results confirm the disappearance of the monomer droplets as a result of a diffusional monomer transfer from monomer droplets to polymer particles in accordance with the mathematical model.

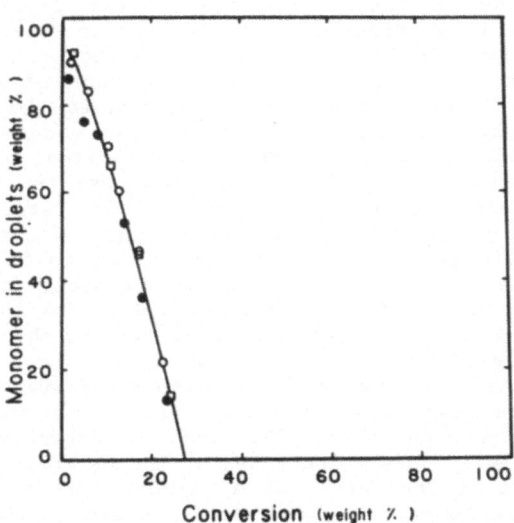

Figure 6: Amount of the monomers in the monomer droplets for the conventional emulsion copolymerization of 50:50 molar ratio VAc-BuA. Points, experimental results; continuous line, prediction of the model.

A close inspection of Figure 3 reveals the existence of a minimum in the amount of both monomers in the polymer particles (a maximum for the monomer droplets) at low conversions during the copolymerization of miniemulsions. Even though hexadecane is present in the droplets entered by radicals, the monomer/polymer ratio is larger than that allowed thermodynamically and monomer is lost to the uninitiated monomer droplets. It will be shown later that when the amount of hexadecane is increased, the minimum in the distribution curve of the monomers in the particles disappears as the swellability of the active droplets is increased. This effect is not observable in the conventional process because particles are formed either in surfactant micelles or more likely by a homogeneous nucleation mechanism.

The distribution of the monomers in the aqueous phase in Figure 5 shows that more VAc is present in the aqueous phase than BuA, reflecting accurately the experimentally determined fact. It is also shown that, for the particular hexadecane concentration used in the simulations (8 mM), there is almost no difference between the two processes in terms of the distribution of the monomers in the aqueous phase.

4.2 Effect of the Hexadecane Concentration

Several simulations were carried out in which the hexadecane concentration was varied from 8 mM to 80 mM in an attempt to quantify the effect of the hexadecane on the distribution of the monomers during the miniemulsion copolymerization of 50:50 VAc-BuA. For clarity, the percentage of the initial amount of both monomers in the monomer droplets and polymer particles, defined as $100x(N_{1,q}+N_{2,q})/(N_{1,0}+N_{2,0})$, is shown in Figure 7 and the percentage of the initial amount of VAc and BuA in the aqueous phase is shown in Figure 8. An interesting effect is observed at the beginning of the polymerization when the monomer droplets are entered by the oligoradicals. It is seen in Figure 7 that the minimum at low conversions in the curve representing the amount of monomer in the polymer particles is reduced as the concentration of hexadecane is increased. Upon entry of oligoradicals in the monomer droplets and formation of copolymer in them, if the monomer/polymer ratio in them is greater than that dictated by the equilibrium swelling thermodynamics, monomer is transferred to non-initiated monomer droplets. As the amount of hexadecane is increased in the monomer droplets their equilibrium monomer/polymer ratio is increased and the amount to be transported to non-initiated droplets is reduced. When the hexadecane concentration is high enough all the monomer initially contained in the initiated monomer droplets is retained. This effect indicates that in the early stages of the polymerization hexadecane acts as a swelling promoter.

Later on however, it can also be seen in Figure 7 that as the concentration of hexadecane is increased the total amount of monomer (VAc+BuA) in the polymer particles is reduced. Conversely, the amount of monomer in the monomer droplets increases as the hexadecane concentration is increased. This is the result of a reduction in the difference in free energy of mixing between monomer droplets and polymer particles as the hexadecane concentration is increased, and is manifested by a reduction in the equilibrium swelling concentration of the monomers in the polymer particles. The hexadecane can be considered, for most of the reaction, as a "monomer retaining agent" in the monomer droplets. This effect is similar to that found by Ugelstad et al.[26] and Azad et al.[27] for the swelling of polymer particles with organic phases containing a low molecular weight water-insoluble compound.

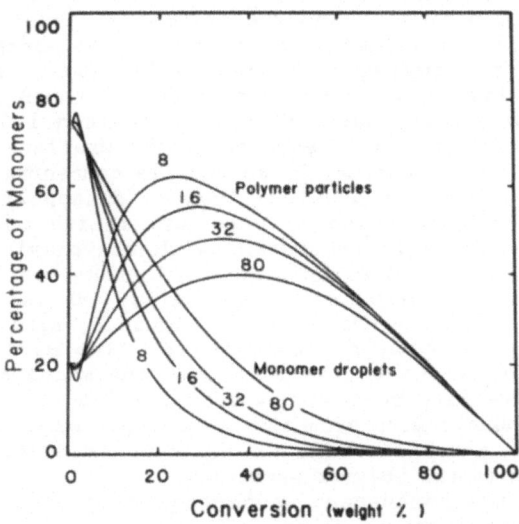

Figure 7: Simulation of the total amount of monomer in polymer particles and monomer droplets for the miniemulsion copolymerization of 50:50 molar ratio VAc-BuA. Numbers in the curves indicate the concentration of hexadecane in mM.

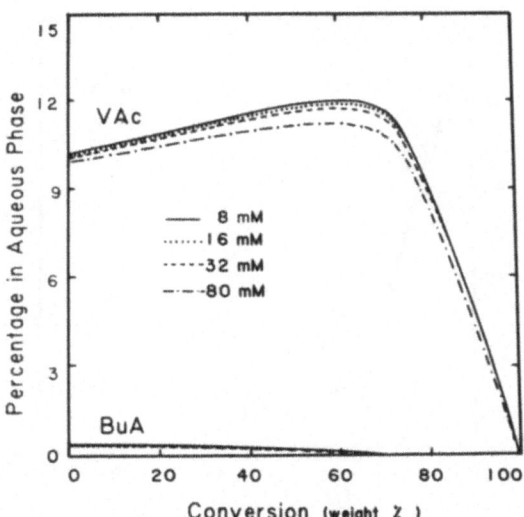

Figure 8: Simulation of the amount of VAc and BuA in the aqueous phase during the miniemulsion copolymerization of 50:50 molar ratio VAc-BuA, for different hexadecane concentrations.

Three polymerizations were carried out at 60°C where hexadecane in various amounts was added to the monomer phase of 50:50 molar ratio VAc-BuA conventional emulsions. The surfactant, sodium hexadecyl sulfate, was in a concentration of 1 mM and the initiator, ammonium persulfate, was in a concentration of 1.1 mM. The amount of the monomers as monomer droplets was monitored during the copolymerization. The results are shown in Figure 9. Run A contained no hexadecane, run B contained 0.54% hexadecane in the monomer phase and run C contained 1.63% hexadecane. Figure 9 shows that the amount of monomer in the droplets increases when hexadecane is added to the monomer phase and its concentration increased. Also, the point at which the droplets seem to disappear is shifted to higher conversion as hexadecane is used and as its concentration is increased. This behavior is consistent with the previously discussed "monomer retaining effect" of hexadecane. This behavior is predicted by the mathematical model when the condition $\phi_{H,p} = 0$ is imposed. The predictions, shown as continuous lines in Figure 9, agree well in trend and magnitude with the experimental results. The deviations at high conversion are probably due to the fact that the model only takes into consideration a convective monomer transport, while at high conversions the monomers are probably more effectively transported through collisions as a result of the destabilization of the monomer droplets due to the surfactant migration to the polymer particles.

The effect of the hexadecane concentration on the amounts of the monomers in the aqueous phase, shown in Figure 8, is a reduction in the water solubility of both monomers as the hexadecane concentration is increased. This prediction was substantiated experimentally and is shown in Table 2. The water solubility of VAc is reduced as the amount of hexadecane added to the monomer is increased. However, the reduction in the solubilities of the monomers should be very small for the usual smaller amounts of cosurfactant employed in the miniemulsion processes.

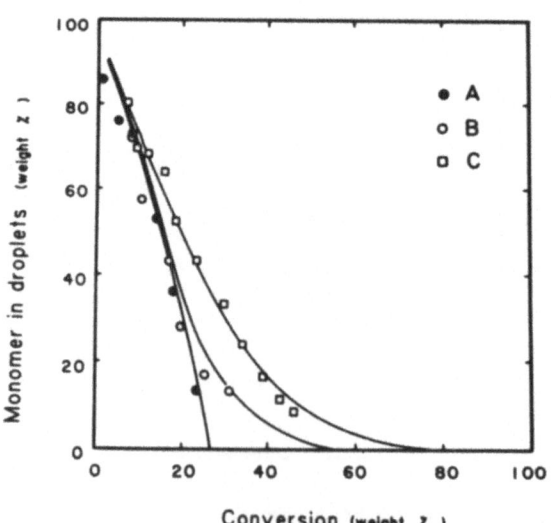

Figure 9: *Total amount of the monomers in the monomer droplets for experiments A (no HD), B (8 mM HD) and C (24 mM HD). Points, experimental results. Continuous lines, predictions of the model.*

Table 2   Water-solubility of VAc in the Presence of Various Amounts of
Hexadecane Determined by Gas Chromatography

$\phi_{H,d}$	S_{VAc} (g. VAc/100 g. H_2O)
0.000	2.25
0.305	2.19
0.580	2.05
0.791	1.79

4.3 Effect of the Size of the Monomer Droplets

The effect of the initial size of the monomer droplets on the
monomer distribution between various phases was also simulated. Figure
10 shows three simulations where the initial size of the monomer droplets
acting as reservoirs was varied from 100 to 500 nm in radius, while the
initial radius of the active droplets (initiated ones) was kept constant
to 100 nm in radius. This would be the case of the polymerization of a
miniemulsion with bimodal distribution.

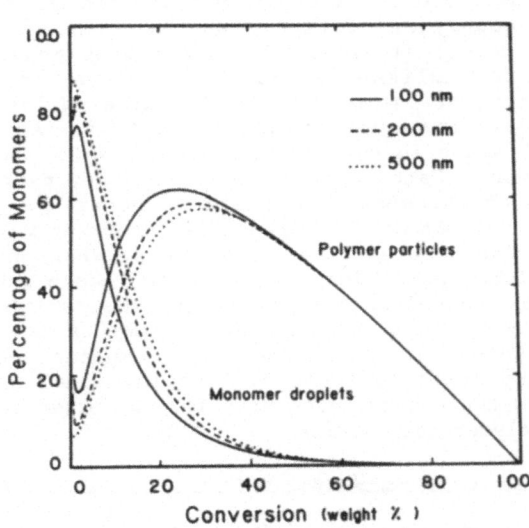

Figure 10:   Simulation of the total amount of monomer in the polymer
particles and monomer droplets for the miniemulsion copolymerization of
50:50 molar ratio VAc-BuA. Radius of the reservoir droplets: 100 nm
continuous line; 200 nm broken line; 500 nm dotted line.

The simulations show that as the initial radius of the monomer droplets acting as monomer reservoirs is increased, the total amount of monomer in the polymer particles is reduced. This is the result of a reduction in the difference in free energy of mixing between monomer droplets and polymer particles as the value of the interfacial energy term of the monomer droplets decreases due to the increase in the radius of the monomer droplets. These results confirm the earlier finding by Poehlein et al.[28] where the swelling ratio of polystyrene latexes was increased as the size of the monomer droplets of styrene emulsions, used to swell the latexes, was reduced.

5. THE EVOLUTION OF THE COPOLYMER COMPOSITION
5.1 Effect of the Value of the Mass Transfer Coefficient
The composition of the copolymer formed during the copolymerization is quite insensitive to the value of the mass transfer coefficient for values greater than $5x10^{-5}$ cm/sec. For lower values of the mass transfer coefficient, the monomer transfer rate is greatly reduced. As discussed previously, the copolymerization becomes controlled by the entry of radicals to the monomer droplets for these low values of the mass transfer coefficients and the monomers in the initiated droplets are consumed at a rate faster than they are incorporated. The result is that for same overall conversions, the copolymer becomes richer in VAc monomeric units as the mass transfer coefficient is reduced.

5.2 Differences in Copolymer Composition Between Processes
Figure 11 compares the prediction of the model (solid lines) for the evolution of the copolymer composition during the miniemulsion and conventional emulsion copolymerization of 50:50 molar ratio of VAc and BuA, with the respective experimental results. The agreement is good even though the model predicts a slightly lower content in VAc monomeric units for the copolymer formed in the conventional emulsion copolymerization process at intermediate and high conversions. These deviations arise from slight imprecisions in the prediction of the relative partitioning of the monomers. A source of imprecision may be the use of not too accurate interaction parameters. Another source may be the presence of water in the organic phases, which was not accounted for in this treatment. Water present in the polymer particles may favor the incorporation of the more hydrophilic monomer[19], i.e., VAc in the present case. This effect may be especially important in the case of the conventional emulsion copolymerization process, where no hexadecane is present in the polymer particles, which could reduce substantially the amount of water and therefore the favorable incorporation of VAc. The presence of water could thus increase the content of VAc monomeric units in the copolymer formed in the conventional emulsion copolymerization process. It is important to indicate that the model predicts, to the extent of the assumptions associated with it, the experimentally observed differences in copolymer composition.

5.3 Effect of the Monomer/Water Ratio on the Composition of the Copolymer
Figure 12 shows the results of four simulations which were carried out to determine the effect of the monomer/water ratio on the composition of the copolymer formed during the miniemulsion copolymerization of 50:50 molar ratio VAc-BuA monomer mixture. The monomer/water ratios have been converted to percent solids content of the latex once polymerized. As the solid content of the latex decreased from 50% to 5% (decreased

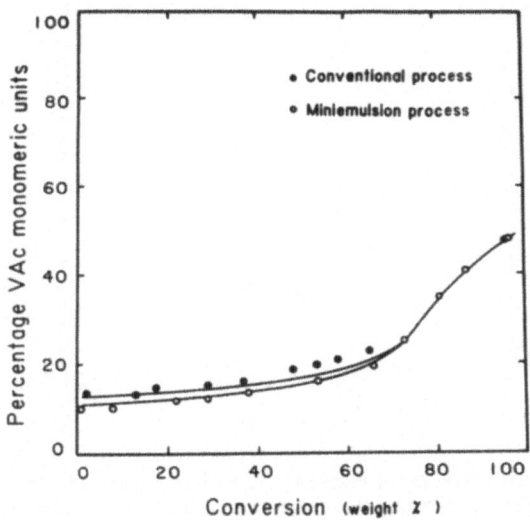

Figure 11: Evolution of the copolymer composition for the miniemulsion and conventional emulsion copolymerization of 50:50 molar ratio VAc-BuA. Points, experimental results. Continuous lines, predictions of the model.

monomer/water ratio) the amount of VAc incorporated to the polymer, before BuA was totally consumed, decreased. This effect is the result of the preferred solubilization in water of VAc over BuA as the amount of water present during the polymerization is increased. Guillot[29,30] reported a similar effect for the conventional emulsion copolymerization of styrene and acrylonitrile. As the monomer/water ratio was decreased the copolymer became richer in the less water soluble monomer (styrene). The effect of the monomer/water ratio in the styrene-acrylonitrile system was more pronounced owing to the higher water solubility of acrylonitrile relative to VAc.

6. **SUMMARY**
    An interparticle monomer transfer model has been presented which can be applied to determine the rate of monomer transport and monomer distribution during the copolymerization of conventional emulsions and miniemulsions. Equilibrium swelling thermodynamics were used in the model to calculate the magnitude of the driving force for the mass transfer to occur. Conditions have been shown for the copolymerization process to become monomer transport controlled.
    The mathematical model showed that a different distribution of the comonomers between phases is expected depending on the process used to carry out the copolymerization. It predicted for the conventional emulsion copolymerization process the disappearance of the monomer droplets, as found experimentally, while it showed that when hexadecane is present in the monomer droplets, they will not disappear as a result of a convective monomer transfer process. Moreover, it showed that

100

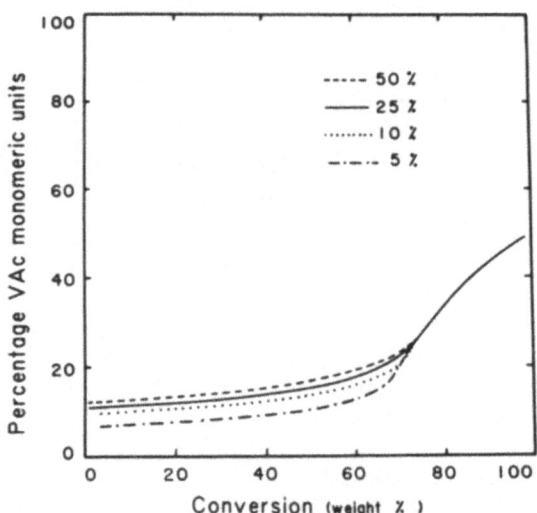

Figure 12: Effect of the polymerizable solid content on the composition of the copolymer formed during the miniemulsion copolymerization of 50:50 molar ratio VAc-BuA monomer mixture.

hexadecane plays two roles during the miniemulsion copolymerization process. Initially, it increases the capability of the monomer droplets to retain monomer once polymerization is initiated in the monomer droplets. On the other hand, once polymer particles are formed, the presence of hexadecane in uninitiated monomer droplets reduces the equilibrium concentration of the monomers in the polymer particles.

On the effect of the size of the monomer droplets on the comonomer distribution the model confirmed the experimental and theoretical finding of Poehlein et al.[28] that the dispersion of monomers in small droplets increases the swellability of polymer latexes.

7.   ACKNOWLEDGEMENT
    The financial support by the National Science Foundation (Grant No. CBT-8513542) and the partial support to J. Delgado by the Basque Government is gratefully appreciated.

NOTATIONS

A_q — area of phase q, cm^2

A_o — initial number of moles of VAc

B_o — initial number of moles of BuA

$C_{ei,q}$ — equilibrium concentration of component i in phase q, $moles/cm^3$

$C_{i,q}$ — concentration of component i in phase q, $moles/cm^3$

D_i — diffusion coefficient of component i, cm^2/sec

H_o — initial number of moles of hexadecane

k_{pi} — propagation rate constant of monomer i, $cm^3/(mole-sec)$

k_1, k_2 — coagulation rate constants, sec^{-1}

$K_{i,q-r}$ — mass transfer coefficient of component i between phases q and r, cm/sec

m_{ij} — ratio of the equivalent number of molecular segments between i and j given by $\bar{V}_i / \bar{V}_j$

m_{ij}^* — experimentally determined value for m_{ij}

$\bar{n}_j$ — average number of radicals j per particle

N_A — Avogadro's number

$N_{i,q}$ — number of moles of component i in phase q

N_{io} — initial number of moles of component i

N_p — total number of particles

N_d — total number of monomer droplets

N_{do} — initial number of monomer droplets

r_d — radius of the monomer droplets, cm

r_p — radius of the polymer particles, cm

$[R\cdot]_{j,a}$ — concentration of radicals j in the aqueous phase, $moles/cm^3$

R_{dp} — rate of change of monomer droplets to polymer particles, sec^{-1}

$R_{p(i,q)}$ — rate of polymerization of monomer i in phase q, $moles/(sec\ cm^3)$

T — temperature ($^\circ K$)

$\overline{V}_i$ — molar volume of i, $cm^3/mole$

V_q — volume of phase q, cm^3

W_o — initial number of moles of water

X_i — fractional conversion of monomer i

$X_{i,q}$ — fractional conversion of monomer i in phase q

γ_d — interfacial tension of the monomer droplets, dyne/cm

γ_p — interfacial tension of the polymer particles dyne/cm

$\phi_{i,p}$ — volume fraction of component i in phase q

χ_{ij} — Flory-Huggins interaction parameter between components i and j

χ_{ij}^* — experimentally determined χ_{ij}

ρ_D — rate of radical absorption per droplet, sec^{-1}

REFERENCES

1. El-Aasser MS: "Preparation of Latexes by the Miniemulsion Process", International Conference: Polymer Latex II, London, (May 1985).
2. Ugelstad J, El-Aasser MS and Vanderhoff JW: *Polym. Letters* 11 503 (1973).
3. Chou YJ, El-Aasser MS and Vanderhoff JW: *J. Dispersion Sci. Tech.* 1 129 (1980).
4. Grimm WL, Min TI, El-Aasser MS and Vanderhoff JW: *J. Colloid Interface Sci.* 94 531 (1983).
5. Lack CD, El-Aasser MS, Vanderhoff JW and Fowkes FM; (ed) Shaw DO: "Macro- and Microemulsions: Theory and Practice", *ACS Series* No. 272 p. 272 (1985).
6. Brouwer WM, El-Aasser MS and Vanderhoff JW: *Colloids and Surfaces* 21 69 (1986).
7. Ugelstad J, Hansen FK and Lange S: *Makromol. Chem.* 175 507 (1974).
8. Hansen FK and Ugelstad J: *J. Polym. Sci., Polym. Chem. Ed.* 17 3069 (1979).
9. Azad ARM, Ugelstad J, Fitch RM and Hansen FK; (eds) Piirma I and Gardon JL: "Emulsion Polymerization" ACS Symp. Ser. No. 24 p. 1 (1976).
10. Chamberlain BJ, Napper DH and Gilbert RG: *J. Chem. Soc., Faraday Trans. I* 78 591 (1982).
11. Choi YT, El-Aasser MS, Sudol ED and Vanderhoff JW: *J. Polym. Sci., Polym. Chem. Ed.* 23 2973 (1985).
12. Delgado J, El-Aasser MS, Silebi CA and Vanderhoff JW: *J. Polym. Sci., Polym. Chem. Ed. Part A* 24 861 (1986).
13. Brooks BW: *Br. Polym. J.,* 3 269 (1971).
14. Brooks BW: *Br. Polym. J.,* 2 197 (1970).
15. Ugelstad J, Mork PC, Kagerud K Herder, Ellingsen T and Berge A: *Adv. Colloid Interface Sci.,* 13 101 (1980).
16. Smoluchowski MV: *Z. Phys. Chem.* 92 129 (1918).
17. Guillot J: *Acta Polym.,* 32 592 (1981).
18. Morton M, Kaizerman S and Altier MW: *J. Colloid Sci.,* 9 300 (1954).
19. Tseng CM, El-Aasser MS and Vanderhoff JW; (ed) Provder T: "Computer Applications in Applied Polymer Science", ACS Symp. Ser. No. 197 197 (1982).
20. Flory PJ: *Principles of Polymer Chemistry*, Cornell University Press, New York (1953).
21. Ugelstad J, Mork PC, Mfutakamba HR, Soleimany E, Nordhuus I, Nustad K, Schmid R, Berge A, Ellingsen T and Aune O; (eds) Poehlein GW, Ottewill RH, Goodwin JW: "Science and Technology of Polymer Colloids", NATO ASI Ser. Vol. I, (1983) p. 51
22. Delgado J: Ph.D. Dissertation, Lehigh University (1986).
23. Scheisser WE: "Differential System Simulator, version 2", Lehigh University (1976).
24. El-Aasser MS, Delgado J, Tseng CM and Vanderhoff JW: unpublished results.
25. Rodriguez V, Delgado J, El-Aasser MS, Silebi CA and Vanderhoff JW: *Graduate Research Progress Reports*, Emulsion Polymers Institute, Lehigh University 26 116 (1986)
26. Ugelstad J, Mork PC, Berge A, Ellingsen T and Khan AA; (ed) Piirma I: "Emulsion Polymerization", Academic Press, NY p. 383 (1982).

27. Azad ARM, Fitch RM and Nomura N: *Org. Coat. Plast. Chem. Prep.* 43 537 (1980).
28. Jansson LH, Wellons MC and Poehlein GW: *J. Polym. Sci. Polym. Lett. Ed.* 21 937 (1983).
29. Guillot J: *Makromol. Chem. Suppl.* 10/11 165 (1985).
30. Guillot J: *Makromol. Chem. Suppl.* 10/11 235 (1985).

RHEOLOGY OF LATEX SYSTEMS AND CONCENTRATED DISPERSIONS

RHEOLOGY OF LATEX SYSTEMS AND CONCENTRATED DISPERSIONS

POSITION PAPER

1. INTRODUCTION

As in the past, polymer latexes will be the experimental systems most often selected to model the rheology of polymer colloids. The major part of the research effort during the next decade will be directed towards concentrated latexes and systems containing dissolved polymer. These will include stable, flocculated and coagulated systems, as well as the effects of flow field conditions on latex stability. It is clear that two complementary lines of approach will receive continuing attention. Firstly there is the microrheological approach, which emphasizes the inter-relation of macroscopic rheological properties with such particle properties as size, charge and adsorbed layer characteristics. The work will require advances in both theory and experiment. Secondly, there is a continual need for physical and mathematical models of the macroscopic behavior itself, in order to provide suitable constitutive equations for use in both laboratory and industry.

While it is clear that the bulk of the work will be centered around shear flow, attention should also be given to extensional flow, since material application and processing often involve significant extensional components. Advances have occurred recently in the experimentalist's ability to measure extensional flows in systems of high fluidity. As this experimental base is extended, we may expect a resulting expansion of the theoretical base and of our understanding of the factors which govern extensional behavior.

2. Experimental Studies
2.1. Viscoelastic Properties

Whereas most laboratories have hitherto concentrated on the continuous shear rheology of polymer colloids, viscoelastic properties are now becoming of wider interest. It is envisaged that increasing attention will be paid to this aspect of their behavior, through both dynamic measurements and by following the response to transient applications of stress or strain.

Modern instrumentation allows dynamic moduli to be studied over a frequency range of 0.0001 Hz to 10 Hz, with strains as low as 0.001. Strains of this low magnitude are often required if a linear response is to be maintained, especially with aggregated latexes. Many gelled systems have characteristic relaxation frequencies within this range. In conjunction with structural studies of concentrated latex systems, measurements of their storage and loss moduli should be carried out systematically over a wide frequency range. The effects of particle charge and concentration, electrolyte concentration, and free and bound polymer concentration should produce marked variation of rheological properties. Study of the viscoelastic behavior of non-aqueous systems over a wide temperature range should also provide valuable insights into the importance of factors such as the motion of stabilizer chains and anchor groups on the rheological properties of the latexes.

At present, most instruments are limited to a frequency sweep which terminates between 10 and 100 Hz. Higher frequencies can be reached by shear-wave propagation, but currently this is limited to frequencies of ca. 200 Hz. Experimental work should be carried out to extend this frequency range, in order to give a reliable estimate of the high-frequency limit of the storage modulus.

A new breed of controlled-stress rheometers allow creep to be studied at shear rates down to 10^{-6} reciprocal seconds. This is often low enough to measure the low-stress limit to the viscosity for weakly gelled systems. However, very little attention has been given to the detailed form of the creep curve, and almost none to the creep recovery curve. These measurements should make it possible to characterize polymer colloid systems by means of their relaxation and retardation spectra. Measurement of the stress relaxation modulus will also become of increasing use in characterizing weakly gelled polymer colloid systems; experimental characterization of the relaxation spectra should in turn lead to further theoretical developments.

2.2 Deformations Other Than Shear

Enqiries into the flow response of dispersions have centered principally on simple shear viscometric flows, where the particle structures that form are controlled in part by the flow fields. In recognizing this interconnection between structure and flow fields, one expects that non-viscometric flows, such as converging or elongational flows, may lead to responses that are quite unlike those that are observed in viscometric flows. In stable, concentrated suspensions, particles typically move in layers which parallel the surfaces of constant shear, but it is likely that non-viscometric flows will force layers of particles to interact and accommodate one another in ways that differ from those required by simple shear flow. Some flows may suppress the formation of layers completely. When this happens, material functions such as the viscosity may exhibit unusual responses.

While some important flow fields will be difficult to achieve for the same reasons that arise with continuum fluids, further complications may result from the non-continuum nature of suspension systems. Elongational flows with free boundaries provide a case in point. With these flows, perturbations in the boundary conditions arising from the discontinuous nature of a suspension may lead to gross instabilities in the flow. Even short of this point, conditions can arise wherein the particles are forced through the interface, thereby roughening the surface. Effects such as these are worthy of study in their own right.

2.3. Effects of External Fields

The behavior of colloidal dispersions when subjected to external fields other than shear fields promises to be increasingly important. Especially interesting is the behavior in gravitational or centrifugal fields, and in electric or magnetic fields superimposed on shear fields.

Gravitational or centrifugal fields will produce non-uniform particle distributions throughout the suspensions. This effect can be exploited in determining particle size and shape distributions. Little or no experimental data are available for well characterized systems in which the relative density between the particle and suspending medium is significantly different from unity. Measurements over a range of relative density values for a range of flow conditions (Peclet numbers) are desirable. Centrifugal fields are often superimposed on shear flow in separation processes in equipment such as hydrocyclones; a detailed study could be of considerable interest in this area.

More important for polymer colloids than gravity fields are their interaction with electric and magnetic fields over a broad range of volume fractions. Recent theories show that electric fields superimposed on shear fields can provide information on particle shape distributions for anisometric particles. Very little experimental work has been done in this area. Since non-spherical polymer colloid particles can now be synthesized, an experimental program to check the theoretical work could be initiated. At high volume fractions, electrically induced flocculation could drastically alter the rheological properties of systems by, for example, inducing liquid/solid-like transitions which can be exploited in a variety of applications. The same concepts apply to the effects of magnetic fields on magnetic polymer colloid particles, which are increasingly produced for new and and important applications. There is currently a great need for experiments on the rheological effects of electric and magnetic fields on well characterized dispersions of electrically and magnetically anisotropic particles. Of special interest will be the measurement of the dynamic complex viscosity over a wide range of conditions and the relationship of these results to the structure in the dispersion.

Finally, another phenomenon that lends itself to experimental study is electrically or magnetically induced flows.

2.4. Time-Dependent Thixotropic and Viscoelastic Effects

With a thixotropic polymer colloid, the shear stress decreases with increasing time of shear at constant shear rate. This effect is commonly studied by accelerating the shear rate uniformly to a predetermined value and then decelerating back to zero. The area of the hysteresis loop between the up and down curves, on a graph of shear stress versus shear rate, is used as a measure of the degree of thixotropy. This measure has little fundamental significance, depending as it does upon the history of the sample, the shear acceleration and the maximum shear rate. It gives no information as to whether the mechanical work done on the sample has been partly stored elastically, or entirely dissipated as heat or converted to a chemical form which cannot be recovered mechanically. This distinction is more critical for viscoelastic materials than for those thixotropic materials which are entirely dissipative.

Research is needed to develop test methods that will cover ranges of both time and shear rate sufficiently broad to establish the true characteristics of the material. These methods should include techniques for reproducing the material's rest state, small-amplitude oscillatory strain as used in dynamic testing, and transient tests such as step-changes in shear rate. Instrumental response must be adequate to record the growth and possible overshoot of shear stress. A wide range of reduced times (Deborah numbers) should be studied by modifying variables such the viscosity of the suspending medium and the temperature of the experiment.

2.5. Interaction of Radiation with Flowing Systems

Diffraction of light is an effective tool for monitoring structure in highly ordered suspensions of spheres, and the utility of this method is far from exhausted. Other techniques are being developed, however, to overcome the limitations of light diffraction, which include the effects of multiple scattering, the need for a high degree of ordering, and limitations imposed by the wavelength of light on the size and spacing of the particles under study. Radiation of other wavelengths can be used to overcome the latter restriction, and neutron scattering provides a viable alternative which reduces the effects of multiple scattering. Methods must be developed, however, which can be used when structures exist which are not rigorous enough to be detected by diffraction techniques. With flocculated suspensions, angular-dependent scattering can be used to gain information about the size and structure of the flocs as a function of the the variables affecting the system. Experimental measurement of the pair distribution function provides an important alternative technique.

Unfortunately, however, most methods can be readily applied only to suspensions of monosized spheres. Ultimately, these methods must be broadened and others developed which will allow one to probe the structure formed in suspensions containing non-spherical particles and particles of more than one size. The major new work needed here is theoretical modeling of the experimental data; the experimental techniques are already at hand.

2.6. Dielectric Spectroscopy

This is the measurement of the complex electrical impedance as a function of the frequency of an applied electrical field. Dielectric spectroscopy when applied to latex dispersions is capable of providing information about processes that affect the interfacial dielectric properties. The wide range of frequencies (ranging from 1 mHz 100 MHz) that can be used in these measurements makes it possible to study the dynamics of processes ranging from particle motion to deformations of electrical double layers and of adsorbed layers at interfaces. The results of the experiments carried out to date on suspensions indicate the high sensitivity of the technique in probing changes in the micro-scale structure of dispersions subjected to a shear field. Although a complete interpretation of the measurements in terms of dispersion structure needs further theoretical work, the method can be conveniently used to study the dynamics of the structural changes that affect the rheological behavior of the dispersions.

Shear flocculation of polymer colloids with different stabilization mechanisms (electrostatic, steric etc.) provides interesting systems for dielectric spectroscopy. The study of concentrated polymer colloids subjected to shear fields should give information on the relaxation times of the different processes associated with changes in the structure of the dispersion. This should be a particularly fertile ground for theoretical development as well as experimental study.

2.7. Experimental Characterization of Certain Dispersions

2.7.1. Sterically Stabilized Latexes. When the dispersion medium is of low molecular weight, these systems can behave as dispersions of non-interacting or "hard" spheres whose radii include the thickness of the stabilizing layer. This thickness depends on the geometry and stiffness of the medium-soluble portion of the stabilizer, and possibly also on shear stress. A series of dispersions should be prepared with homologous soluble moieties of different molecular weights, and another series with similar geometries but different rigidities of the soluble sections. If possible, the surface concentration of stabilizer should be an in-dependent variable for the experiments. Both the rheological behavior and the thickness of the stabilizing layer should be studied as a function of particle size, concentration and shear stress.

In polymeric media, whether polymer melts or polymer solutions, the large size of the molecules produces steric-elastic stabilization. It is important to relate quantitatively the molecular weights at which this occurs (as shown by a rise in the low-shear limiting viscosity) to both particle size and concentration and to the temperature.

2.7.2. <u>Flocculated Systems</u>. Particle flocs can occur through destabilization of a fluid dispersion or through break-up of a gel. Rheological behavior will be a function of the geometry and the mechanical properties of the floc, which depend in turn on the force which produced flocculation. Three floc-producing mechanisms are: (1) removal or neutralization of the stabilizing layer; (2) depletion flocculation; and (3) bridging by associative thickeners. In addition, we have two-phase systems such as latices whose high effective volume fractions induce ordering through a hard-sphere condensation. There is a marked need to identify differences and similarities in the rheological behavior of these various two-phase systems. Steady-state, dynamic, and transient data are needed to guide both macroscopic modeling and further development of microrheological theories.

2.8. <u>Electroviscous Effects</u>

A reasonably good theoretical framework exists for the treatment of the primary and secondary electroviscous effects in dispersions of spherical particles up to moderate volume fractions. In the past, however, lack of precise experimental methods to characterize interfacial electrical properties has been a major obstacle to progress in the area of moderately concentrated systems. In addition to conventional static techniques such as potentiometric and conductometric titrations and electrophoretic mobility measurement, dynamic measurements such as dielectric spectroscopy and ultrasonic potential determinations should be utilized in the electrical characterization. A study of the complete viscoelastic properties of well characterized polymer colloids should be carried out to obtain information on the effects of particle number, surface charge (and hence counterion numbers as well as electrostatic potential), and electrolyte concentration on the rheological properties of the systems. An additional area worthy of both experimental work and theoretical studies of electroviscous effects is their extension to non-spherical particles. Some recent experimental data on rotary diffusion coefficients have shown significant electrolyte effects in semi-dilute suspensions. Theoretical work in this area should contribute importantly to our understanding of the non-equilibrium statistical mechanics of suspensions.

2.9. <u>Characterization of Suspensions in Non-Newtonian Fluids</u>

Suspensions which are chosen for study in the laboratory generally consist of particles dispersed in a rheologically simple fluid such as water. The complex flow response which occurs with these systems can be attributed to particle-particle and particle-fluid interactions. There are, however, a number of colloidal systems, such as filled elastomers and rubber-reinforced polymers, which consist of particles dispersed in rheologically complex fluids, and our understanding of flowing dispersions will not be complete until these systems have been characterized. Factors which generally influence the flow response of dispersions in Newtonian liquids will certainly come into play with these more complex suspensions. In some cases the responses caused by these factors may simply build upon the the rheologically complex behavior of the suspending fluid, for example as a relative viscosity, but more complex interactions are possible and these must also be characterized and understood through careful study.

2.10. <u>Characterization of Non-Spherical Particles</u>

Recently it has become possible to synthesize monodisperse non-spherical polymer colloids. These these could be used in many ways, for example as fillers to change the rheology of many systems and as model particles for studying anisotropic suspensions. The shape of asymmetric particles can be determined from rheological and rheo-optical measurements. Except for very dilute systems of spheroidal and some ellipsoidal particles, very few experiments have been performed with well characterized anisotropic systems. Useful experiments that could be attempted are measurements of orientation distribution function in various flows using rheoptical techniques. This knowledge should provide a starting point in developing theoretical treatments of the rheology of such systems and the testing of those theories.

2.11. Complete Rheological Studies on Prototypical Systems
Dispersions show a variety of different rheological behavior under different conditions of stress, strain, stress rate and strain rate. Earlier sections of this chapter described specific experimental techniques that can be used. In a classical material (such as a Hookean solid or a Newtonian liquid) the various data obtained using different techniques can all be interpreted in terms of the same material parameters. In polymer colloid systems, however, this is not the case. The behavior under one testing condition cannot, in general, be interpreted and used to predict behavior under more complex flow situations. It is necessary to study a material under a wide and comprehensive range of test parameters in order that its full range of rheological response may be determined. To be more specific, a constitutive equation may contain several material parameters or constants. No single test technique will yield all of them; they have to be determined using different testing techniques over a wide range of test parameters. The available range of test techniques are as follows:

a. Controlled shear stress experiments where the test parameter is stress, ranging from zero to a high value. The following categories of experiment can be undertaken:

*   Creep below the yield stress and static yield stress measurements.

*   Creeping flow above the yield stress, with measurement of flow curves and calculation of a dynamic yield stress.

*   Determination of the complex modulus by oscillatory stress applied at various amplitudes and frequencies.

*   Behavior under sudden stress reversal.

b. Controlled shear strain experiments from zero to large strains such as:

*   Stress relaxation measurements.

*   Dynamic experiments as a function of strain amplitude and frequency.

c. Controlled shear strain rate experiments which range from zero to 10^6 reciprocal seconds. For example:

*   Stress growth following strain rate application, giving flow curves

which permit determination of a dynamic yield stress.

* Stress decay on cessation of steady flow.

* The effect of rapid shear rate reversal.

Attempts should be made to monitor growth of normal stresses, as well as the normal stress coefficients under steady shear and the decay of normal stresses on cessation of flow. It should be recognized, however, that normal stresses may be very low in polymer colloid systems except at very high volume fractions. An alternative approach may be the measurement of the relaxation spectrum using dynamic or transient testing.

A corresponding set of experiments should be carried out under controlled elongational stress, elongational strain and elongational strain rate. Dynamic Young's moduli under transient tests and elongational viscosities under steady straining are the rheological functions to be sought.

Test results should be interpreted in terms of constitutive equations derived from theory, and from macroscopic or phenomenological modeling. Appropriate constitutive equations may incorporate parameters such as temperature, phase composition, and other material parameters (particle size, size distribution and shape). Experiments performed over a range of such material parameters will provide validation of the constitutive equations.

At present very few polymer colloid systems have been fully studied in this way. It is not known which characteristics will typify the various types of system (stable, flocculated, two-phase, etc.). It is therefore important that well characterized systems of each different type be prepared and studied.

2.12. Wall Effects (width and geometry of the flow channel)

The usual formulation of constitutive equations or viscosity of a disperse system describes the bulk property of an infinite expanse of the suspension. It is known experimentally that, if the flow channel is less than one or two hundred particle diameters in width, the measured viscosity depends on the diameter-to-width ratio. Traditionally this is interpreted in terms of a wall slip-velocity concept. It is also known experimentally that, as the dispersed phase concentration increases and as the flow velocity increases (i.e., as the Reynold's Number becomes large), the wall effect can no longer be explained in terms of this classical concept. In important flow processes such as atomization, coating, and flow through blood vessels and other body channels, dispersions are made to flow through channels with a width of only a few particle diameters. It is therefore important to consider wall effects from a theoretical standpoint and to check the conclusions by careful experimentation using well characterized systems such as polymer colloids.

3. Theoretical Studies
3.1. Concentrated Dispersions of Anisometric Particles

Anisotropic polymer colloid particles include fused doublets of spherical particles as well as rods and platelets of biological origin. These can be sufficiently represented as bodies of revolution; it is unlikely that further lowering of symmetry (e.g. to tapes, triaxial ellipsoids or unequal spherical doublets) will introduce important differences in rheological behavior. The microrheological description of a dispersion of axisymmetric particles now includes, in addition to the quantities that characterize spheres, an orientation distribution function with two angles as arguments. At low volume fractions, this function should be sought by solving a diffusion equation in orientation space. Increasing the concentration to the semi-dilute range leads to severe steric interactions between the particles. These have been successfully treated by considering the transverse angular mobility to be greatly reduced, a concept analogous to the repetition of concentrated polymer molecular systems. In general, the essential features of the orientation distribution are embodied in an order-parameter tensor.

Theory needs to be extended from the lowest shear rates, where the shear and normal stresses are linear and quadratic respectively, to the shear-thinning region and beyond. Here it appears that the use of a single characteristic time associated with one rate process will be inadequate, so that a theory should produce a spectrum or distribution of characteristic times. Approximations to make the theory tractable need to be chosen carefully so that no features of rheological behavior are obscured. Most noteworthy in this regard is the negative primary normal stress difference that has been observed over a restricted range of higher shear rates in both particulate and liquid-crystalline materials.

3.2. Microscopic Theories

3.2.1. Non-Equilibrium Statistical Mechanics.
In weak flows, colloidal forces exert their maximum influence on the rheology of dispersions, and in fact the effects reflect the corresponding thermodynamic non-idealities. For single-phase systems, equilibrium and non-equilibrium statistical mechanics as developed for dense fluids, and the non-equilibrium analog for colloidal systems which includes hydrodynamic effects, can provide powerful techniques for predicting thermodynamic properties such as osmotic pressure as well as the weak flow limits of the viscometric functions, i.e., the zero shear viscosity and the high-frequency limit to the frequency-dependent storage modulus. This approach yields both the equilibrium and non-equilibrium structure factors, and therefore is amenable to experimental tests through both rheological and scattering experiments.

For stable ordered latexes the linear viscoelastic behavior has been estimated using a lattice model. Extension to a wider range of conditions and to non-linear effects may prove to be possible. However, more significant advance may prove possible through adaption of the results from the theory of dense fluids. Approximations for the high-frequency limit of the shear modulus for both repulsive and weakly attractive systems follow from the colloid pair-potential and the calculation of the appropriate perturbed hard-sphere structure. Creep behavior below the yield stress is more difficult to predict, requiring approximations for the rate of self-diffusion in addition to the potential.

A more rigorous theory must include multi-body interactions through the potential of mean force. To date this has only been developed for dispersions of hard spheres, with hydrodynamic interactions approximated by pairwise addition, and for stable charged spheres at low ionic strengths and volume fractions, for which hydrodynamic interactions can be neglected. These results capture some of the important effects of thermodynamic interactions, but show that improved models of the hydrodynamic interactions are necessary for quantitative predictions. The statistical mechanical techniques to represent other types of interparticle potentials may already be available.

Future work for stable disordered systems can address a number of interesting and important questions once satisfactory hydrodynamic models have been developed:

*   Electroviscous effects in electrostatically stabilized dispersions.

*   Effects of steric stabilizer layers.

*   Bimodal and generally polydisperse systems of hard spheres.

*   Effects of weak interactions through the sticky sphere model.

*   Multicomponent dispersions of hard, sticky and permeable spheres as models for associative thickener systems.

*   Spheres polarized by an external field.

The current approach is limited to structures that are close to equilibrium, to a single disordered phase, and to spherical particles. Generalization to stronger flows may be possible, though those may be handled more easily by the hydrodynamic calculations for ordered layers or by computer simulations as described below. Alternative, though somewhat similar, treatments for non-spherical particles are described in a later section. The extension to ordered or glassy phases should also be possible.

Systems which undergo equilibrium phase separations due to weak attractions represent another important class of materials with interesting rheological behavior. Recent work demonstrates the success of equilibrium statistical mechanics in predicting phase diagrams. During flow, however, these macroscopic phases break up into a heterogeneous mixture. This represents a first-order departure from equilibrium which requires a different theoretical approach.

3.2.2. <u>Ordering in Sheared Suspensions</u>. Several important studies indicate that stable colloidal dispersions under simple shear follow a flow mechanism in which layers of hexagonally packed particles slide over one another. In some systems, disordering may be observed at high shear rates. Theoretical effort is needed to enable the flow to be described and to determine the conditions under which such flows are stable. One place to start is to look at the flow of systems at high Peclet numbers, where Brownian motion can be neglected. A recent two-dimensional model of sheared aligned cylinders could be extended to the flow of a layer of spheres and subsequently to an array of spheres. A computer simulation of a two-dimensional system of disks also showed the transition to ordered flow and subsequently a disordering stage. Use could be made of recently developed theories describing the flow of spatially periodic structures. Existing theory for cylinders suggests that such systems can be shear-thickening, because of restructuring of the cylinders into fewer more densely packed arrays. However, much more theoretical work is needed, especially in regard to the stability of the proposed periodic structures.

It is possible in principle to introduce fields other than hydrodynamic into the theory. Probably the simplest to consider would be the DLVO-type of repulsive force, or a steric repulsion in the case of polymer-coated particles, which would prevent particles from making close contact. Inclusion of the effects of external fields will lead to non-uniform distributions, thus destroying the global order in the suspension. In such cases, the flow could perhaps be treated as having a gradually varying global structure, locally resembling a regular ordering.

3.2.3. <u>Flocculated Systems</u>. Flocculated dispersions pose more difficult theoretical problems because of the non-equilibrium and history-dependent nature of the structure at rest. In the absence of governing equations, structures must be assumed and the consequences tested by comparison with experiment.

With strongly flocculated dispersions, recent characterization of fractal structures has opened the way to percolation and scaling theories capable of correlating volume fraction dependence of the elastic response to small-amplitude deformations. Considerable work remains, however, in relating the fractal dimension to the mechanical prehistory (i.e., the degree of shear during flocculation), as well as relating the interparticle potential to the magnitude of the modulus.

With weakly flocculated dispersions at moderate volume fractions, the structures may primarily reflect hard-sphere repulsions. As suggested above, non-equilibrium theory can then be applied to predict the linear viscoelastic properties.

In either case the predictions represent approximations which must be tested against experiments or computer simulations.

3.2.4. <u>Computer Simulations</u>. The success in dynamically simulating the thermodynamic and transport properties of simple molecular systems immediately suggests the possibilities of similar studies of colloidal dispersions. The presence of a viscous fluid damps Brownian fluctuations, converting the process from that of collisions among molecules to stochastic displacements of particles. The formalism for integrating the equations of motion on the diffusion time scale is well established. In some cases, it may prove feasible to integrate Langevin's equations in lieu of the full equations of motion. Recent work suggests that pairwise additive forces may adequately approximate many-body hydrodynamic interactions.

Though promising, the hydrodynamic approximation presents difficulties, since the numerical computations in three dimensions become prohibitively time-consuming, and hence extremely expensive. Tractable operations must be developed before a wide range of interesting concentrated systems can be addressed.

With flocculated dispersions, existing hydrodynamics should suffice. Hence the creation, deformation and rupture of networks appear to be tractable simulations. In this case, the simulations can address important problems outside the scope of other approaches.

4. Conclusion

Developments of the past decade have greatly increased our capability to study the rheology of polymer colloids. From the experimental side we have seen advances in rheometric instrumentation, in preparation and characterization of model systems, and in use of complementary techniques such as diffraction and dielectric spectroscopy to elucidate the structure of flowing dispersions. From the theoretical side we see more powerful and rigorous application of nonequilibrium statistical mechanics to flowing systems, new and insightful modeling concepts such as repetition and structured flow, and increasing capability of computers to solve complex equations and to simulate rheological experiments. With such methods at our disposal, we can confidently anticipate that the decade to come will greatly advance our understanding of the rheology of polymer colloids.

5. **Panel**

Discussion Leader:
    I. M. Krieger (Case Western Reserve University, USA)

Secretary:
    J. Goodwin (University of Bristol, United Kingdom)

Participants:
    C. E. Chaffey (University of Toronto, Canada)
    D. C-H. Cheng (Warren Spring Laboratory, United Kingdom)
    R. L. Hoffman (Monsanto Company, USA)
    D. S. Jayasuriya (S. C. Johnson & Son, Inc., USA)
    M. Ojalvo (National Science Foundation, USA)
    D. Quemada (Universite Paris VII, France)
    W. B. Russel (Princeton University, USA)
    C. Silebi (Lehigh University, USA)
    T. van de Ven (McGill University, Canada)

POLYMER COLLOIDS AS MODEL SYSTEMS FOR STUDYING RHEOLOGY OF DISPERSIONS

Irvin M. Krieger

Case Western Reserve University
Cleveland, Ohio 44106, USA

1. INTRODUCTION

Interest in the rheological behavior of polymer colloids is both practical and theoretical. Many commercial plastics are originally polymerized as dispersions in aqueous or nonaqueous media. The resulting polymer colloids may be coagulated directly to form bulk polymer, or they may be blended with pigments, fillers and/or soluble binders to form coatings, adhesives, sealants, foams, etc. Rheology is important in both polymerization and subsequent processing. Practical systems range from simple dispersions of spherical particles to formulations which contain particles of different kinds together with stabilizing polymers and/or surfactants.

Polymer colloids provide dispersions which can be very well characterized and which correspond closely to idealized models. They permit us to confront theory with experiment, under conditions where the two should agree well. Any disagreement must then be attributed to flaws in the theory or to faulty experiment, rather than to discrepancies between experimental system and idealized model. We need data of this kind if we are to improve our experiments and refine our theories.

Polymer colloids are formed under the influence of interfacial tension. For this reason the sphere, which is the most tractable particle shape from the viewpoint of theory, is also the easiest shape to achieve. Furthermore, methods are available to produce concentrated dispersions of spheres which are highly uniform in size, with diameters ranging from 0.1 to 10 micrometers.

Particle interactions can also be controlled and varied in polymer colloids. In aqueous systems, changing surface charge and electrolyte environment can control the strength and range of interparticle Coulombic forces. In both aqueous and nonaqueous systems, we can vary thickness and nature of stabilizing layers. Availability of polymers of narrow molecular weight range permits study of dispersions of spherical polymer colloids in well characterized polymeric media and in polymer-containing solutions.

To this date, relatively little experimental work has been done with well characterized nonspherical polymer colloids. Nonspherical particles of defined shapes are less available than are spheres, and are more difficult to characterize. Despite the increased complexity of their mathematical models, however, some progress has been made in treating nonspherical particles theoretically.

2. RIGID SPHERES IN NEWTONIAN MEDIA

Of all colloidal dispersions, the simplest to treat theoretically is the dispersion of uniform rigid spheres in a Newtonian medium. Here the interaction potential is infinite for center-to-center distances less than one diameter, and zero for greater separations. Only continuum properties of the medium need be considered - principally the viscosity η_o and the density ρ_o. At ordinary temperatures, Brownian motion plays an important role throughout the entire colloidal size range.

Before even considering the fluid-dynamic equations for this system, important insight can be gained through dimensional analysis[1]. A list of nine system variables which must enter into the rheology-structure relationship is given in Table 1. Since the dimensionality of each variable can be expressed in terms of the three fundamental units of mass, length and time (M,L,T), the desired equation must be expressible as a functional relationship among six independent dimensionless groups. One such set is shown in Table 2. Other dimensionless variables which can be constructed as products of these six may be very convenient in special contexts. Examples are the shear strain $\gamma - \dot{\gamma}t$ and the reduced shear stress $\sigma_r - \sigma a^3/(kT)$, where $\sigma - \eta\dot{\gamma}$ is the shear stress.

TABLE 1. Variables in the Rheology-Structure Relation for Rigid-Sphere Dispersions

Variable	Symbol	Dimensions
Viscosity of dispersion	η	$ML^{-1}T^{-1}$
Viscosity of medium	η_o	$ML^{-1}T^{-1}$
Density of medium	ρ_o	ML^{-3}
Density of particle	ρ_p	ML^{-3}
Radius of particle	a	L
Number density of particles	n	L^{-3}
Shear rate	$\dot{\gamma}$	T^{-1}
Elapsed time	t	T
Thermal energy	kT	ML^2T^{-2}

TABLE 2. Dimensionless Variables in the Structure-Rheology Relation for Rigid-Sphere Dispersions

Name of Group	Symbol	Group
Relative viscosity	η_r	η/η_o
Volume fraction	ϕ	$4\pi n a^3/3$
Relative density	ρr	ρ/ρ_o
Peclet number	Pe	$\eta_o\dot{\gamma}a^3/(kT)$
Reduced time	t_r	$kTt/(\eta_o a^3)$
Internal Reynolds number	Re	$a^2\dot{\gamma}\rho_o/\eta_o$

The dimensional analysis tells us that there must exist a master equation for uniform rigid-sphere dispersions:

$$\eta - f(\phi, Pe, \rho_r, t_r, Re) \tag{1}$$

Some important implications of the dimensional analysis have been tested. Under conditions of slow (Re → 0), steady (t_r → ∞) flow of a neutrally buoyant (ρ_r → 1) dispersion, Equation 1 reduces to:

$$\eta_r - f(\phi,Pe,1,\infty,0) - f(\phi,Pe) \tag{2}$$

Or, equivalently

$$\eta_r - g(\phi,\sigma_r) \tag{3}$$

The validity of Equation 3 has been confirmed for various particle diameters below 0.6 micrometers (i) for aqueous latices by Krieger and Wood[2], (ii) for redispersions of crosslinked polystyrene in polar organic media of various viscosities by Papir and Krieger[3], (iii) for sterically stabilized dispersions of poly(methylmethacrylate) in silicone media of different viscosities by Choi and Krieger[4,5], and (iv) for silica dispersions in cyclohexane by de Kruif, van Iersel, Vrij and Russel[6].

Returning to Equation 1, it would be interesting to study transient behavior, where the limitation t_r → ∞ can be removed. A diffusion time t^* = $\eta_o a^3/(kT)$ should characterize the time-dependent thixotropic transformation. For aqueous latex particles of diameter around 0.25 micrometers, t^* is of the order of a millisecond, which is smaller than the response times of most instruments now in use. To measure thixotropic response times, it is therefore necessary to improve instrument response or to increase the time constant of the system.

In view of the applicability of the master equation

$$\eta_r - f(\phi,\sigma_r,t_r) \tag{4}$$

where $t_r - t/t^*$, increase of t^* is an attractive possibility. Although t^* depends strongly on particle radius, a can be increased by only a factor of two before interparticle van der Waals forces become too important to ignore. An alternative would be to increase η_o. Use of polymeric thickeners is not appropriate, unfortunately, because of adsorption of the polymer onto the particles. Oligomeric media such as polybutenes and silicones should be acceptable so long as the molecular size of the oligomers remains negligible compared to the diameter of the polymer colloid particles. The most attractive route would be to use oligomers of low molecular weight at reduced temperatures, where their viscosity is augmented.

No rheological effects have as yet been measured which depend on either ρ_r or Re. The reduced density ρ_r should play a role in establishing concentration gradients under gravitational and centrifugal forces. Lin, Peery and Schowalter[7] have shown that the internal Reynolds number Re defines conditions where inertial forces on the particles can be regarded as negligible relative to viscous forces.

Chaffey[8] has investigated the consequences of including other parameters in the rheological equation of state, such as particle deformability or anisotropy and interparticle forces of various kinds. He

suggests suitable dimensionless variables which should facilitate
correlations in systems where these parameters affect the rheology of the
dispersion.

3. ELECTROVISCOUS EFFECTS

The "first electroviscous effect" is the influence of
electrical forces on the intrinsic viscosity. It produces only a small
increment in intrinsic viscosity over the Einstein value of 5/2, and is
very difficult to measure precisely. The subject has been well studied
theoretically for spherical particles[8]. Its magnitude depends on surface
charge (or zeta potential) and on the Debye radius of the ionic
atmosphere. With the methods now available to control both of these
variables, it would seem that the time is ripe for a definitive
experimental test of available theory. Such an attempt was recently made
by McDonogh and Hunter[9], with inconclusive results.

The "second electroviscous effect", that due to interparticle
Coulombic forces, is much larger, however, and therefore easily measured.
Krieger and Eguiluz[10] varied electrolyte concentration and valency as
well as ϕ in a deionized monodisperse polystyrene latex, but did not vary
particle diameter or surface charge. Their deionized latices showed an
apparent yield stress, which decreased and eventually vanished with
electrolyte addition. High-shear limiting viscosity was independent of
electrolyte content, however. A striking observation was that reduction
of viscosity depends on a dimensionless group which is the ratio of
charge density of added counterions to particle charge density. The
second electroviscous effect therefore depends on the equivalent
concentration of counterions, but not on their valency type. This
implies that most of the added counterions become part of the compact
double layer, until their concentration exceeds that which would
neutralize the charge on the particles.

Goodwin et al.[12,13], and later Buscall[14], determined shear moduli in
deionized latices by measuring propagation velocity of shear waves. Both
the presence of rigidity and the existence of an apparent yield stress
are reminiscent of the behavior of crystalline solids. The behavior of
these systems below their apparent yield stresses merits further study,
and should benefit from experiments and theory developed for crystalline
solids. The analogy is brought closer by verification through optical
diffraction measurements of long-range ordering in deionized latices[15,16].
Apparently a transition occurs with electrolyte addition, from deionized
systems which show apparent yield stresses to fluid systems with finite
low-shear viscosities. Is this a gradual change, or does it represent a
first-order phase transition? Optical diffraction studies suggest the
latter; Hachisu and Kobayashi[17] even determined a phase diagram
experimentally.

The question is further compounded by measurements by Buscall et
al.[17] of creeping flow at stresses well below the apparent yield stress.
They found limiting Newtonian viscosities of very high magnitudes.
Furthermore, they succeeded in estimating low-shear viscosities from the
Ree-Eyring transition-state theory of liquid viscosity[18], using an
appropriate double-layer potential for the activation barrier. It would
appear that creeping Newtonian flow is normal whenever a shear stress is
applied to a crystalline solid, but that such flow is detectable only
when the applied stress is an appreciable fraction of the apparent yield
stress. This leaves open the question of a change in rheological regime
when enough electrolyte has been added to cross over into the two-phase
region of the phase diagram.

Benzing and Russel[19] have developed a self-consistent field theory of the second electroviscous effect, and compared its predictions with experiment. Russel[20], in a study of the rheological effects of particle interactions, reports better success in dealing with ordered polymer colloids than with the disordered state.

4. STERICALLY STABILIZED DISPERSIONS

We might expect sterically stabilized polymer colloids to behave like dispersions of rigid particles. For spherical particles, then, master Equations 1 through 4 should be applicable. It will be necessary, however, to include the layer of stabilizing polymer as part of the effective volume fraction ϕ. If ϕ_o is the volume fraction of the polymer colloid itself and δ the thickness of the stabilizing layer, then

$$\phi = \phi_o[1+(\delta/a)^3] \tag{5}$$

Choi and Krieger[4,5] prepared poly(methylmethacrylate) dispersions of various uniform diameters in hexane, using an ABA triblock stabilizer with poly(dimethylsiloxane) as the soluble A-block and polystyrene as the anchoring B-block. These were then redispersed in silicone fluids. The observed non-Newtonian viscosities failed to follow Equation 2 when ϕ_o was used as the volume fraction. When an adsorbed layer thickness δ of 7.0 nanometers (determined from the intrinsic viscosity) was used in Equation 5, however, excellent correlation was obtained, with curves corresponding closely to those observed with hard-sphere dispersions. Low-shear limiting viscosities were somewhat higher, however, possibly indicating some compaction of the stabilizing layer during shear flow. Willey and Macosko[21] also were able to use Equation 3 to correlate their data on sterically stabilized poly(vinylchloride) particles dispersed in various organic media. Their η_r values were higher than those observed in the other examples, possibly due to swelling of the particles.

5. DISPERSIONS IN POLYMERIC MEDIA

Polymer colloids may consist of particles dispersed in liquid polymers or in solutions of polymers in solvents of low molecular weight. To the extent that these media can be treated as continua of given ρ_o and η_o, these systems should obey the master equation (Equation 1). However, complications arise for polymer solutions unless solvent, polymer and particle surface are energetically similar, i.e., chemically homologous or at least closely matched in cohesive energy density. And for both polymer solutions and liquid polymeric media, complications arise when the size of the polymer molecules is appreciable relative to the size of the dispersed particles.

In many cases, soluble polymer is added to stabilize or destabilize the dispersion. The efects on colloid stability of energy differences among solvent, dissolved polymer and particle surface have been well discussed by Buscall and Ottewill[22]. However, the rheological consequences have not been adequately investigated. Yet polymeric solutes are often added to aqueous or nonaqueous polymer colloids specifically in order to modify their rheological behavior. Other polymer colloids, for example aqueous latices prepared with ionic comonomer, contain dissolved polymer as a necessary consequence of the method of polymerization.

A few examples can be cited of systems where solvent, dissolved polymer and particle surface are energetically very similar. Buscall and

Ottewill[22] mention aqueous latices stabilized by nonionic surfactants containing poly(ethylene oxide) fragments, with added poly(ethylene oxide) solute. A nonaqueous system suggested by Choi and Krieger would use low-molecular-weight poly(dimethylsiloxane) as solvent, high-molecular-weight poly(dimethylsiloxane) as solute and a diblock or triblock stabilizer whose soluble block is poly(dimethylsiloxane). (The commercially available silicones provide poly(dimethylsiloxane)s of nearly uniform molecular weight, ranging from low to high degrees of polymerization). These systems, where interactions are almost entirely entropic, can serve as a baseline for the study of more general systems whose rheological behavior is complicated by the presence of significant energy differences among solvent, dissolved polymer and particle surface.

One effect which should be demonstrable even in purely entropic systems is depletion flocculation, which is closely related to the "creaming" process used commercially to concentrate natural and synthetic latices. Polymeric solutes whose molecules are large in spatial extent must adopt compact configurations if they are to accommodate to the small interparticle void spaces in concentrated polymer colloids. When sufficient solvent is on hand, an alternative configuration is available. Solvent can fill the void spaces in a compacted (flocculated) dispersion which contains all of the particles, while the soluble polymer forms a separate solution phase. If, as in natural rubber, the particles are less dense than the solvent, the colloidal dispersion rises to the top (creams).

Of greater rheological interest is a phenomenon, christened "steric-elastic stabilization" by Napper and Smitham[23,24], which can occur when solvent is absent (or if the amount on hand is not enough to support depletion flocculation). Molecules of the liquid polymer must then occupy the interparticle voids in a *space-filling* configuration. This requires them to adopt configurations much more compact than those they would take on in the unconfined liquid polymer. The compressive action is effectively provided by the particles, and the reaction is an elastic force of expansion directed against the particle surfaces. The result is equivalent to an interparticle repulsive force. The rheological consequences of steric-elastic stabilization are therefore very similar to those produced by the second electroviscous effect.

Choi and Krieger[4,5] redispersed four different particle sizes (core diameters 190, 280, 410 and 625 nanometers) of their sterically stabilized PMMA spheres in three different silicone fluids (nominal viscosities 10,50 and 200 centistokes). They characterized the rheological behavior of the 12 disperse systems, listed in Table 3, as "gelling" (G) or "fluid" (F), according to whether or not they showed yield stresses. Figure 1 shows the difference in flow properties between fluid and gelling dispersions. Regardless of temperature, all the dispersions in the 10-cSt fluid showed fluid behavior at all stresses, while all the dispersions in the 200-cSt fluid showed gelling behavior. In the 50-cSt fluid, the two larger particle sizes gave fluid dispersions, while the two smaller sizes gave gelling dispersions.

TABLE 3. Model Sterically Stabilized Dispersions (Four particle sizes in three silicone oils of different viscosities)

Oil ⇒ Particle ⇓	10 cSt	50 cSt	200 cSt
85 nm	F	F	G
141 nm	F	F	G
204 nm	F	G	G
310 nm	F	G	G

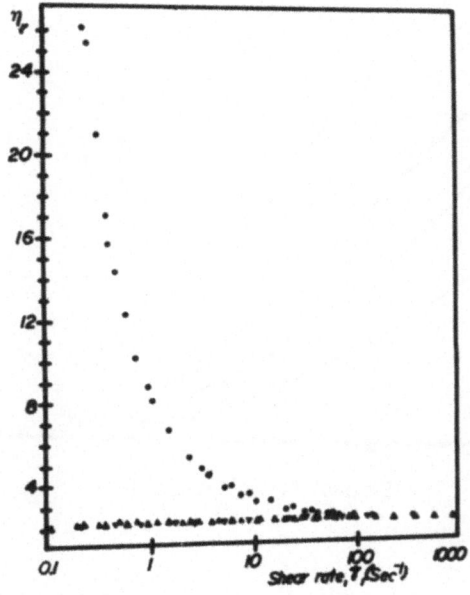

Figure 1: "Fluid" and "gelled" dispersions at ϕ = 20%.

Molecular weights of the three silicone fluids were 1300, 4400 and 11,000. It is clear that the gelling dispersions correspond to situations where molecular sizes of the liquid polymer molecules are appreciable relative to interparticle spacing, conditions conducive to steric-elastic stabilization. Since elastic forces increase with temperature, viscosity of the gelling dispersions should likewise increase. Figure 2 shows the steep increase of viscosity with temperature of one of the gelling dispersions. On the other hand, rheological behavior of the six fluid dispersions, shown in Figure 3, followed the corresponding states principle of Equation 3. This implies that the sizes of their molecules are small enough to allow the silicone polymers to behave as continuous fluids in these dispersions.

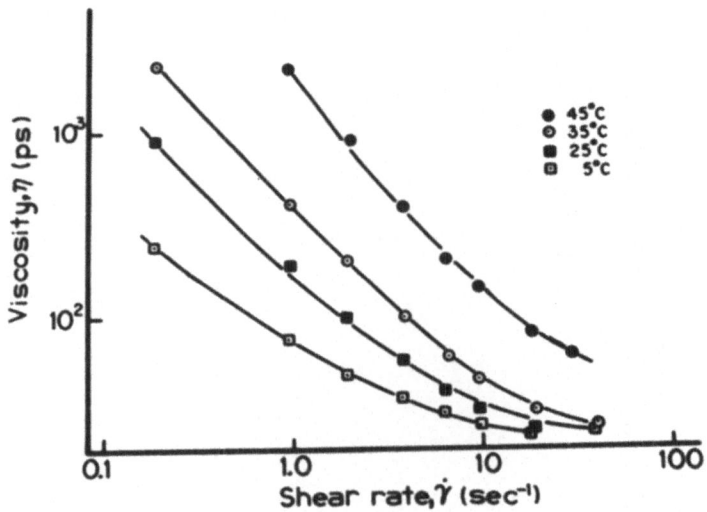

Figure 2: Viscosity increase with temperature of a gelling dispersion. Particle radius 85 nm, medium viscosity 50 cSt, volume fraction 40.

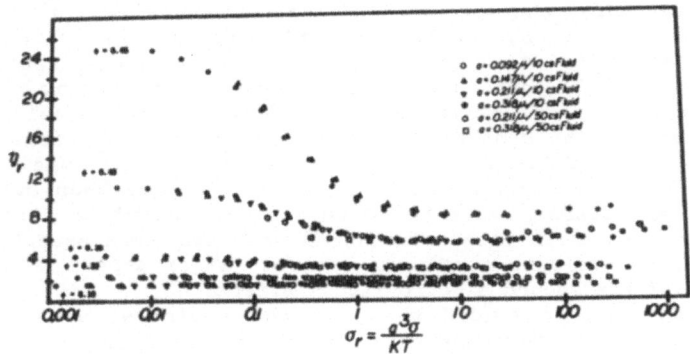

Figure 3. Superposition of dimensionless viscosity data for the six sets of fluid dispersions.

The possible rheological consequences of using polymeric media are therefore rich and varied, and deserve further study.

1. In the simplest case, the added polymer merely modifies the properties of the medium. This should increase the viscosity of the dispersion as well as its thixotropic relaxation time, and perhaps introduce an elastic component.

2. By adsorbing to form a monolayer on the particle, the polymer confers steric stabilization and augments the volume fraction of the disperse phase.

3. When the sizes of the dissolved polymer molecules are appreciable relative to interparticle spacing, destabilizing effects due to bridging or to depletion flocculation come into play. The rheology of such associatively thickened dispersions has itself attracted much atention[26].

4. Finally, when size and concentration of the polymer become so high that its molecules must be compressed in order to accommodate themselves to the interparticle void spaces, elastic stabilization sets in, accompanied by gelation and inverse temperature effects.

6. THE YIELD STRESS CONTROVERSY

The concept of a yield stress implies that at steady stresses below a certain critical value σ_y the system does not flow, but that above σ_y it "yields" and flows like a liquid. Two general methods exist for measuring σ_y: (1) the *stress method*, where stress is increased from zero and the σ-value noted where continuous flow is first detectable; and (2) the *strain-rate method*, which extrapolates $\dot{\gamma}$ vs. σ curves back to intersect the σ-axis at σ_y. Most rheological instruments for characterizing non-Newtonian fluids are rotational instruments whose independent variable is the rotational speed. It is not surprising, therefore, that the strain-rate method is the more widely used for measuring σ_y. The values obtained, however, can vary widely depending on the range over which data were obtained and the method of extrapolation employed. Furthermore, the very action of continuous shearing can destroy the structure which is responsible for rigidity of the system to flow below σ_y. Unless this structure is capable of reforming rapidly, the strain-rate method will tend to give erroneously low yield stresses.

A new generation of rotational rheometers has been developed which permit the torque to be controllably increased from zero. Some of these are sufficiently sensitive to detect rotation at very small angular velocities. With these instruments, creeping flow has been detected in many systems at stresses well below their apparent yield stresses (as determined by the strain rate method). Cited above[13] are measurements of Buscall et al. on deionized polymer latices, which appear to gel under the influence of the second electroviscous effect.

It may well be contended that, in the strictest sense, a yield stress does not exist. Nevertheless, the yield stress concept has proven useful in characterizing flow behavior of many important systems, including some polymer colloids. If we are to retain the concept, the yield stress may have to be redefined for given applications, perhaps by specifying a shear rate below which the system is considered not to flow. In cases where stress and strain-rate methods give two different values, we may need to distinguish between the yield stresses of static and flowing systems.

7. ORDERING DURING SHEAR FLOW

In a classical investigation of the mechanism of rheological dilatancy, Hoffman[26,27] sheared a uniform poly(vinylchloride) dispersion in a rotational viscometer which was equipped with transparent shearing surfaces. Simultaneous optical diffraction observations clearly showed that, as the viscosity decreased with increasing shear rate, the dispersion formed hexagonally ordered sheets of particles whose planes were normal to the velocity gradient. As shear rate increased further, order increased and interparticle spacing decreased. Eventually, shear-thickening set in, gradually in dispersions of volume fraction below 52%, but catastrophically at higher volume fractions. At the same time, the diffraction effects disappeared, indicating instability of the ordered structure. Recent measurements of diffraction of light by sheared suspensions by Van de Ven and Tomita[28] confirm Hoffman's observation of ordering during shear flow.

Choi's measurements[4,5] on his uniform sterically stabilized "fluid" dispersions showed gradual onset of shear thickening. He found that shear thickening was much more pronounced at reduced temperatures. Choi's studies did not extend to the high volume fractions where Hoffman observed catastrophic instability. Whereas Hoffman explained the instability of ordered flow in terms of the DLVO potential, Choi's explanation was based on a competition between shear rate and particle diffusion.

Hoffman puts forth a mechanism for shear flow of uniform dispersions which covers the entire shear rate range, from the low-shear Newtonian regime to the onset of shear thickening and even beyond. As illustrated by Choi in Figure 4, the low-shear Newtonian regime (I) corresponds to flow of a particle configuration characteristic of the static dispersion. The high-shear Newtonian regime (III) corresponds to fully developed hexagonally ordered particle sheets flowing smoothly past one another. Shear-thinning regime (II) is the transition as the ordered structure develops. Finally, in shear-thickening regime (IV), the ordered flow becomes unstable.

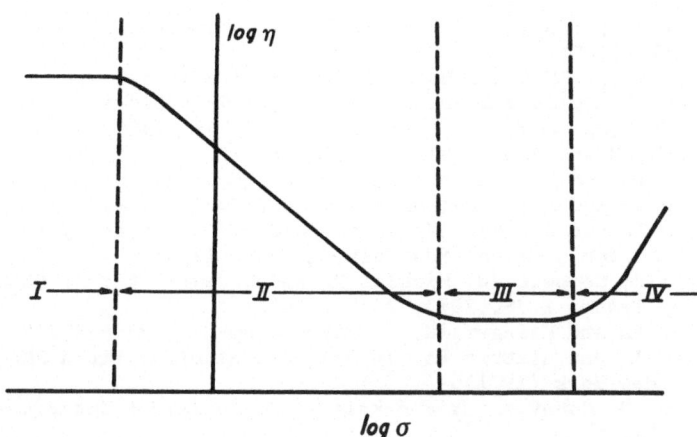

Figure 4: Schematic graph of the four flow regimes: (I) Newtonian flow of randomly ordered dispersion; (II) Shear thinning as layering develops; (III) Fully developed layered flow; (IV) Shear thickening due to instability of layered flow.

Appealing as we may find this global explanation, it raises questions as yet unanswered and challenges as yet unmet. Regimes very similar to I, II and III are observed with heterodisperse polymer colloids and with solutions and melts of macromolecules. Are heterodisperse systems capable of organizing into ordered sheets as a consequence of shear flow? We should be able to derive the functional dependence of η_r on ϕ in regime III based on a layered-flow mechanism. And we should also be able to derive the dependency of ϕ_r on σ_r in regime (III) based on shear-induced onset of ordering.

REFERENCES

1. Krieger IM: *Trans Soc. Rheology* 7, 101 (1963).
2. Woods ME and Krieger IM: *J. Colloid & Interface Sci.* 34, 269 (1970).
3. Papir YS and Krieger IM: *J. Colloid & Interface Sci.* 34, 126 (1970).
4. Choi GN: Ph.D. Thesis, Case Western Reserve University (1983).
5. Choi GN and Krieger IM: *J. Colloid & Interface Sci.* 113, 94, 101 (1986).
6. de Kruif CG; van Iersel EME; Frij A and Russel WB: *J. Chem. Phys.* 83, 4717 (1985).
7. Lin CJ; Peery JH and Schowalter WR: *J. Fluid Mech.* 44, 1 (1970).
8. Chaffey CE: *Colloid & Polymer Sci.* 255, 691 (1977).
9. Booth F: *Proc. Roy. Soc. (London)* A203, 533 (1950).
10. McDonogh RW and Hunter RJ: *J. Rheology* 27, 189 (1983).
11. Krieger IM and Eguiluz M: *Trans. Soc. Rheology* 20, 29 (1976).
12. Goodwin JW and Smith RW: *Disc. Faraday Soc.* 57, 126 (1974).
13. Goodwin JW and Khidher AM: in *Colloid and Interface Science*, Kerker M(ed), Academic Press, New York, IV 529 (1976).
14. Buscall R; Goodwin JW; Hawkins MW and Ottewill RH:. *J. Chem. Soc. Faraday Trans. I* 78, 2873, 2889 (1982).
15. Hiltner PA and Krieger IM: *J. Phys. Chem.* 73, 2386 (1969).
16. Krieger IM and Hiltner PA: in *Polymer Colloids*, Fitch RM(ed); Plenum Press, New York (1971).
17. Hachisu, S; Kobayashi Y and Kose A: *J. Colloid & Interface Sci.* 42, 509 (1981).
18. Ree T and Eyring H: *J. Appl. Phys.* 26, 793, 800 (1955).
19. Benzing DW and Russel WB: *J. Colloid & Interface Sci.* 83, 167, 178 (1981).
20. Russel WB: *J. Rheology* 24, 287 (1980).
21. Willey SJ and Macosko CW: *J. Rheology* 22, 381 (1978).
22. Buscall R and Ottewill R: in *Polymer Colloids*.
23. Smitham JB and Napper DH: *J. Colloid & Interface Sci.* 54, 467 (1976).
24. Smitham JB and Napper DH: *J. Chem. Soc. Faraday Trans. I* 72, 2425 (1976).
25. Hunter RJ: *Adv. Colloid & Interface Sci.* 17, 197 (1982).
26. Hoffman RL: *J. Colloid & Interface Sci.* 46, 491 (1974).
27. Hoffman RL: *Adv. Colloid & Interface Sci.* 17, (1982).
28. Tomita M; Takano K and van de Ven TGM: *J. Colloid & Interface Sci.* 92, 367 (1983).

THEORETICAL APPROACHES TO THE RHEOLOGY OF CONCENTRATED DISPERSIONS

W.B. RUSSEL

Department of Chemical Engineering
Princeton University
Princeton, New Jersey

1. INTRODUCTION

Suspensions of submicron particles in a liquid respond to flow in a variety of ways depending on the size, concentration, and shape of the particles and the nature and magnitude of the interparticle potentials. The most dramatic phenomena occur when one of the interparticle forces dominates, e.g., strong van der Waals forces for aqueous latices at high ionic strengths[1] or carbon black particles in mineral oil[2], long range electrostatic repulsions for colloidal crystals[3,4], and the interactions between adsorbed polymer layers in sterically stabilized suspensions near closest packing[5,6]. Even with hard sphere interactions though, the rheology is significantly shear-thinning at moderate concentrations[7,8]. In each case the non-Newtonian phenomena, whether elastic or pseudoplastic, derive from many-body interactions involving both hydrodynamic and interparticle forces.

The goal of our work is a quantitative theoretical understanding of the rheology of colloidal suspensions of industrial relevance. To this end we focus on well-defined systems for which the dominant forces can be identified and experimental data is available. For these we seek to predict specific rheological parameters such as the low shear viscosity and the high frequency modulus and dynamic viscosity for stable dispersions and the elasticity and plastic flow for flocculated suspensions. A complete theory requires knowledge of the interactions, both hydrodynamic and thermodynamic; the resulting structure, at rest and under flow; and proper averaging of this information to obtain the bulk stresses[9,10]. Inevitably the complex many-body interactions in the concentrated suspensions of interest preclude an exact treatment short of computer simulation[11], so suitable approximations must be identified. Since the approach differs markedly between stable systems, for which the rest state corresponds to thermodynamic equilibrium, and flocculated systems far from equilibrium, we address the two classes separately in the following.

Reprinted with permission from <u>Powder Technology</u>.

2. A NON-EQUILIBRIUM THEORY FOR STABLE DISPERSIONS

2.1. Introduction

This approach capitalizes on statistical mechanical descriptions of dense fluids at equilibrium to extend exact non-equilibrium treatments of pair interactions to finite concentrations. The former provides a rest state which accommodates many-body thermodynamic interactions accurately, if not completely rigorously. The latter requires the incorporation of many-body hydrodynamics and the reduction of an N-body conservation equation to the pair level. Neither are possible exactly, but reasonable approximations are emerging.

For charged spheres at low ionic strengths and dilute concentrations Ohtuski[12] justifiably ignored the hydrodynamic interactions. For small perturbations from equilibrium the superposition approximation for the N-body distribution function led to an integro-differential equation for the non-equilibrium pair distribution function. Numerical solutions then determined the structure and the associated rheological parameters for steady shear and small amplitude oscillations. Russel and Gast[13] subsequently addressed concentrated suspensions of hard spheres through a pairwise additive approximation for the hydrodynamics. Neglecting the integral terms representing explicit three-body couplings still left many-body thermodynamic interactions in the form of the potential of mean-force acting on a pair. The results for the low shear limiting viscosity and the high frequency modulus and dynamic viscosity identified this long range thermodynamic interaction as the sources of the high viscosities, shear thinning, and elasticity which appear at volume fractions $\phi > 0.25$. Our current efforts strive to improve the hydrodynamics through the pairwise additive force approximation[11] to treat a variety of concentrated suspensions quantitatively. Below we review the nature of the theory and the results for hard spheres.

2.2. Formulation

The theory is developed in configuration space in terms of the probability $P_N d\underline{x}_1 \ldots d\underline{x}_N$ of finding N spheres at positions $\underline{x}_1, \ldots, \underline{x}_N$ within a volume V. The conservation equation takes the form

$$\frac{\partial P_N}{\partial t} + \sum_{i=1}^{N} \nabla_i \cdot \underline{j}_i = 0 \tag{1}$$

with the fluxes

$$\underline{j}_i = P_N \left(\underline{U}_i + \sum_{j=1}^{N} \underline{\omega}_{ij} \cdot \underline{F}_j \right) \tag{2}$$

arising from both the applied flow through $\underline{U}_i$ and translation due to the forces

$$\underline{F}_i = -\nabla_i V - kT \nabla_i \ln P_N. \tag{3}$$

generated by gradients in the interparticle potential V and thermodynamic potential $kT \ln P_N$.

For flow weak relative to Brownian motion $Pe = a^2 \gamma / D_o \ll 1$ for spheres with radius a and diffusivity D_o subjected to shear rate γ. This permits the probability density to be expanded about equilibrium and the conservation equation to be is integrated over N-2 spheres. The pair distribution function then takes the form

$$P_2 = n^2 g(r) \{1 - \frac{a^2}{2D_o} \frac{\mathbf{r \cdot E \cdot r}}{r^2} \exp(i\omega t) f(r)\} \tag{4}$$

with n the number density of spheres, $\mathbf{r}$ the center-to-center vector, $g(r)$ the equilibrium radial distribution function, $\mathbf{E}$ the rate of strain tensor, and ω the frequency of the oscillating strain field. The governing equation, without three-body integrals, reduces to

$$-i \frac{a^2\omega}{2D_o} f + \frac{1}{r^2} \frac{d}{dr} r^2 G \frac{df}{dr} - G \frac{d}{dr} \frac{V_{mf}}{kT} \frac{df}{dr} - \frac{6H}{r^2} f = -W + r(1-A)\frac{d}{dr} \frac{V_{mf}}{kT} \tag{5}$$

with the boundary conditions

$$\lim_{r \to 2} r(1-A) + G \frac{df}{dr} = 0$$

$$\lim_{r \to \infty} f = 0. \tag{6}$$

In these equations the potential of mean force $V_{mf} = -kT\ln g(r)$ plays the same role as the pair potential in a dilute theory. The hydrodynamic functions G, H, and A here and J and W below are known from solutions to the Stokes equations for two interacting spheres.

For weak flows the viscous, interparticle, and Brownian forces contribute separately to the bulk stress as

$$\Sigma^H = 2\mu_o \mathbf{E} \exp(i\omega t) \{1 + \frac{5}{2}\phi + \frac{5}{2}\phi^2 + \frac{15}{2}\phi^2 \int_2^\infty J(r)g(r)r^2 dr\}$$

$$\Sigma^I = \frac{9}{20} \phi^2 \mu_o \mathbf{E} \exp(i\omega t) \int_2^\infty r^3(1-A(r)) \frac{dg}{dr}(r)f(r)dr \tag{7}$$

$$\Sigma^B = \frac{9}{20} \phi^2 \mu_o \mathbf{E} \exp(i\omega t) \int_2^\infty r^2 W(r)g(r)f(r)dr$$

with μ_o the fluid viscosity. For a steady shear flow $\omega = 0$ so f is real and all three stresses contribute to the low shear limiting viscosity η_o. But in the high frequency limit the time dependent and forcing terms balance in (5) making f imaginary. Consequently, the stress assumes the form

$$\Sigma^H + \Sigma^I + \Sigma^B = 2(\eta'_\infty - i \frac{G'_\infty}{\omega})\mathbf{E} \exp(i\omega t) \tag{8}$$

with the viscous stress determining the dynamic viscosity η'_∞ and the interparticle and Brownian forces producing G'_∞.

2.3. Results

For hard spheres at finite volume fractions the equilibrium pair distribution function g(r) develops an oscillatory structure due to interactions of the pair with other spheres. Under flow the resulting volume fraction dependent potential of mean force generates a complementary structure in f(r) at both zero and high frequencies (Figure 1). When f is in phase with dg/dr the contribution to the stress from the potential of mean force is positive and increases with volume fraction. Consequently the low shear viscosity (Figure 2) and the high frequency modulus (Figure 3) increase rapidly when three-body interactions become important.

The difference between η_0 and η_∞' reflects the contribution of the potential of mean force to the low shear viscosity. For Pe>>1 viscous stresses would render these thermodynamic contributions insignificant, suggesting that $\eta(\gamma\rightarrow\infty)\sim\eta_\infty'$. Thus the thermodynamic stresses $\Sigma^I+\Sigma^B$ provide a mechanism for shear thinning even in hard sphere suspensions.

The data of Krieger[7] and de Kruif, et al.[8] for aqueous polystyrene latices and silica spheres in cyclohexane, respectively, demonstrate the theory to be qualitatively, but not quantitatively, correct (Figure 2). The divergence of the low shear viscosity at volume fractions of $\phi\sim0.63$ and the shear thinning for $\phi>0.25$ are at least strongly suggested by the predictions for η_0 and $\eta_0-\eta_\infty'$, repectively. The predicted values err in magnitude significantly though, presumably because of the approximate hydrodynamics.

3. APPLICATION OF PERCOLATION THEORIES TO FLOCCULATED NETWORKS

3.1. Introduction

In contrast to the fluidlike response of stable systems, flocculated suspensions respond elastically to small deformations with moduli which depend strongly on the volume fraction of particles. As with the stable systems the elasticity arises solely from the non-hydrodynamic interactions. For aggregation resulting from van der Waals attraction, deformation of the network changes the relative positions of the particles, thereby increasing the total potential energy of the system and producing an interparticle force. If the particles actually fuse, deformation of the network strains the particles themselves, reflecting their internal elasticity.

Given the appropriate force law and the exact positions of the particles, one could in principle calculate the elastic deformation of the network. In practice, however, the large number of particles in the network and the absence of detailed information on the structure makes this unappealing. Hence, as with the stable systems, we week an approach which captures the statistical characteristics of the structure and averages the stresses to determine the effective modulus.

The primary difficulty lies in the non-equilibrium nature of the structure which implies a dependence on the mode of formation. Recent work on Brownian flocculation of very dilute, unstable dispersions[14,15] detects flocs containing N particles in a radius R such that $N\sim N_0(R/a)^D$ with D~1.75-2.0, i.e. less than the dimension of physical space. These correspond to a scale invariant, or fractal microstructure with the length scales and mean volume fraction varying radially within the floc as $\phi_{fl}\sim\phi_0(r/2a)^{D-3}$. Other measurements on flocs broken down by a strong shear flow[16], again at infinite dilution, indicate a similar structure

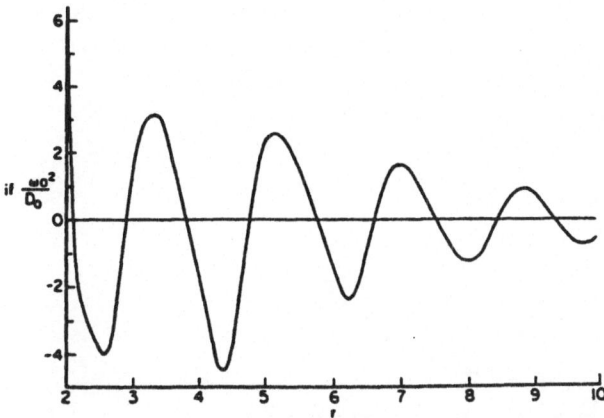

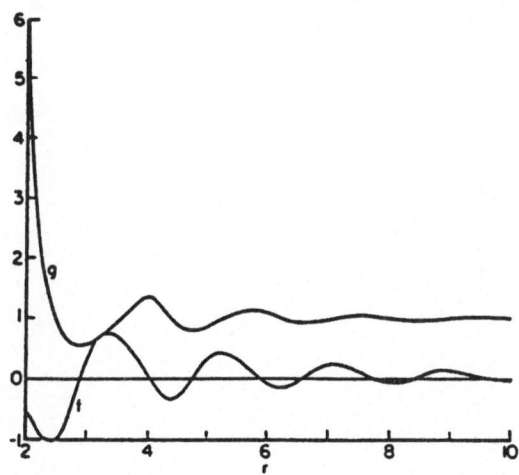

Figure 1: Radial dependence of pair distribution functions for hard spheres in weak flows at $\phi = 0.50$[13]: a) steady shear; b) small amplitude oscillations.

136

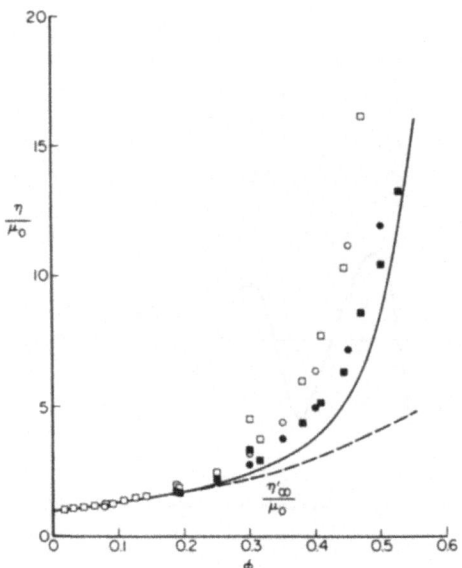

Figure 2: *Comparison of predictions for hard spheres[13] for the low shear limiting viscosity (———) and the high frequency dynamic viscosity (----) with data for aqueous latices.*

	Pe = 0	Pe = ∞
aqueous latices[7]	O	●
silica spheres in cyclohexane[8]	□	■

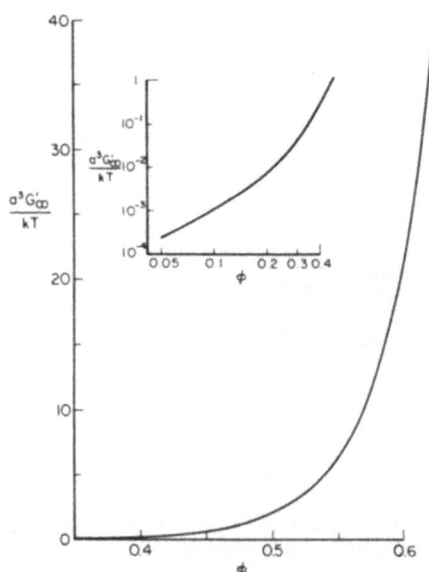

Figure 3: *Predictions of the high frequency modulus for hard spheres[13].*

but with $D=2.5$. In either case growth at a finite volume fraction ϕ, in the absence of gravity, will eventually lead to an infinite network with fractal domains of size $R/2a=(3\phi_o/D\phi)^{1/(3-D)}$.

Simulations of colloidal aggregation have successfully explained the fractal dimensions noted above as the consequence of cluster-cluster aggregation with no subsequent rearrangements[17]. But the extensive computations required appear unattractive for studies of the elasticity of flocculated networks. Instead we have modelled the randomness of the network through percolation theory and calculated the resulting moduli through an effective medium theory[18].

The percolation approach generates the structure by placing sites randomly on a lattice with bonds connecting adjacent filled sites. In the absence of the strong correlations caused by the attractions between colloidal particles, an infinite network forms only above a critical fraction p_c of occupied sites. Below this threshold only finite clusters exist, while just above p_c the probability of a site being part of the infinite cluster increases as $(p-p_c)^m$ with $m=0.3-0.4$ and $p_c=0.312$ for a simple cubic lattice[18]. The infinite cluster alone has fractal dimension $D=2.49$[19]. Presumably the introduction of correlations between sites would alter both, but no results are available.

Transport properties of percolation networks can be calculated exactly by explicit numerical solution of balance laws for each configuration generated in a simulation or approximated analytically by examining a single site embedded in an effective medium. Numerical results for the conductivity K establish that near threshold

$$K - K_o \left\{ \begin{array}{ll} 0 & p_c>p \\ a(p-p_c)^n & p>p_c \end{array} \right. \tag{9}$$

with K_o the conductivity of a single bond, since conduction requires a continuous infinite path from site to site. For a simple cubic lattice $n=1.5\pm0.2$ and $a=1.37$[18].

At higher site fractions the effective medium approximation becomes accurate, determining the conductivity as[18]

$$K - K_o\{p - 1.52p(1-p) + O(1-p)^2\} \tag{10}$$

Figure 4 summarizes these results. The renormalized effective medium approximation of Sahemi, et al.[20], provides a similar picture.

In general though, the elasticity of a network differs qualitatively from the conductivity. While conductors carry a scalar, current, the bonds in the elastic network transmit force, a vector. For example, for the simple cubic lattice the infinite network at $p=p_c$ conducts but does not support an elastic stress unless the bonds between particles transmit torques. With central forces alone the bonds rotate freely, allowing the lattice to collapse under simple shear. The formulation of a model for the elastic network thus requires either a lattice with more nearest neighbors or an angular dependent interparticle potential. Consequently, the elasticity threshold can be higher than for conductivity and the scaling above threshold different[21].

3.2 The Effective Medium Approximation

The effective medium approximation focuses on a single site and replaces the remainder of the partially filled network with a homogeneous

138

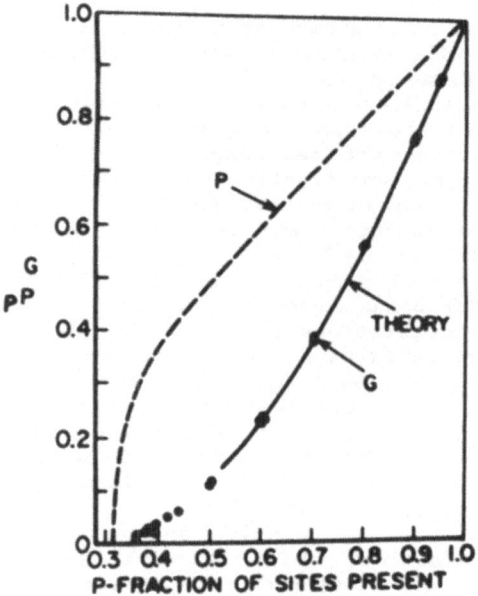

Figure 4: Results for site percolation on a simple cubic lattice[18]*: pP, the fraction of sites in the infinite cluster; K, conductivity.*

network in which the bonds have uniform but unknown force constants. An applied bulk strain $\underline{\varepsilon}$ is then transmitted through the effective medium to bonds surrounding the site which is occupied with probability p and vacant with probability (1-p). The force constants for the effective medium or, equivalently, the elastic moduli then follow from the self-consistency condition that the average strain in the bonds surrounding the site equal the applied strain, i.e.

$$\underline{\varepsilon} = p\underline{\varepsilon}^{occupied} + (1-p)\underline{\varepsilon}^{empty} \tag{11}$$

Solution of the boundary value problem for the homogeneous network containing a single inhomogeneous site requires the Green's functions for the deformation of the homogeneous network due to a point force. Proper superposition of these point forces then produces strain fields which satisfy the balance laws throughout the inhomogeneous network. The calculations are described in detail elsewhere[20-23].

3.3. Triangular Networks with Central and Bond Bending Forces

To examine the effects of lattice geometry and interaction potential we[23] applied the effective medium approximation to two-dimensional triangular lattices with the force and torque on the ith site proportional to the relative displacements and rotations of the surrounding sites. Two constants, G_o and kG_o, characterize the magnitudes of the central and bond bending forces, respectively.

Figure 5 displays the results for Poisson's ratio ν and the bulk modulus $K' = 2G'(1+\nu)/3(1-2\nu)$ as functions of the site fraction p and the ratio k of bond bending to central force constants. For reference the

exact threshold for conductivity is 0.50. Several points are worth
noting: (1) The elasticity threshold decreases to that for conductivity
when k>0. (2) Poisson's ratio decreases to zero at threshold for k>0. (3)
For k=1 ν=0 so the bulk modulus and conductivity become equal with
K'=K-2G'/3. (4) The site fraction at threshold is quite high, i.e. the
lattice is over half full. The last fact is responsible for the accuracy
of the effective medium for these problems[21].

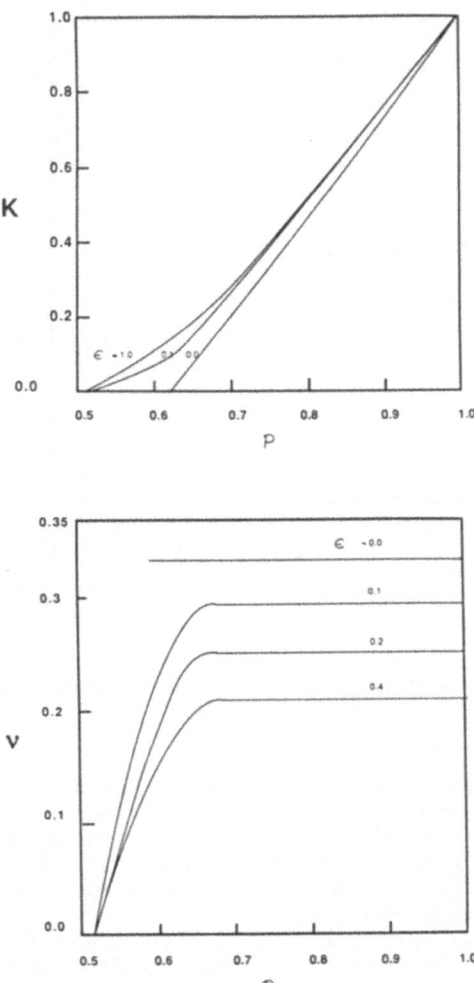

*Figure 5: Predictions from effective medium theory for site percolation
on a two-dimensional triangular lattice as a function of the strength of
bond-bending force[23]: a) bulk modulus; b) Poisson's ratio.*

3.4. Application to Flocculated Dispersions

Data available for the shear modulus and conductivity show several features relevant to these predictions. First, measurements of Mewis and coworkers[2] for carbon blacks in mineral oil indicate G'/K=constant independent of volume fraction (Figure 6). Second, those data and two sets obtained with aqueous polystyrene latices (Figure 7) exhibit detectable values for G' at $\phi \sim 0.01$-0.03. Third, the shear moduli depend very strongly on volume fraction.

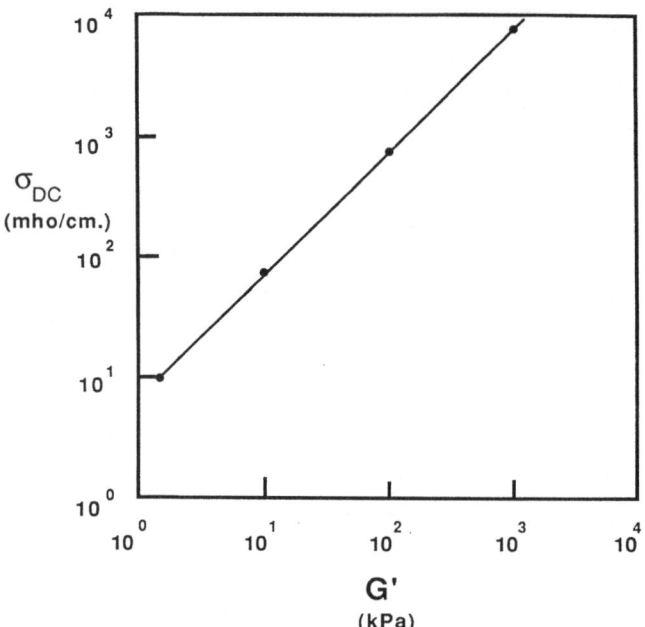

Figure 6: Conductivity and shear modulus for carbon blacks dispersed in mineral oil[2].

The juxtaposition of these data with the predictions of Figure 5 leads us to two conclusions: (1) Despite expectations to the contrary based on interaction potentials between spheres, the bond bending forces appear to be comparable to the central forces. (2) The randomly filled lattice characteristic of the percolation problem does not adequately model a flocculated network with $\phi \sim p$.

The first indicates that the results cited above for conduction on simple cubic lattices, a much easier problem, apply to the shear modulus. The latter suggests that either correlations must be included explicitly in the percolation model or a more appropriate relationship between the site fraction and the volume fraction developed.

We take the latter tack[23]. One failure of the analogy between the flocculated network and the partially filled lattice is the presence of individual sites and clusters not attached to the infinite cluster. Hence we strip the lattice of these by defining the volume fraction as

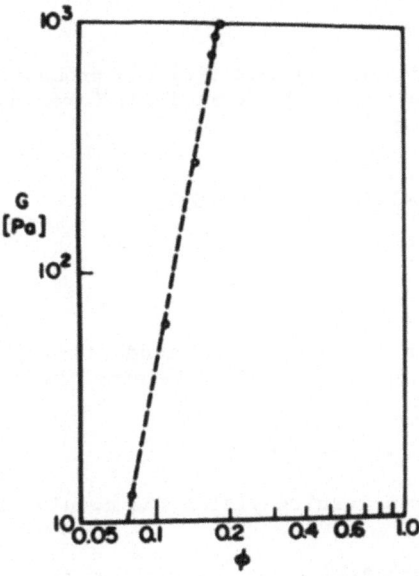

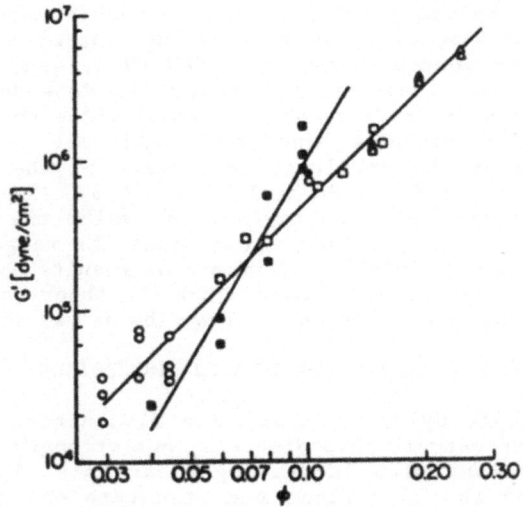

Figure 7: Data for the shear modulus for aqueous polystyrene latices:
a) a = 0.26 μm, ω = 300 Hz[29]; b) a = 0.03 μm, ω = 0[30].

$$\frac{\phi}{\phi_m} = pP(p) \qquad (12)$$

with $\phi_m = 0.52$ for the simple cubic lattice and $P(p)$ the probability that a site resides in the infinite cluster. Kirkpatrick's[18] results for the simple cubic lattice indicate that

$$\phi = \phi_m \begin{cases} 0.5(5.32p-1.66)^m & p<0.5 \\ p & 0.5<p \end{cases} \qquad (13)$$

with $0.3<m<0.4$ and $\phi_m = 0.52$.

In addition the numerical results for conductivity[18] and the renormalized effective medium approximation[20] determine the conductivity and, hence, the shear modulus as

$$G' = G_0(1.45p-0.45)^n \qquad (14)$$

with $1.3<n<1.7$. Combining the two results yields the modulus in terms of the volume fraction as

$$G' = G_0 \begin{cases} (0.273)^n(2\phi/\phi_m)^{n/m} & \phi<0.5\phi_m \\ (1.45\phi/\phi_m-0.45)^n & 0.5\phi_m<\phi \end{cases} \qquad (15)$$

Thus at low volume fractions, i.e. $\phi<0.26$, the modulus should vary as ϕ^s with $3.25<s<5.67$ before turning over to a weaker dependence for $\phi>0.26$. The threshold disappears since the spanning cluster has a fractal dimension less than the spatial dimension, $D=2.49<3$, and, therefore, fills space at any volume fraction. The power law dependence at low volume fractions reflects the ability of additional sites to stiffen the system by securing dangling chains to the elastic backbone.

Figure 8 illustrates the predicted dependence of the modulus on volume fraction for $n=1.7$ and $m=0.36$ such that $s=4.67$. The functional form resembles that of the data in Figure 7 for which the equivalent values of s vary from 2.5 to 5.3. Note though that the range in s from the theory arises from uncertainty in the numerical results while that in the data must reflect different structures. Thus the theory captures the general features of the phenomena but cannot describe as yet the details.

4. A SELF-CONSISTENT FIELD MODEL FOR FLOCCULATED DISPERSIONS
4.1. Introduction

Application of a steady shear flow to a flocculated suspension necessarily ruptures the network into discrete, but strongly interacting flocs whose size decreases with increasing shear rate. The internal structure of each floc should reflect the structure of the original network, with some rearrangement due to the viscous stresses. If the volume fraction of particles varies spatially within the floc then decreasing the floc size changes the effective volume fraction of the dispersion, thereby altering the stresses and making the rheology non-Newtonian.

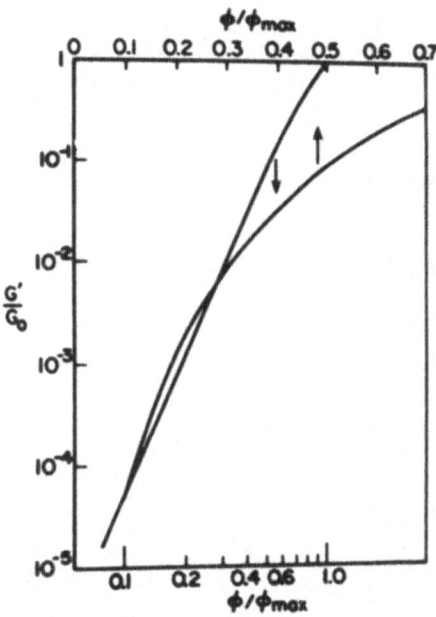

Figure 8: Predicted values of shear modulus from (15) with n - 1.7 and m = 0.36.

The simplified treatment developed here is based on our recent model for the breakup of isolated flocs subjected to a shear flow[28]. This mean-field theory views the flocs as porous spheres of radius R with radially varying internal volume fraction ϕ_{fl}. In accord with recent results for both Brownian flocculation[14] and the breakup of flocs in shear[16] we assume $\phi_{fl}=\phi_o(r/2a)^{D-3}$ corresponding to a fractal structure with dimension D varying from 1.7 to 3.0 depending on the history of the floc. The internal flow is governed by Brinkman's equations[24] with local permeability $k(\phi_{fl})$ and local effective viscosity $\mu(\phi_{fl})$ which depend on volume fraction[25] and hence on position.

Internal flow transmits forces to the solid skeleton of the floc. The resulting elastic deformation follows from an equilibrium equation which depends on the shear modulus G' and Poisson's ratio ν of the floc and includes the viscous drag as a body force. The modulus varies with local volume fraction as discussed in the previous section while Poisson's ratio is held constant. We assume the floc to rupture when the strain energy due to this deformation exceeds a critical value locally, borrowing the von Mises criterion for yielding of brittle solids[26]. The model then assumes a monodisperse suspension of flocs of the maximum coherent size.

Interactions are approximated by embedding the floc in an effective medium (Figure 9). The viscosity jumps from the internal value μ to that of the pure fluid μ_o at radius R and then to μ^* for the effective medium at radius δR with the Stokes equations governing the fluid motion in both regions. For hard spheres at dilute concentrations $\delta=2$ to reflect the excluded volume which prevents closer approach. At higher concentrations, however, three-body and higher order interactions increase the number density of spheres near contact, suggesting $\delta<2$. With flocs some

144

interpenetration is inevitable, leaving δ an unknown but important parameter. Once the full velocity and stress fields are known, setting the average local stress in the suspension equal to that of the effective medium determines μ^*.

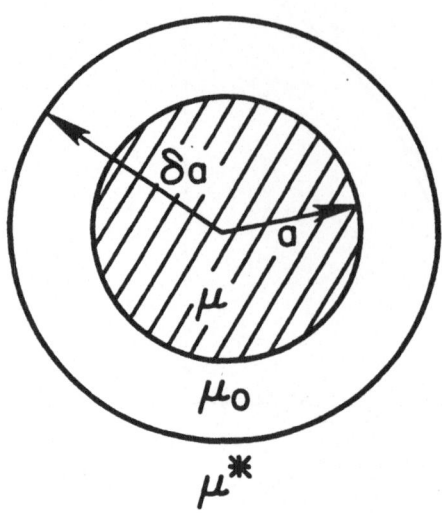

Figure 9:   Unit cell for self-consistent field model.

The approach has flaws since mean-field theories are not strictly valid for fractals[27]. Also the present brief treatment avoids solving the field equations by assembling existing solutions from the self-consistent field model for hard spheres[24] and the breakup of porous flocs at infinite dilution[28]. Nevertheless the theory predicts features of the rheology commonly observed with flocculated systems, suggesting the approach has some merit.

4.2. Mathematical Formulation
The fluid flow is incompressible throughout but governed by different momentum equations in the three regions:

$\delta R < r < \infty$
$$0 = -\nabla p + \mu^* \nabla^2 \underline{u}$$

$R < r < \delta R$
$$0 = -\nabla p + \mu_o \nabla^2 \underline{u}$$

$0 < r < R$
$$0 = -\nabla p + \mu \nabla^2 \underline{u} - \frac{\mu}{k} \underline{u}$$

(16)

At the interfaces the velocities and normal tractions must be continuous while

$$\underline{u} \to \underline{E} \cdot \underline{r} \text{ as } r \to \infty.$$

The deformation of the solid matrix of the floc must satisfy

$$\nabla \cdot \underline{\sigma} + \frac{\mu}{k} \underline{u} = 0$$

with

$$\underline{\sigma} = 2G'\underline{\varepsilon} + \frac{2G'\nu}{1-2\nu} \text{tr}(\underline{\varepsilon})\underline{\delta} \qquad (17)$$

and $\underline{\varepsilon}$ the strain. In the absence of solid-solid contact between flocs $\underline{\sigma} \cdot \underline{n} = 0$ at $r = R$. The von Mises criterion assumes the material to yield when $\underline{\sigma}_d : \underline{\sigma}_d > 2S^2$ locally.

The constitutive relations for the coefficients in these equations can be assembled from several sources. For flocs with low internal volume fractions the permeability is affected only by interactions with nearest neighbors so that[28] $\mu = \mu_o$ and

$$\frac{\mu}{k} = C \frac{9\mu_o \phi_{f1}}{2a^2} \qquad (18)$$

with $C < 1$ representing a shielding coefficient.

The experimental data and the theory cited above suggest a power law dependence of the shear modulus on volume fraction of the form

$$G = G_o \phi_{f1}^n \qquad (19)$$

Similarly

$$S = S_o \phi_{f1}^n \qquad (20)$$

implies that yield occurs at a constant strain so that $S_o = G_o \varepsilon_{max}$.

The formulation is completed by equating the average stress in the suspension to that in the effective medium, i.e.

$$\langle \underline{\sigma} \rangle = 2\mu_o \underline{E} + n \int_0^R (\underline{\sigma} - 2\mu_o \underline{e}) d\underline{r} = 2\mu^* \underline{E} \qquad (21)$$

Here $\underline{\sigma}$ is the viscous stress in the fluid, $\underline{e}$ the local rate of strain, and n the number density of flocs. Note that the overall volume fraction of particles in the dispersion is related to n through

$$\phi = 4\pi n \int_0^R r^2 \phi_{f1} dr \qquad (22)$$

4.3. Approximate Results for Nondraining Flocs

Results available in the literature provide an indication of the rheology expected from this model. Brinkman[24] first constructed the self-consistent field theory for hard spheres with $\delta=1$, obtaining in our notation for a simple shear flow

$$\sigma = (\mu_o\gamma)/(1-2.5\phi_{eff}) \tag{23}$$

with

$$\phi_{eff} = \frac{4\pi R^3}{3}n$$

$$= \frac{D}{3}\left[\frac{R}{2a}\right]^{3-D}\frac{\phi}{\phi_o} \tag{24}$$

In the non-draining limit our theory for the breakup of isolated flocs[28] related the maximum floc size to the stress through

$$\frac{S_{0.}}{\sigma} = \frac{45}{14}(R/a)^{n(3-D)} \tag{25}$$

Since the theory fails when $R\sim a$ we add a minimum floc size such that

$$\frac{S_0}{\sigma} = \frac{45}{14}\left[(R/a)^{n(3-d)}\right]-1 \tag{26}$$

Combining these results relates the stress to the shear rate through

$$\frac{\sigma}{S_o} - \frac{5D2^D\phi}{48\phi_o}\frac{\sigma}{S_o}\left[1+\frac{14}{45}\frac{S_o}{\sigma}\right]^{1/n} = \frac{\mu_o\gamma}{S_o} \tag{27}$$

This constitutive equation for σ vs. γ has several interesting features. For example, in the weak flow limit, i.e. $\mu_o\gamma/S_o \to 0$

$$\frac{\sigma}{S_o} = \frac{14}{45}\left[\frac{5D2^D}{48}\frac{\phi}{\phi_o}\right]^n \tag{28}$$

indicating creep at a constant stress while the flocs break down. Note that the magnitude of this stress is predicted to correlate with the shear modulus. Alternatively, for $S_o/\sigma \to 0$

$$\sigma = \frac{1}{1-B\phi/\phi_o}\left\{\mu_o\gamma+\frac{7D2^D}{216n}S_o\frac{\phi}{\phi_o}\right\} \tag{29}$$

with $B=5D2^D/48$, corresponding to Bingham plastic behavior. In this limit the plastic viscosity $\eta_{pl}=\mu_o/(1-B\phi/\phi_o)$ arises solely from hydrodynamic effects, i.e. is independent of S_o characterizing the strength of the interparticle contacts, while the residual or Bingham stress, corresponding to the constant term, varies linearly with S_o.

The model contains a number of parameters. Some, such as the fluid viscosity and the particle radius, should be known. Others, such as S_o and n, might be deduced from independent measurements of the shear modulus. The parameter D, motivated by the fractal character of flocs studied at dilute concentrations, as well as ϕ_o might be determined by static light scattering if the system is transparent.

To illustrate the behavior predicted at intermediate shear rates we set $\phi_o = -2^{D-3}D/3$, so that $\phi_{eff} = \phi$ for R=a. Then

$$\frac{\sigma}{S_0} - \frac{5}{2}\phi \frac{\sigma}{S_0} \left\{ 1 + \frac{14}{45} \frac{S_0}{\sigma} \right\}^{1/n} - \frac{\mu_0\gamma}{S_0} \tag{30}$$

so that as $\mu_0\gamma/S_0 \to 0$

$$\frac{\sigma}{S_0} - \frac{14}{45} \left[\frac{5}{2}\phi \right]^n \tag{31}$$

and as $\mu_0\gamma/S_0 \to \infty$

$$\frac{\sigma}{S_0} - \frac{1}{1-2.5\phi} \left\{ \frac{\mu_0\gamma}{S_0} + \frac{7\phi}{9n} \right\} \tag{32}$$

Figure 10 illustrates the smooth transition between these limits for several volume fractions.

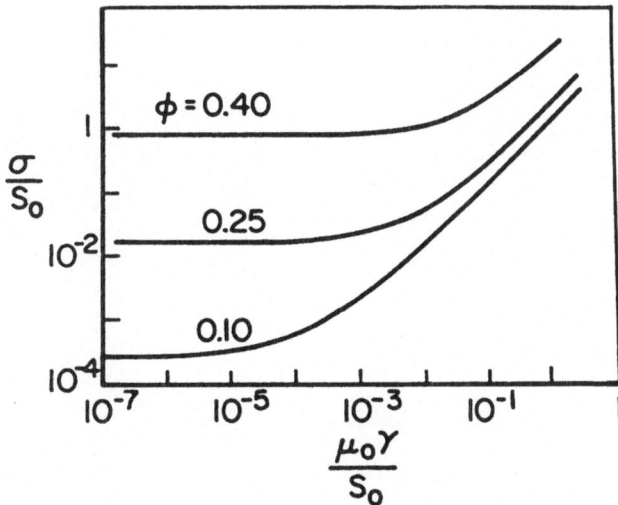

Figure 10: *Stress-rate of strain relations predicted by self-consistent field model with $\phi_0 = 2^{D-3}D/3$ and n = 4.5.*

148

5. SUMMARY

For stable suspensions the non-equilibrium statistical mechanics formalism promises a reasonable description of weak flows for which the structure departs only slightly from equilibrium. Our next step, improving the approximation of the many-body hydrodynamic interactions and evaluating the integral terms neglected here, will be tested against the hard sphere data before moving onto interparticle potentials appropriate for electrostatic, steric, and weak van der Waals forces. For strong flows there as yet appears no alternative to simulations.

Flocculated suspensions still pose considerable difficulties. The percolation theory captures some features of the elasticity of the rest state, but does not reflect appropriately the variation in network structure, i.e. fractal dimension, arising from different histories. Introduction of relevant correlations or the actual physical processes producing the structure may require computer simulations. The self-consistent field treatment of flowing systems appears to explain the constant stress region observed with many flocculated systems at low shear rates and the subsequent shear thinning. The theory is as yet incomplete and untested though.

6. ACKNOWLEDGEMENTS

This work was supported by the International Fine Particles Research Institute and the National Science Foundation through grant CPE-8212327.

REFERENCES

1. Hunter RJ: *Adv. Coll. Inter. Sci.* **17** 197 (1982).
2. Mewis J: *Suspension Rheology.* IFPRI Report: FRR 0601, October 1, 1985.
3. Buscall R; Goodwin JW; Hawkins MW; and Ottewill RH: *J. Chem. Soc. Far. Trans. I* 78, 2983, 2889 (1982).
4. Lindsay HM and Chaikin PM: *J. Chem. Phys.* 76, 377 (1982).
5. Hoffman RL: *Adv. Coll. Inter. Sci.* 17, 161 (1982).
6. Willey SJ and Macosko CW: *J. Rheol.* 22
7. Krieger IM: *Adv. Coll. Inter. Sci.* 3, 111 (1972).
8. de Kruif CG; van Iersel EMF; Vrij A; and Russel WB: *J. Chem. Phys.* 83, 4717 (1985).
9. Batchelor GK: *J. Fluid Mech.* 41, 545 (1970).
10. Batchelor GK: *J. Fluid Mech.* 83, 97 (1977).
11. Bossis G and Brady JF: *J. Chem. Phys.* 80, 5141 (1984); *J. Fluid Mech.* 155, 105 (1985).
12. Ohtsuki T: *Physica A* 108, 441 (1981).
13. Russel WB and Gast AP: *J. Chem. Phys.* 84, 1815 (1986).
14. White DA and Huang JS: in *Kinetics of Aggregation and Gelation* (eds., Family P and Landau DP). Elsevier, 1984, p. 19.
15. Schaeffer DW; Martin JE; Wiltzius P; and Cannell DS: *Phys. Rev. Lett.* 52, 2371 (1984).
16. Sonntag RC and Russel WB: *J. Coll. Inter. Sci.* 113 399 (1986).
17. Family F and Landau DP (eds): *Kinetics of Aggregation and Gelation.* Elsevier, 1984.
18. Kirkpatrick S: *Rev. Mod. Phys.* 45, 574 (1973).
19. Havlin S; in *Kinetics of Aggregation and Gelation* (eds., Family F and Landau DP). Elsevier, 1984, p. 145.
20. Sahimi M; Scriven LE; and Davis HT: *J. Phys. C.: Solid State Phys.* 17, 1941 (1984).
21. Feng S and Sen PN: *Phys. Lett.* 52, 261 (1984).
22. Feng S; Thorpe MF and Gaboczi E: *Phys. Rev. B* 31, 276 (1985).
23. Mall S and Russel WB: *J. Rheol.* (accepted).
24. Brinkman HC: *Appl. Sci. Res.* **A1**, 27 (1947).
25. Kim S and Russel WB: *J. Fluid Mech.* 154, 269 (1985).
26. Adler PM and Mills PM: *J. Rheol.* 23, 25 (1979).
27. Cates ME and Witten TA: *Phys. Rev. Lett.* (submitted).
28. Sonntag RC and Russel WB: *J. Coll. Inter. Sci.* (in press) (1986).
29. Buscall R: *Colloids and Surfaces* 5, 269 (1982).

STRUCTURE FORMATION IN FLOWING SUSPENSIONS

Richard L. Hoffman

Monsanto Company
Springfield, Massachsusetts 01151

1. INTRODUCTION

Following the discovery of particle structure in flowing suspensions in the early 1970's[1,2] several different types of suspensions have been examined carefully, and it has become evident from this work that a number of different structures can occur. Dilute suspensions of monosized polystyrene particles in water comprise one system that has been studied[3-11], and a progression of structures have been observed as the shear stress increases. Without any shear, particles tend to order in a three-dimensional structure, but under increasing levels of shear the ordering shifts to two-dimensional structures, and then one-dimensional structures which finally break up into an amorphous structure at high levels of shear. With more concentrated suspensions of similar composition, a different progression of structures is reported and some authors indicate that order is maintained at all levels of shear[12] while others report a nearly amorphous structure at high levels of shear[13].

Suspensions of monosized polyvinyl chloride spheres in fluids such as di-2-ethylhexyl phthalate have also been examined over a wide range of concentrations, and a number of structural changes have been charted[14]. These systems appear to be stabilized principally by steric stabilization[15], and as a result they provide an interesting contrast to the suspensions of polystyrene spheres which are stabilized by charge alone. Changes in structure have been documented as a function of the concentration of particles, the level of shear, the flow fields utilized and the age of the sample. With these systems an important link occurs between a change in structure and the onset of dilatant or discontinuous viscosity behavior[1,2].

Other studies with dilute and concentrated suspensions of large monosized glass and polystyrene spheres have led to the conclusion that structures also form when these systems flow[16-19]. Since the particles in these systems are much greater than one micrometer in size, colloid forces of interaction are expected to be negligible. In the discussion which follows, we review the results of all these studies, and consider some of the challenges that remain in this growing area of interest.

2. STRUCTURE IN DILUTE SUSPENSIONS OF POLYSTYRENE SPHERES

Much of the work published on dilute suspensions of polystyrene spheres has been done by Clark, Ackerson and coworkers[3-8]. They worked with sets of monosized particles which ranged in size from roughly 0.1 to 0.25 μm in diameter, and the concentration of particles was nominally 0.1 percent by weight. Strong forces of repulsion between the particles were established by treating these aqueous suspensions with ion exchange resins. Early experiments were done with either a rocking cuvette or a concentric cylinder cell in Couette flow, but more recent experiments have been done in torsional flow with a rotating disk cell. Some

152

conclusions obtained with the rather ambiguous velocity profiles in the rocking cuvette have now been abandoned in favor of more recent results obtained in the Couette and torsional flows.

In these studies Clark, Ackerson and coworkers observed a progression of structures which occurred as the rate of shear was increased. At equilibrium without shear, a body centered cubic (bcc) crystal structure was observed, but at low levels of shear the crystallities oriented so that the planes of densest packing were parallel to planes of constant shear, and axes of closest packing pointed in the direction of flow. With notation commonly applied to crystal structures, this means that planes having Miller indices of (110) were parallel to planes of constant shear while axes along the [111] direction pointed in the direction of flow. This arrangement is pictured in Figure 1 where planes of constant shear are parallel to the plane of the paper, and the direction of flow is shown by the axis labeled $\underline{V}$. The axis labeled $\underline{v}$ is normal to planes of constant shear; $\underline{e}$ is perpendicular to $\underline{V}$ and $\underline{v}$ following the convention of Clark and Ackerson.

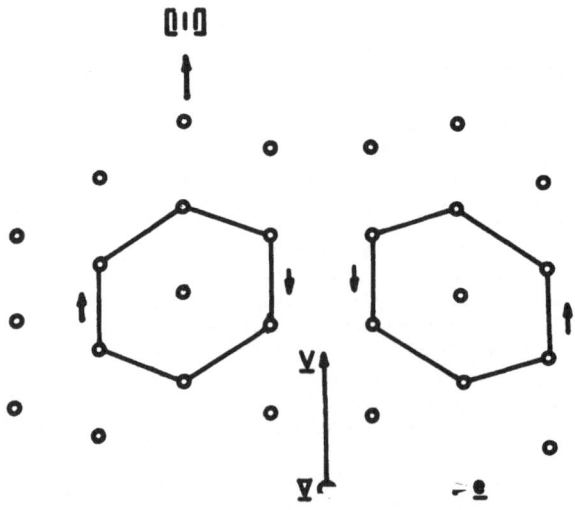

Figure 1: One layer of a bcc packing showing twin structures. Planes of constant shear are parallel to the paper.

Two twin structures are possible as shown by the arrangement of particles depicted on the left and on the right of the $\underline{V}$ axis. The difference between the twins is highlighted by lines showing the distorted hexagon pattern in each of the twins. Light scattering data taken at low shear led Clark, Ackerson and coworkers to conclude that deformation causes a transition from three to two-dimensional order with increasing shear. At small but finite shear rates the twins oscillated as indicated by the four small arrows next to the hexagons and exchanged identities periodically. Further increases in shear, to 10 sec^{-1}, led to a distorted two-dimensional hexagonal close packed structure (2D-hcp) oriented roughly midway between the two twins. This structure may result when hydrodynamic forces prevent a switch from one twin to the other in the time scale required. At this point, two-dimensional layers of the distorted hexagonal close packed structure appeared to slide over one another in the direction of flow. As the shear rate was raised to even higher levels, this sliding layer structure became unstable and degenerated into strings of particles extending parallel to the direction of flow as shown in Figure 2. Yet higher levels of shear ultimately led to a generally random packing of the spheres or amorphous structure. This general progression of structures from three-dimensional to amorphous with increasing levels of shear can probably be associated with a transition in the forces dominating the system. At very low levels of shear the repulsive forces between spheres will dominate and a high level of order will prevail. But at higher levels of shear the hydrodynamic forces will predominate, and in these dilute suspensions this dominance apparently leads to an amorphous structure. Intermediate structures are likely to be associated with varying intermediate ratios of these two opposing forces.

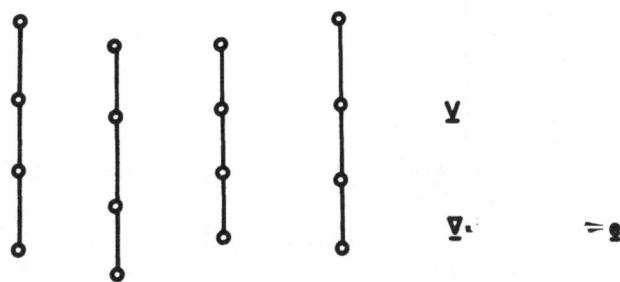

Figure 2: Strings of a few particles running parallel to the direction of flow, $\underline{V}$, have some mean distance of spacing along $\underline{e}$, but no interstring positional correlation along $\underline{V}$.

154

Pieranski, Joanicot and Pansu have also worked with dilute
suspensions of polystyrene spheres in aqueous media of low ionic
strength[9-11]. As reported by Pansu[11] these systems contained particles
nominally 0.1 μm in diameter at the level of a few percent. Couette and
torsional flow were both used to examine structure in these suspensions,
and the stress-strain behavior of these suspensions was measured in the
Couette system as shown in Figure 3. In this figure, ω is the angular
velocity of the inner cylinder and Γ is the torque.

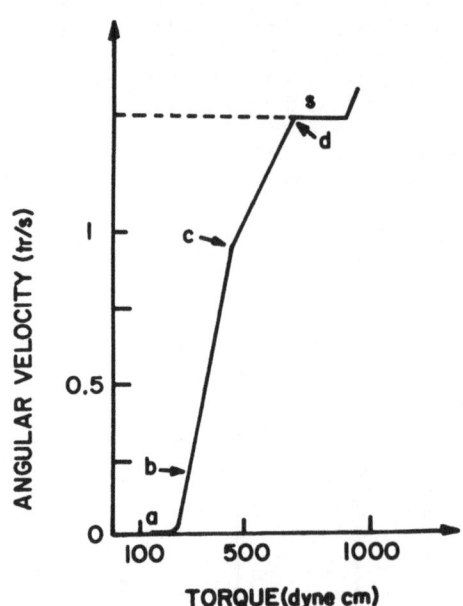

*Figure 3: Stress-strain plot obtained by Joanicot and Pieranski with
their Couette viscometer has four different regions of response. (a-b)
plastic flow of the colloidal crystal; (b-c) coexistence of a plastic
phase and a melted phase; (c-d) flow of a completely melted suspension;
d) onset of Taylor instabilities with the return of crystalline regions
where the flow converges between the vortices (toroidal rings).*

Under low levels of shear, Pieranski and coworkers report that the
particles order into the same oriented bcc structure described by Clark,
Ackerson and coworkers. Observing structure while their suspensions were
deformed in the Couette rheometer, Joanicot and Pieranski established
several different regions of behavior with increasing rates of shear.
Plastic flow of the oriented bcc structure occurred over the region from
(a) to (b) in Figure 3, until it began to "melt" at the inner cylinder of
the rheometer at point (b). From (b) to (c) a melted phase and a crystal
phase coexisted, presumably as the interface between the phases moved
from the inner to the outer wall of the viscometer. Then from (c) to (d)
the suspension flowed as a completely melted system until Taylor
instabilities set in at point (d).

Much of what Pieranski, Joanicot and Pansu report concurs with the observations made by Clark, Ackerson and coworkers, but the sharp transition between a crystal phase and a "melted" phase does not seem to fit with the more gradual transition of structures reported by the latter authors. Reasons for this difference are not clear, and it is likely that further experimental work will be necessary to resolve this matter.

With the occurrence of Taylor instabilities at the higher levels of shear, Joanicot and Pieranski found interesting interactions between the structure and flow. Taylor instabilities led to toroidal rings in which the flow followed the path traced by the arrows in Figure 4. As a result, narrow ring-shaped crystalline regions reappeared where the flow converged between the toroidal rings. Recrystallization occurred in these regions because the shear rate dropped below the critical level required for melting to occur. Following the cessation of shear, the suspension also crystallized into a bcc structure throughout the toroidal rings. In these rings the orientation of the bcc structure rigorously followed the prior flow patterns. That is, the (110) planes of the packing were coincident with surfaces of constant shear, and these crystal planes tended to orient with [111] rows parallel to the direction of the flow. Knowing this one is tempted to suggest that a completely amorphous structure did not occur even when the suspension was in the "melted" state. It is possible, however, that reorientation occurred over the finite but short time span required for the shear to drop to zero.

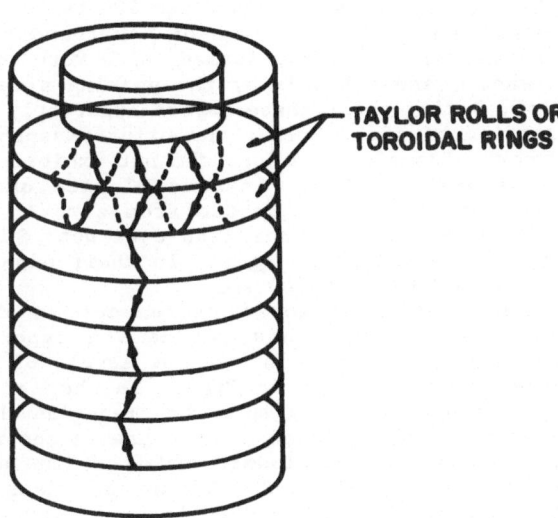

TAYLOR ROLLS OR
TOROIDAL RINGS

Figure 4: Pattern of Taylor instabilities in the Couette viscometer.

Joanicot and Pieranski also found that orientation of the bcc structure in the toroidal rings had a profound effect on the direction of motion in the rings when flow was reinitiated. Reinitiating the flow caused the crystals to "melt" again, but the direction of flow in the Taylor rolls was intimately tied to relative values of the angular velocity, ω, of the inner cylinder in each of two successive runs. Under certain conditions the direction of flow in the Taylor rolls reversed. Reasons for this relate to changes in the vectorial direction of the flow in the rolls with changes in ω, and the fact that there are two directions in an oriented bcc (110) plane which will allow for flow in the direction of closest packed rows. See Figure 1 and note that twins are not required. One of the closest packed rows, marked [111] on Figure 1, will promote flow in the same direction as before while the other will induce flow which is wound in the opposite direction around the toroidal rings. Which direction prevails is determined by the closest packed row that comes nearest to matching the orientation of the flow path required by the Taylor rolls.

3. STRUCTURE IN SUSPENSIONS OF POLYSTYRENE SPHERES AT HIGHER CONCENTRATIONS

Tomita and van de Ven studied aqueous suspensions of monosized polystyrene spheres in which the volume fraction of the particles ranged from 0.05 to 0.45[12]. Particles with two different diameters, 0.128 and 0.497 μm, were available for this work. Strong forces of repulsion between the particles were realized by deionizing the latices with ion exchange resins before the experiments were begun, and torsional flow between parallel glass plates was used to study particle structure during shear. The spacing between the plates ranged from 100 to 300 μm.

With these suspensions Tomita and van de Ven found that layers of hexagonally close packed particles were formed which were coincident with planes of constant shear, and one axis of the packing always pointed in the direction of flow. During shear these layers apparently slid over one another in the direction of flow. Unlike the dilute suspensions studied by the previous authors, these suspensions did not exhibit shear induced melting at higher levels of shear. Instead, Tomita and van de Ven observed that as the shear rate increased from zero to intermediate levels, the diffraction patterns became less distinct. Subsequently at higher shear rates they became clear again. In their experiments shear rates spanned a range from 200 to 10,000 sec.$^{-1}$.

Using two different diffraction measurements to determine the spacing between particles, Tomita and van de Ven reported that the spacing changes with the level of shear. At low levels of shear the 2D-hcp was almost equilateral as shown in Figure 5a, but at intermediate levels, the packing became distorted within the planes as shown in Figure 5b. Further increases to high levels of shear caused a compression of the packing within the layers to a more closely packed 2D-hcp structure which was nearly equilateral once again. See Figure 5c. Mass conservation requires that the spacing between the layers increase when the compression occurs within the layers. The lower the volume fraction of particles, the greater was the increase in spacing, and the higher the volume fraction of particles, the higher was the shear rate at which the spacing increased.

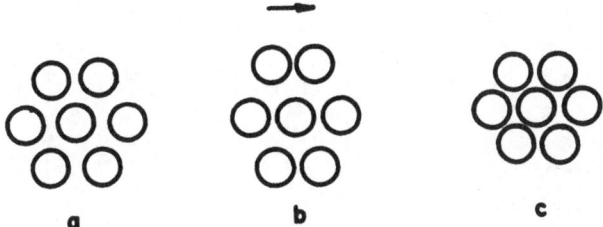

Figure 5: Changes in the spacing of spheres within the layers of hexagonally close packed particles as a function of the level of shear: (a) low shear ($<10^2$ sec.$^{-1}$); (b) intermediate shear ($\sim10^3$ sec.$^{-1}$); (c) high shear (6×10^3 sec.$^{-1}$). The levels of shear are for a particle loading of 26% by volume. The direction of flow is shown by the arrow.

Quite recently Ackerson, Hayter, Clark and Cotter[13] have reported results from a study of aqueous, charge stabilized suspensions of intermediate concentration containing polystyrene spheres 0.109 μm in diameter. The solids level was nominally 14% by weight. The sample cell used in this study was a Couette flow device in which the outer cylinder was rotated to generate the shear fields. The outer cylinder was nominally 75 mm in diameter with a gap of 1.5 mm between the inner and outer walls. Neutron scattering was used to follow changes in structure as shear rates were increased from zero to levels as high as 1720 sec.$^{-1}$.

Although these suspensions appear to be quite similar to the ones studied by Tomita and van de Ven, Ackerson and coworkers did not see any evidence for particles reordering into more dense layers at high levels of shear. They only saw evidence for shear induced melting with increasing levels of shear. (Higher levels of shear may have been necessary to see reordering). They did find, however, that low levels of shear oriented the particles in the manner described by Tomita and van de Ven. Layers of particles packed in a 2D-hcp structure were parallel to surfaces of constant shear and one axis of the packing pointed in the direction of flow. Without any flow these layers were apparently stacked in the third dimension in close packed, face centered cubic (fcc) twin structures. Two layers of this fcc packing are shown in Figure 6a. The plane of the paper is parallel to planes of constant shear, and the direction of flow is shown by $\underline{V}$. One layer of the 2D-hcp particles (closed circles) with Miller indices of (111) in the fcc structure is shown over another (111) layer (open circles).

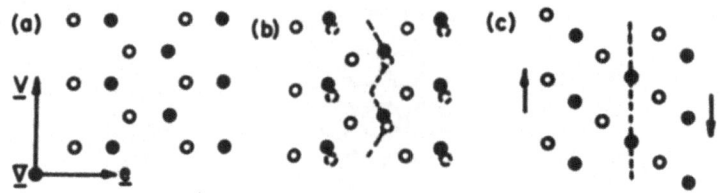

Figure 6: Particle structures as a function of shear as reported by Ackerson and coworkers for aqueous suspensions of polystyrene spheres at ~14 Wt. % solids.

Increasing the level of shear to 30 sec.$^{-1}$, Ackerson and coworkers obtained a diffraction pattern which led them to suggest that (111) layers slip over one another in a zig-zag fashion from one fcc registration site to the next as shown by the dashed line. The dashed circles show the equilibrium positions of the solid circles given in Figure 6a.

Still higher levels of shear, 160 sec.$^{-1}$, gave a diffraction pattern which indicates that particle placement in the 2D-hcp layers was distorted. Oscillatory in-plane shear distortions indicated by the arrows in Figure 6c are thought to occur as a means of minimizing the resistance to flow. Note that the placement of spheres in Figure 6c is distorted relative to the placement of the spheres in Figure 6a. Evidence for a tendency toward more random placement of spheres or amorphous structure also appears in the diffraction pattern.

Finally at a shear rate of 400 sec.$^{-1}$, the diffraction pattern obtained gives evidence for a largely random or amorphous structure although the scattering intensity varies in a way which indicates that the particles tend to align in the direction of flow. This result stands in contrast to the observations of Tomita and van de Ven who saw evidence for particles reordering into more dense layers at high levels of shear. A number of factors could be responsible for these differences. One is that higher levels of shear may have been necessary in the work of Ackerson and coworkers to see the return of order. On the other hand, lack of full shear induced melting in Tomita and van de Ven's work may relate to the fact that light did not pass through their samples but was scattered back from particles near the surface of one of the glass plates. If structure in the bulk of the sample is different from structure near the restraining boundaries of the walls then their results would be biased towards structure near the walls. Particles structure could also be influenced by the spacing between the walls of the cell. Tomita and van de Ven had a spacing of 100 to 300 μm while the spacing was at least 1 mm in all the other studies discussed in detail up to this point.

4. STRUCTURE IN SUSPENSIONS OF POLYVINYL CHLORIDE SPHERES

Hoffman has studied suspensions of monosized polyvinyl chloride spheres in oils such as di-2-ethylhexyl phthalate[1,2,14]. Sets of monosized particles ranging from 0.416 to 1.25 μm have been used in this work. A distinguishing feature of these suspensions is that they exhibit dilatant and discontinuous viscosity behavior over a wide range of particle concentrations from roughly 18 to 60% by volume. An example of this flow behavior is given in Figures 7 & 8 with data on suspensions of polyvinyl chloride (PVC) in di-2-ethylhexyl phthalate (DOP). A change in structure is intimately tied to this behavior[1]. Results obtained with these suspensions also provide an interesting contrast to the work on aqueous suspensions of polystyrene spheres because these suspensions appear to be stabilized principally by steric stabilization[15] although charge stabilization could also be involved[20].

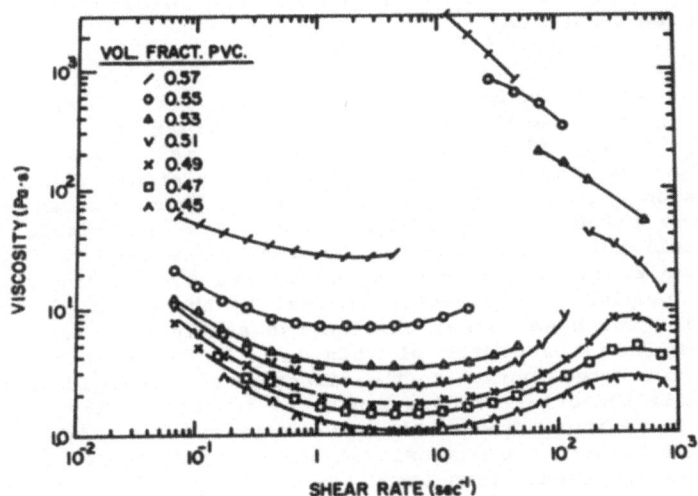

Figure 7: Dilatant and discontinuous flow behavior as a function of particle concentration in suspensions containing 1.25 μm polyvinyl chloride spheres in di-2-ethylhexyl phthalate. The data were taken in cone and plate flow at 25°C.

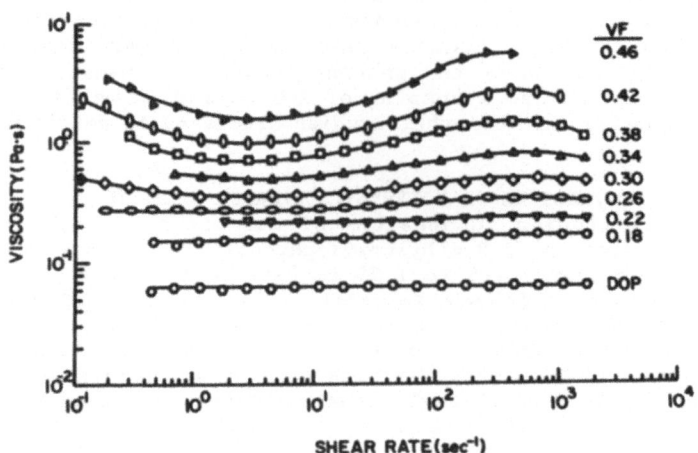

Figure 8: Dilatant and discontinuous flow behavior as a function of particle concentration in suspensions containing 1.25 μm polyvinyl chloride spheres in di-2-ethylhexyl phthalate. The data were taken in cone and plate flow at 25°C. VF signifies the volume fraction of the particles in the suspension.

For his recent studies of structure as a function of particle concentration, shear level, and sample age, Hoffman used torsional flow and simple shearing flow, both between parallel glass plates[14]. In torsional flow the spacing between the plates was either 25 or 60 μm. The results obtained in either case were the same. Thicker samples gave the same patterns although they became fainter as the sample thickness increased until multiple scattering finally obliterated the diffraction patterns completely. Structure was monitored by passing white light through the samples in a direction normal to the glass plates, and in torsional flow the beam passed through the sample at a radial position roughly 2 cm from the center of rotation. Diffraction patterns were projected onto a screen normal to the direction of the incident beam.

Unlike all the other studies mentioned in this review, Hoffman used white light to monitor structure because it provides a very easy means of distinguishing between two-dimensional and three-dimensional structures. With a two-dimensional structure (or grating) all wavelengths are diffracted without discrimination so that each diffraction spot will show a continuous spectrum of colors extending in the radial direction from the center of the diffraction pattern. A three-dimensional grating, however, is selective in that only certain discrete wavelengths will give diffraction spots[21,22]. As a result, separate monochromatic spots will be observed with a three-dimensional grating. Diffraction patterns obtained from two-dimensional gratings which are stacked one on top of another in a random manner will have the same form as the diffraction pattern from a single layer as long as the direction of the axes does not vary[1].

Throughout his work, Hoffman has only obtained diffraction patterns characteristic of two-dimensional ordering. With samples containing 0.54 volume fraction PVC in DOP, several different diffraction patterns were obtained at various levels of shear. Samples which were allowed to sit for a few seconds after being sheared in simple shearing flow below the discontinuity shear rate (~8 sec.$^{-1}$) gave the six arm star shown in Figure 9a. The full spectrum occurs in each of the arms with the colors running from blue where the solid dots are to red where the x's are. The rest of the field is dark. (The central spot for the undiffracted beam has been omitted from this and all other diffraction patterns shown in Figure 9). This pattern indicates that the particles were ordered into layers of two-dimensional hexagonal packing which were parallel to planes of constant shear. With the arrows pointing in the direction of flow, this diffraction pattern also tells us that one axis of the packing pointed in the direction of flow. The ordering which gives this pattern is shown in Figure 9b.

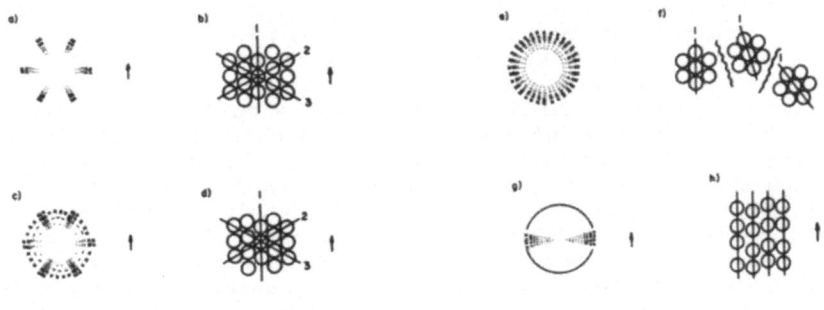

Figure 9: *Particle ordering and diffraction patterns for sheared suspensions of polyvinyl chloride spheres in di-2-ethyl hexyl phthalate. In each diffraction pattern, (a), (c), (e), and (g), the full spectrum occurs with colors running from blue where the solid dots are to red where the x's are. Arrows show the direction of flow.*

During shear, at a level below the viscosity discontinuity shear rate, the diffraction pattern is a bit different as shown in Figure 9c. The two horizontal arms are weaker than the other four, and there is some weak bridging of color between the arms. The packing that yields the two less intense arms in this diffraction pattern is shown in Figure 9d. It is distinguished from the packing shown in Figure 9b by a difference in the rigor of the packing along the three axes. Particles located along the axis pointing in the flow direction (labeled "1" in Figure 9d) are more uniformly placed than they are in the axial directions labeled "2" and "3". This occurs as a result of slight displacements of the "1" rows relative to one another in the flow direction. The weak bridging of colors between the arms of the star is caused by some deviation from the condition that one axis of the packing always points in the direction of flow.

Moving to shear rates just above the discontinuity shear rate with the same sample, one finds an abrupt change to the diffraction pattern shown in Figure 9e. In this pattern, uniform rings of color are seen in place of the six arm star, and the central portion of the pattern inside the rings is brightened by white light. This diffraction pattern is caused by a more random placement of the particles in which the layers of particles still tend to exist and the particles still tend to pack in a two-dimensional hexagonal packing within these layers, but the orientation of axes of the packing varies randomly from one place to the next as shown in Figure 9f. These differences in the two-dimensional hexagonal packing are quite evident in photomicrographs of the particles taken after shear. This abrupt change in structure apparently occurs because particles begin jamming into one another between the layers thereby destroying the long range order. Most of the randomization of particle placement occurs within the layers, however, because of the restraints imposed by the high volume fraction of solids, i.e. 0.54. Distances between particles are of the order of 0.04 times the particle diameter for the packing that is observed at low shear.

Dropping the solids level to a volume fraction of 0.46 and shearing the sample at a low level of shear (3.7 sec.$^{-1}$), Hoffman obtained a diffraction pattern very similar in form to the one shown in Figure 9c, except that the intensity of the arms dropped relative to the intensity of the rings of color bridging between the arms. This information led him to conclude that the ordering is basically the same as the ordering shown in Figure 9d, except that there is more deviation from the condition that one axis of the 2D-hcp points in the direction of flow. Increasing the level of shear to 105 sec.$^{-1}$ which is just short of the peak in the dilatant viscosity region of the curve, Hoffman obtained a white disk for his scattering pattern. This scattering pattern indicates that the ordering of the particles was random. It appears, however, that the bias towards layers was still there because within five seconds after shear the white disk reverted to the ring pattern shown in Figure 9e, but without the white center. Apparently the particles were randomly placed in rough layers, and once flow ceased the repulsive forces between the particles were able to move them into the local cells of hexagonal packing shown in Figure 9f.

Reducing the solids loading to even lower levels (38 and 22%), and testing the samples at a low level of shear, Hoffman obtained yet another diffraction pattern shown in Figure 9g. This pattern was basically a white disk with a band extending across the disk at right angles to the direction of flow. In the most distinct case observed, the band looked like one set of arms from the star pattern. At a solids level of 38% there was faint evidence of rings of color, but all traces of the rings were gone at 22%. This pattern is probably caused by a tendency for the particles to align in the direction of flow as shown in Figure 9h. The white disk surrounding the band indicates that the particles were otherwise randomly ordered. After shear, these patterns reverted to the ring pattern shown in Figure 9e, but without a white center. Once again this change in the pattern after shear indicates a bias during shear toward the layering of particles in planes parallel to planes of constant shear.

Finally at a solids level of 8%, Hoffman obtained white disk patterns both during and after shear. This indicates that the particles were randomly placed under both of these conditions.

5. STRUCTURE IN SUSPENSIONS OF LARGE GLASS AND POLYSTYRENE BEADS

Evidence of structure has also been observed in suspensions of glass beads in various suspending fluids. Patzold[16,17] studied suspensions of monosized glass spheres in mineral oil and glucose/water solutions. The size of his various sets of spheres ranged from 14 to 79 μm. Because of the size of the beads and the fluids involved, one expects Brownian motion to be suppressed and the effect of colloid chemical forces of interaction to be minimal. Under these conditions, the flow response obtained will depend upon the flow field superimposed upon the suspension and the hydrodynamics of the flow about the particles.

Viewing suspensions at a high solids level after shear, Patzold found that his spheres were ordered in a two-dimensional hexagonal packing. Information on packing in the third dimension was not obtained. As with the other suspensions mentioned in this review, Patzold found that the two-dimensional hexagonal packing was parallel to planes of constant shear, and one axis of the packing pointed in the direction of flow.

In another study concerned with suspensions of large particles, Gadala-Maria and coworkers have recently looked for structure in dilute and concentrated suspensions of neutrally buoyant polystyrene spheres in silicone oils[18,19]. The size of their spheres ranged from 40 to 50 μm in diameter, and the suspensions were tested in a clear plastic Couette viscometer.

The outside diameter of the inner cylinder was 9.929 cm and the inside diameter of the outer cylinder was 10.193 cm thereby giving a gap to particle diameter ratio of roughly 29. Flow was driven by rotating the outer cylinder of their Couette device.

To monitor the placement of the particles, the suspensions were illuminated with white light from a fiber optics lamp, and images of the particles in the plane of the suspension perpendicular to the axis of rotation were viewed with a camera. Observations were generally restricted to the surface plane of the suspension except in the most dilute suspension where it was possible to view particles slightly below the surface. Gadala-Maria and coworkers believe that end effects are small. Data obtained from the photographs were used to calculate pair distribution functions for particles about a reference particle.

Using this procedure to collect data on suspensions containing spheres at 1 to 5% by volume, Gadala-Maria and coworkers found that doublets tended to be aligned in the direction of flow, but alignment of larger groups of particles, e.g. triplets, was not evident. This result seems to be consistent with the fact that an isolated doublet of spheres will rotate with a variable angular velocity which favors alignment in the direction of flow[23,24].

In more concentrated systems, e.g. with spheres at a level of 50% by volume, their data indicate that particles tend not only to align in the direction of flow, but also they tend to have two other favored positions around a sphere. With a reference axis running through the center of a sphere and in the direction of the gradient in shear, the latter two positions occur roughly 40 degrees away from this axis and on those sides of the sphere from which other spheres approach it in the flow.

These observations give evidence for structure in dilute and concentrated suspensions which are dominated by hydrodynamic interactions. Thus one must conclude that the assumption of an isotropic distribution of spheres is incorrect even for these relatively simple systems. An isotropic distribution of particles may only occur when the dominant force in a flowing suspension is Brownian.

6. SUMMARY

The occurrence of particle structures in flowing suspensions is now recognized as an important aspect of suspension rheology. Systems which display a high degree of ordering include charge stabilized suspensions of polystyrene spheres in aqueous media and sterically stabilized suspensions of polyvinyl chloride spheres in oil-like fluids called plasticizers. In studies with these suspensions, the particles have ranged in size from 0.1 to 2.0 μm, and strong forces of repulsion between the particles are a key factor in structure formation.

There is evidence, however, for structure formation when the colloid chemical forces of interaction are expected to be negligible. Ordering has been observed, for example, in suspensions of large monosized glass and polystyrene spheres in various suspending fluids. In these suspensions hydrodynamic forces should be dominant.

Further work is needed to establish the full extent of structure formation in flowing suspensions systems. This work will help build our understanding of the phenomenon.

REFERENCES

1. Hoffman RL: *Trans. Soc. Rheol.*, **16**, 155 (1972).
2. Hoffman RL: *J. Colloid Interface Sci.*, **46**, 491 (1974).
3. Clark NA and Ackerson BJ: *Phys. Rev. Lett.*, **44**, 1005 (1980).
4. Ackerson BJ and Clark NA: *Phys. Rev. Lett.*, **46**, 123, 1981).
5. Ackerson BJ and Clark NA: *NATO Adv. Study Inst. Ser.* Ser. B 73, 781 (1981).
6. Hanley HJM; Rainwater JC; Clark NA and Ackerson BJ: *J. Chem. Phys.* 79 4448 (1983).
7. Ackerson BJ and Clark NA: *Physica*, **118A**, 221 (1983).
8. Ackerson BJ and Clark NA: *Physical Rev.*, **30**, 906 (1984).
9. Pieranski P: *Contemp. Phys.*, **24**, 25 (1983).
10. Joanicot M and Pieranski P: *J. Physique Lett.*, **46**, L-91 (1985).
11. Pansu B: *Viscoelastic Properties of Colloidal Crystals*, presented at the European Mechanics Colloquium #191 on The Physics of Dispersions of Small Particles, April 1985, Cambridge University, Cambridge, England.
12. Tomita M and van de Ven TGM: *J. Colloid Interface Sci.*, **99**, 374 (1984).
13. Ackerson BJ; Hayter JB: Clark NA and Cotter L: *J. Chem. Phys.*, **84**, 2344 (1986).
14. Hoffman RL: *Structure Formation in Sheared Suspensions with Dilatant of Discontinuous Flow Behavior*, presented at the 58th Colloid and Surface Science Symposium, June 1984, Carnegie-Mellon University, Pittsburgh; *Flow Induced Microstructure in Concentrated Suspensions*, presented at the European Mechanics Colloquium #191 on The Physics of Dispersions of Small Particles, April 1985, Cambridge University, Cambridge, England.
15. Willey SJ and Macosko CW: *J. Rheol.*, **22**, 525 (1978).
16. Michele J; Patzold R and Donis R: *Rheol. Acta*, **16**, 317 (1977).
17. Patzold R: *Rheol. Acta*, **19**, 322 (1980).
18. Husband DM and Gadala-Maria F: *J. Rheol.*, **31**, 95 (1987).
19. Parsi F and Gadala-Maria F: *Correlation Between Shear Stress and Particle Arrangement in Concentrated Suspensions*, Presented at the 57th Annual Meeting of the Society of Rheology and by private communication.
20. Merinov YA; Guzeev VV; Berezov LV and Krupnova MN: *Colloid J. USSR*, **43**, 392 (1981).
21. Sommerfeld A: *Optics, Lectures on Theoretical Physics*, Vol. IV, Academic Press, New York (1964).
22. Lipson H and Cochran W: *The Crystalline State Vol.III, The Determination of Crystal Structures*, G. Bell & Sons Ltd., London (1957).
23. Bartok W and Mason SG: *J. Colloid Sci.*, **12**, 243 (1957).
24. Darabaner, CL and Mason SG: *Rheol. Acta*, **6**, 273 (1967).

VISCOELASTIC PROPERTIES OF CONCENTRATED STERICALLY STABILIZED LATICES AND
THE EFFECT OF ADDITION OF FREE POLYMER

Th F TADROS

ICI Plant Protection Division
Jealaott's Hill Research Station
Bracknell, Berks. RG12 6EY
United Kingdom

1. INTRODUCTION

Sterically stabilized latices are those containing adsorbed or grafted polymer chains which are strongly solvated by the medium. These heavily solvated grafted or adsorbed chains confer repulsion when the centre-to-centre separation of the particles R becomes less than twice $(a+\delta)$ where a is the particle radius and δ is the adsorbed layer thickness. According to the classical theories of steric interactions (the so called pragmatic theories by Napper[1] two main contributions may be distinguished for repulsion. The first contribution is a mixing term, G_{mix}, which appears as soon as the chains overlap with each other. Under such overlap conditions, the segment density becomes greater than in the rest of the adsorbed layer. This means that the chemical potential of the solvent in the overlap region is lower than that in bulk solution. A simple expression for G_{mix} may be derived using the Flory-Krigbaum theory[2] as for example was given by Fischer[3], i.e.

$$G_{mix} = \frac{2kT\, V_2^2}{V_1}\, v_2^2\, (1/2 - \chi)\, R_{mix}\, (h) \qquad (1)$$

where V_2 and V_1 are the molar volumes of polymer and solvent respectively, v_2 is the number of adsorbed or grafted polymer chains per unit area, χ is the Flory-Huggins chain-solvent interaction parameter, k is the Boltzmann constant and T the absolute temperature. R_{mix} is a geometric function of a, δ and h (the interparticle separation distance) which depends on the form assumed for the segment density distribution ϕ_2. (z) and the interaction region.

The second contribution to steric repulsion arises from the loss of configuration entropy on approach of a second surface[4], the so-called volume restriction (G_{VR}) or elastic G_{el} term,

$$G_{el} = 2kT_2\, \ln\frac{\Omega(h)}{\Omega(\infty)} = 2kT\, v_2\, R_{el} \qquad (2)$$

where $\Omega(h)$ is the number of configurations at distance h and $\Omega(\infty)$ is the number of configurations at infinite distance. R_{el} is a geometric function of a, δ and h which depends on ϕ_2 (z) and the interaction region.

The form of the total interaction free energy (G_T) - distance (h) curve between two flat plates having adsorbed tails or loops is given in

168

Figure 1, after Hesselink et al[5] (for details of the calculation the reader should refer to reference 5). The calculations shown in Figure 1 show that G_M starts at longer distance of separation than G_{VR}. This is understandable since G_{mix} appears as soon as the layers overlap, whereas G_{VR} (or G_{el}) only becomes operative at significant overlap (mostly G_{VR} becomes operative when the distance of separation between the adsorbed layers becomes smaller than one adsorbed layer thickness). Figure 1 also shows the repulsion due to tails and loops and the free energy minimum, G_{min}, which results from the residual van der Waals attraction.

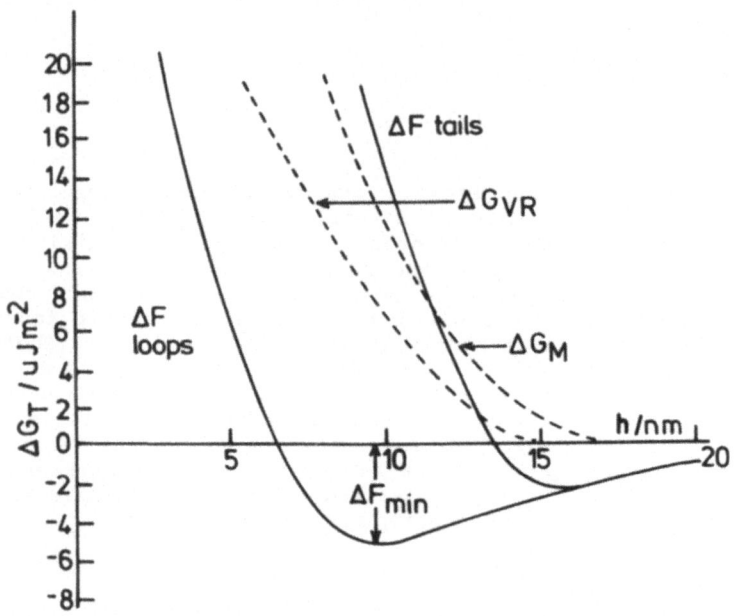

Figure 1: *Total free energy between two flat plates having adsorbed tails or loops. Parameters used in the calculation: Hamaker constant $A = 10^{-20}$; expansion coefficient $\alpha = 1.2$; adsorption density $M = 2 \times 10^{-8}$ gcm^{-2}, molecular weight $M = 6000$; root mean square end to end distance of unperturbed chain $(r^1)_0^{1/2} = 52$ nm; area per chain = 50 nm^2.*

Several methods may be applied for investigation of the steric repulsive force between adsorbed or grafted chains. The earliest techniques were based on measurement of the thickness of the layer using capacitance[6] or light reflectance[7] methods. The latter method was used for poly(vinyl alcohol) films with various molecular weights from which the repulsive force was measured and compared with the theoretical predictions using the Hesselink et al[7] theory. Although the investigations on liquid films give a great deal of information on steric repulsion between polymer layers, they do not represent the real situation in a disperse system consisting of polymer latices (or droplets in an emulsion). The first attempt to measure the repulsive force between particles having adsorbed polymer layers was made by Doroszkowski and Lambourne[8] who measured the compression force for latex particles spread on a heptane/water interface using a surface balance. These measurements give the repulsive force between the particles in two dimensions.

Measurement in three dimensions have been made using a compression cell apparatus by Ottewill and coworkers[9]. Latex dispersions (average diameter 155 nm) in a hydrocarbon medium (dodecane) and containing a stabilizer of poly (12-hydroxystearic acid) were compressed in a specially designed osmotic pressure cell and the change in the volume fraction of the dispersion was measured as a function of the applied pressure, at room temperature. Initially, at a volume fraction of 0.47, osmotic equilibrium occurred slowly. However, at a volume fraction of about 0.55, the resistance to compression increased considerably with small changes of volume fraction and at 0.566 very considerable resistance to compression was experienced. Indeed the system started to behave as a material with a large elastic modulus. Electron microscopy showed the particles to be in a hexogonally close packed array. It was thus possible to calculate the distance between the particle cores and from a knowledge of the bare particle radius, the shell thickness was found to be 7.25 nm. Since the length of the fully extended poly(hydroxystearic acid) is about 9.0 nm, it was suggested that some penetration or compression of these layers must have taken place during the compression process.

It is clear from the above compression experiment that a sterically stabilized dispersion changes from a viscous to an elastic system over a narrow range of volume fraction, at which steric interaction (due to compression and/or penetration becomes significant. It is thus possible, in principle, to use rheological measurements to follow such an interaction in sterically stabilized dispersions. This represents the first part of this short review. For this purpose sterically stabilized latex dispersions of polystyrene with grafted poly(ethylene oxide) were used. Details of the preparation of these latex particles will be given in a forthcoming publication[10]. The average particle diameter of these particles was 350 nm, as measured by photon correlation spectroscopy. This represents the hydrodynamic radius of the particle and grafted poly(ethylene oxide)(PEO). The latter had an average molecular weight of 2000.

The second part of this short review is concerned with the effect of free (non-adsorbing) polymer on the interaction between these sterically stabilized latex dispersions. It has been observed before[11] that dilute polystyrene latex particles carrying terminally-anchored PEO chains dispersed in water, flocculated weakly and reversibly over a certain concentration region of added free PEO. The concentration range at the onset of weak flocculation decreases with increase of molecular weight of the "free" PEO. The origin of the attraction has been discussed in detail by several authors[12-14]. Basically, when the net energy of interaction between the polymer and the surface χ_s (χ_s is the difference in energy between a solvent molecule and the surface and that between a polymer segment and the surface) is smaller than a critical value ($\chi_{s,\ crit}$), the polymer chain tends to be repelled from the surface (the entropy loss when a chain becomes close to the surface is not compensated by an adsorption energy). As a result a "depletion layer" develops at the interface where the segment concentration is lower than in bulk. Overlap of the depletion layers on two approaching particles leads to expulsion of solvent from the overlap region into bulk solution (at a high polymer concentration). In other words, the solvent molecules in the overlap region tend to reduce their chemical potential resulting in flocculation of the dispersion. Thus, depletion flocculation may be considered to be the opposite of steric repulsion in which solvent molecules diffuse from bulk solution into the overlap region (where the chemical potential of the solvent is lower). The magnitude of the depletion flocculation energy

170

is of the order of the osmotic pressure of the free polymer solution, whereas its range is in the region of twice the radius of gyration of the free polymer coil.

The above phenomenon of depletion flocculation when applied to concentrated latex dispersions gives an ideal system for study using rheological techniques and in this review we will report some preliminary results using polystyrene latex dispersion with grafted PEO chains and flocculated by addition of PEO with M_w = 20,000, 35,000 and 90,000. Details of the experimental techniques will be given in a forthcoming publication[15]. However, below a brief summary of the techniques used for studying viscoelastic behaviour will be given.

2. PROCEDURES FOR STUDYING VISCOELASTIC BEHAVIOUR OF CONCENTRATED LATEX DISPERSIONS

There are essentially three procedures for studying the viscoelastic behaviour of concentrated dispersions. The first procedure (transient measurements) consists of performing "static" deformation measurements, i.e. transient measurements whereby a constant deformation or stress is applied to the system and the time variation of stress or strain is monitored. The second procedure dynamic measurements) consists of applying an oscillating deformation and the resulting stress is compared with the strain and the phase difference of the sine waves is measured as a function of frequency. The third procedure is based on shear wave propagation. These measurements are supplemented by steady state shear stress - shear rate measurements, which are of course measurements at relatively larger deformation. Given below is a brief summary of the principle of oscillatory, shear wave propagation and steady state techniques (which have been applied in the present systems).

2.1. Oscillatory Measurements

In these measurements a small amplitude sinusoidal oscillation is applied to a viscoelastic material and the resulting stress compared with the strain. For a viscoelastic material, the stress oscillates at the same frequency but out of phase with the strain. The amplitude ratio of the stress to strain gives the complex modulus, G^*, whereas the two sine waves will have a phase difference, δ, that is used to give the storage (G') and the loss (G'') components[16]. This is illustrated in Figure 2.

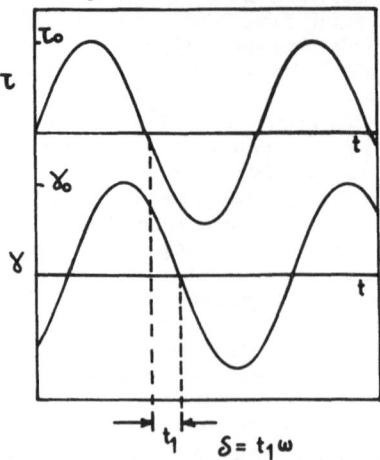

Figure 2: Corresponding stress and strain wave forms.

It can be seen that the in phase component gives the elastic component G' and the out of phase (quadrature component) gives the loss or viscous component G". The values of G', G" and G* are given by,

$$G' = G* \cos\delta \qquad (3a)$$
$$G" = G* \sin\delta \qquad (3b)$$
$$G* = G' + iG" \qquad (3c)$$

A complex viscosity $\eta*$ may be defined and this can be resolved with a real part η' and imaginary part $\eta"$,

$$\eta* = \eta' - i\eta" \qquad (4a)$$
$$\eta' = G"/\omega \qquad (4b)$$
$$\eta" = G'/\omega \qquad (4c)$$

where ω is the frequency in rad s^{-1}.

Experimentally one makes measurements of the complex modulus and its phase lag δ over a wide frequency range. This can be done by using the Bohlin rheometer (Bohlin Rheologie, Lund, Sweden). The general shape of the G', G", η' versus frequency curve (note that all parameters are plotted in dimensionless units G'/G*, G"/G*, $\eta'/\eta*$ and t_r where t_r is the relaxation time) for a Maxwell model is shown in Figure 3[16]. It can be seen that at high frequency, G' reaches a maximum whereas G" approaches zero, i.e. the system behaves as an elastic solid, whereas at low frequency G">G'. Moreover η' reaches a limiting value at low ($\eta'_{\omega \to o} \to \eta_{(o)}$, the residual viscosity) and it decreases rapidly at high frequency.

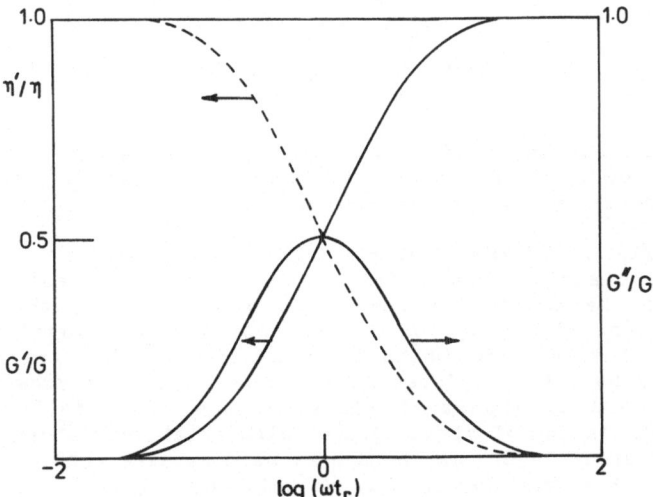

Figure 3: Variation of G'/G, G"/G and η'/η with reduced frequency.

2.2. Shear wave propagation

This procedure consists of initiating a shear wave by a small amplitude shear strain and measuring the time taken for the resulting wave to travel a known distance[16]. This can be carried out using a commercially available instrument, the "Pulse Shearometer" (Rank Brothers, Bottisham, Cambridge, U.K.) from which measurement can be made over a wide range of distances to obtain reliable measurements of wave velocity V. From V, the modulus, G_∞ (which in the pulse shearometer is obtained at a small amplitude of less than 10^{-4}rads^{-1} and high frequency of 200 Hz) is simply given by the equation,

$$G(\infty) = \rho V^2 \tag{5}$$

where ρ is the density of the dispersion.

2.3. Steady state shear stress (τ) - shear rate ($\dot{\gamma}$) measurements

This can be obtained using commercial concentric cylinder or cone and plate viscometers, e.g. Haake Rotovisco (Haake, Germany). The dispersion is placed in the concentric cylinder platens and the shear rate is continuously increased rotating the inner cylinder while measuring the stress at each shear rate value. The stress-strain curves are automatically plotted (usually one performs measurements from low to high shear rates reaching a certain value and then the shear rate is gradually decreased to obtain the downward curve in order to test for any time effects) and the data analyzed using the Bingham or Casson equations, i.e.

$$\tau = \tau_\beta + \eta_{app}\dot{\gamma} \tag{6}$$

$$\tau^{1/2} = \tau_c^{1/2} + \eta_c^{1/2}\dot{\gamma}^{1/2} \tag{7}$$

where τ_β is the extrapolated (Bingham) yield value and η_{app} is the slope of the linear portion of the stress-shear rate curve. τ_c is the Casson yield value whereas η_c is the corresponding viscosity.

3. VISCOELASTIC PROPERTIES OF STABLE LATEX DISPERSIONS

The viscoelastic properties of sterically stabilized latex dispersions is determined by the volume fraction ϕ of the dispersion which determines the range of interaction and the magnitude of the interparticle interaction (which is determined by the steric repulsive force). Viscoelasticity is determined by the balance of Brownian motion, hydrodynamic and interparticle interactions. Sterically stabilized dispersions consisting of rigid spheres with relatively short (and dense) adsorbed or grafted layers may behave as hard spheres with a radius given by $(a+\delta)$ where δ is the adsorbed layer thickness. This is usually tested by plotting the relative viscosity η_r versus volume fraction of the dispersion. As an illustration Figure 4 shows such plots for a polystyrene latex dispersion with grafted PEO chains.

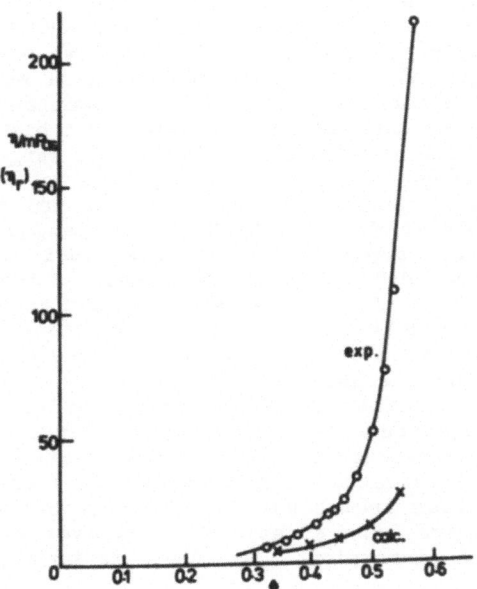

Figure 4: Variation of η_r with ϕ.

The viscosity - volume fraction was also calculated using the Dougherty-Krieger equation(17), i.e.

$$\eta_r = [1-(\phi/\phi_p)]^{-[\eta]\phi_p} \qquad (8)$$

where $[\eta]$ is the intrinsic viscosity (that is equal to 2.5 for hard spheres) and ϕ_p is the maximum packing fraction (0.64 for random packing and 0.74 for hexagonal close-packing). ϕ_p was obtained from plots of $1/\sqrt{\eta_r}$ versus ϕ and extrapolation to zero and this gave a value of 0.7.

It is clear from Figure 4 that the calculated $\eta_{(r)}$ - ϕ curve is below the experimental curve. This is due to two main effects. Firstly, the ϕ values are lower than the effective volume fraction ϕ_{eff} which is that of the particles plus the adsorbed layer.

$$\phi_{eff} = \phi \ (1 + (\delta/a))^3 \qquad (9)$$

Note that if a value of δ of 10nm is taken (a reasonable value for a PEO layer with a molecular weight of 2000), then the experimental curve approaches the theoretical one very closely. However, this is not justified since the adsorbed layer may get compressed at high ϕ values. The second and most obvious discrepancy between the experimental and theoretical curve is due to the relatively "soft" interaction of sterically stabilized dispersions. The Dougherty-Krieger equation (17) is based on perfect hard spheres which of course is not the case with terminally grafted PEO chains on polystyrene latex.

The viscoelastic properties of concentrated sterically stabilized latex dispersions are illustrated in Figure 5 in which G^*, G' and G'' are plotted as a function of frequency ω (in Hz) at various ϕ values. At ϕ = 0.44, $G''>G'$ and $G''\sim G^*$, with all values being relatively low. In other

174

words the dispersion behaves as a viscous fluid. This is not surprising since at such low volume fractions the interparticle separation is greater than twice the adsorbed layer thickness and repulsion is only occasionally felt during a Brownian collision (two-body interaction). Under these conditions, the dispersion behaves as a viscous fluid with little elastic contribution. When the volume fraction is increased to 0.465, G" is still higher than G' but the values are higher than those at $\phi = 0.44$. Here again, the interparticle distance is larger than twice the adsorbed layer thickness. With further increase in ϕ to 0.5, G' now becomes larger than G" and the modulus values are now an order of magnitude higher than at $\phi = 0.465$. Assuming random packing of the particles, the particle separation is of the order of 30 nm which is comparable to twice the adsorbed layer thickness. As the volume fraction is further increased to 0.575, G' now becomes much larger than G" and it approaches G* very closely. Under these conditions, the sterically stabilized latex dispersion behaves as an elastic body. At such volume fractions, the interparticle separation is of the order of 12 nm which is lower than twice the adsorbed layer thickness. Under these conditions strong interaction between the PEO chains occurs as a result of compression and/or interpenetration. Thus, these oscillatory measurements give a great deal of information on interaction of sterically stabilized dispersions. As is clear from Figure 5, the system changes from being more viscous (G">G') to more elastic (G'>G") over a narrow range of volume fraction of the dispersion, i.e. when ϕ is increased from 0.465 to 0.5.

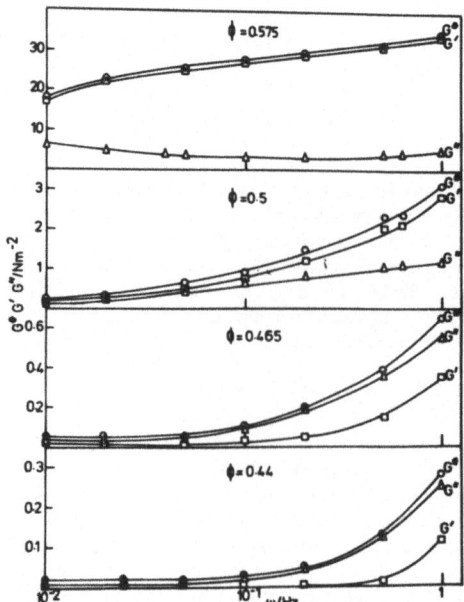

Figure 5: Variation of G*, G' and G" with frequency at various ϕ values.

Within this range of increase of ϕ, it is likely that the chains undergo some compression and interpenetration of the peripheries. Clearly the grafted chains are not of equal size (due to the distribution of molecular weight of PEO) and the longest chains can interpenetrate and some compression may also occur. When ϕ is increased significantly above 0.5, the system becomes predominantly elastic whereby significant interpenetration and compression of the chains occur. Indeed when ϕ is further increased to 0.585 and 0.62, the modulii increase by several orders of magnitude and so does the dynamic viscosity. At such high ϕ values the latex behaves as a near elastic solid.

4. FLOCCULATED LATEX DISPERSIONS

As mentioned in the introduction, the sterically stabilized latex dispersion can be flocculated by addition of free PEO. The viscoelastic properties of concentrated sterically stabilized polystyrene latex dispersions were first investigated by Heath and Tadros[18] who used a physically adsorbed graft copolymer of poly (methyl methacrylate - methacrylic acid) and methoxy - capped poly (ethylene oxide) (M_w - 750). Flocculation in this case was induced by addition of PEO with M_w of 20,000, 35,000 and 90,000. The viscoelastic behaviour was investigated using creep and shear wave propagation measurements. Above a critical volume fraction of free polymer, ϕ_p^+, marked viscoelasticity was observed. However, this system was not ideal since the flocculation could not be reversed when the polymer volume fraction was reduced below ϕ_p^+ after flocculation. It was then suspected that the physically adsorbed graft copolymer may undergo lateral movement and some "bridging" flocculation by the free PEO may have occurred. For that reason another polystyrene dispersion was prepared in which the sterically stabilized (PEO) was grafted to the particles, thus removing the possibility of lateral movement of the chains.

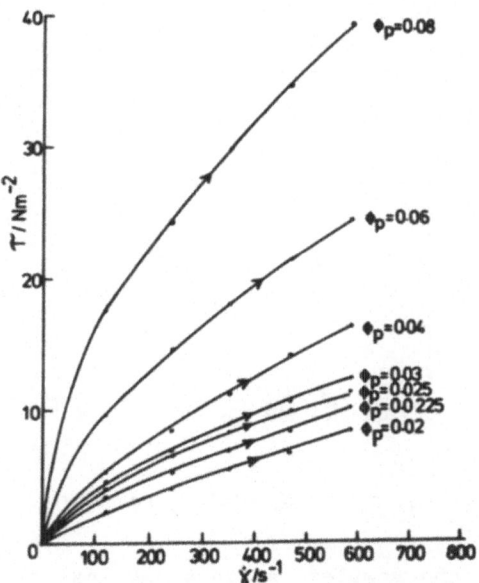

Figure 6: Shear stress-shear rate curves for polystyrene latex dispersions (ϕ - 0.3) at various ϕ values (PEO M_w - 20,000)

The dispersion was then flocculated by addition of PEO with $M_w = 20,000$, 35,000 or 90,000. In this case, the flocculation was reversible in the sense that when ϕ_p was reduced below ϕ_p^+, the system regained its stability. Viscoelasticity in this system was studied using shear wave propagation (pulse shearometer), steady state shear stress - shear rate measurements and oscillatory measurements (using the Bohlin rheometer). As an illustration, Figure 6 shows τ-$\bar{\gamma}$ plots for a polystyrene latex dispersion with $\phi = 0.3$ at various additions of PEO with $M_w = 20,000$. As is clear the system changes from near Newtonian at $\phi_p \leq 0.02$ to pseudoplastic at $\phi_p \geq 0.0225$. As ϕ_p is increased, the pseudoplastic behaviour increases and the extrapolated yield value is increased. This is illustrated in Figure 7 which shows plots of τ_β and τ_c versus ϕ_p. The figure also shows the variation of the high frequency modulus G_∞ with ϕ_p. All rheological parameters show a rapid increase above $\phi_p = 0.02$, which can be taken as the flocculation point ϕ_p^+. Similar results were obtained for PEO 35,000 and 90,000 but in this case the value of ϕ_p^+ was low. For example $\phi_p^+ = 0.01$ with $M_w = 35,000$ and $\phi_p^+ = 0.005$ with $M_w = 90,000$.

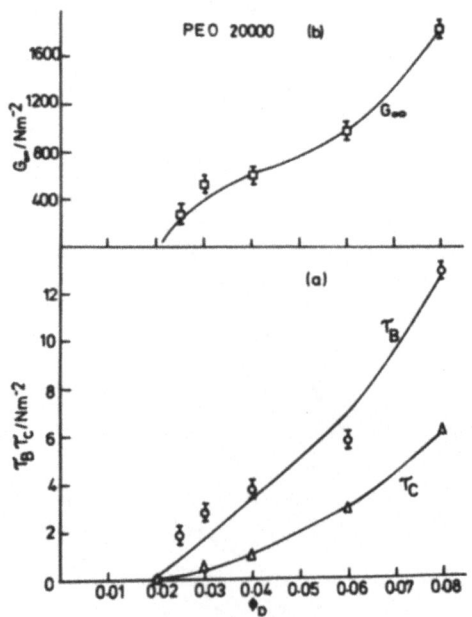

Figure 7: Variation of τ_β, τ_c and G_∞ with ϕ_p for PEO with $M_w = 20,000$

The results for the complex, storage and loss modulii are shown in Figure 8 for PEO 20,000 (and $\phi = 0.3$). Again all modulii show a rapid increase above $\phi_p = 0.02$. Note, however, that G">G' since ϕ is relatively low (0.3).

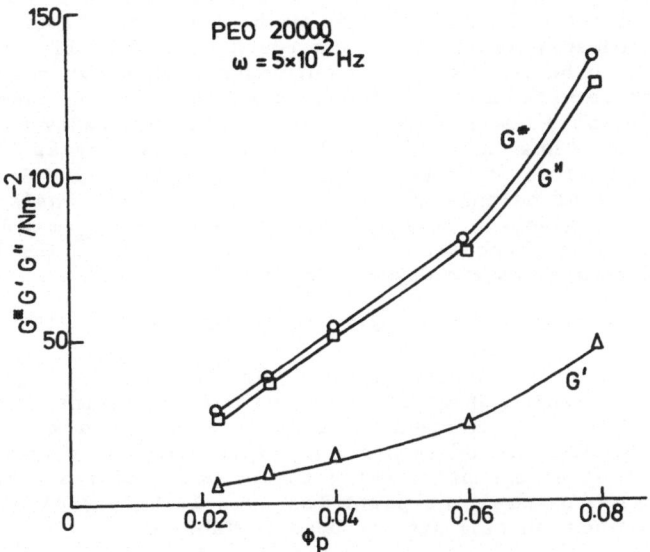

Figure 8: Variation of G*, G' and G" with ϕ_p for PEO with M_w – 20,000

The rheological parameters can be related to interparticle interaction provided some reasonable assumption can be made. For example, the Bingham yield stress, τ_β, can be related to the amount of energy needed to separate the flocs into single units[19,20],

$$\tau_\beta - N \ E_{sep} \qquad (10)$$

where N is the total number of contacts between the particles in flocs and E_{sep} is the energy required to break each contact. The total number of contacts N may be related to the particle volume fraction ϕ and the average number of contacts per particle, n, by[19],

$$N - \frac{1}{2} \ \frac{3\phi n}{4\pi a^3} \qquad (11)$$

combining (10) and (11),

$$\tau_\beta - \frac{3\phi n E_{sep}}{8\pi a^3} \qquad (12)$$

Equation (12) contains two unknowns n and E_{sep}. However, if it is assumed that the flocs are close-packed structures, then n must be between 8(random close-packing) and 12 (hexagonal close packing). Thus, E_{sep} may be calculated from τ_β if one assumes a reasonable value for n.

The latter may be taken to be equal to 8 (a more realistic value since the flocs will probably be randomly packed). Moreover, one has to assume that above the yield point all contacts are broken. This assumption seems justified since the flocculation is weak. Above the flocculation point, the system probably consists of a network of weakly flocculated latex. These flocs can be reversibly broken under high shear resulting in the formation of primary units. Evidence that this is the case is obtained from the relatively small dependence of plastic viscosity on free polymer concentration. Thus at relatively high shear rates most of the flocculated structure is broken down.

The results of calculation of E_{sep} from τ_β on the basis of the above assumptions are given in Table 1. These values of E_{sep} may be equated to the free energy of flocculation due to depletion, G_{dep}. The latter can be calculated from Asakura and Oosawa's equation, i.e.

$$G_{dep}/kT = -(3/2) \, N_p \beta^2 x^2 \; ; \; 0 < x < 1 \tag{13}$$

where N_p is the volume concentration of the polymer ($N_p = (\pi/6)(2R_g)^3 C N_{av}$) where R_g is the radius of gyration of the coil, C is the concentration in mol m^{-3} and N_{av} is the Avogodro number. $\beta = a/R_g$ and $x = [R_g-(h/2)]/R_g$ where h is the distance of separation between the outer surfaces. Clearly when h = 0, i.e. at the point where the polymer coils are "squeezed out" from the region between the particles, x = 1. Values of G_{dep} calculated according to equation (13) are included in Table 1.

TABLE 1. Summary of the results of E_{sep} calculated from the experimental values and G_{dep} calculated using Asakura and Oosawa's equation.

ϕ_p		E_{sep}/kT	G_{dep}/kT
(a) PEO M_w = 20,000			
0.025	2.0	9.1	25.3
0.03	2.8	12.7	30.3
0.04	3.8	17.3	40.5
0.06	5.8	26.4	60.7
0.08	13.1	15.6	80.9
(b) PEO M_w = 35,000			
0.015	2.3	10.5	28.6
0.02	4.4	20.0	38.1
0.03	7.0	31.9	57.1
0.04	11.7	33.2	76.2
(c) PEO M_w = 90,000			
0.01	1.2	5.5	55.0
0.015	2.8	12.7	82.0
0.02	4.4	20.0	110.1
0.025	5.9	26.9	137.6

It can be seen from the data shown in Table 1 that the calculated G_{dep} values are significantly higher than E_{sep} obtained from the experimental r_β values. The discrepancy is higher the higher the molecular weight of the free polymer. This is not surprising since the G_{dep} values are based on a hard sphere model, which is unrealistic for the present system consisting of sterically stabilized particles with PEO chains (tails). Recently Vincent et al[21] developed a model for depletion flocculation of "soft" spheres. Calculations of G_{dep} based on this model for the present system will be given in a forthcoming publication[22].

REFERENCES

1. Napper DH: Polymeric Stabilization of Colloidal Dispersions, *Academic Press, London (1983)*.
2. Flory PJ and Krigbaum WR: *J. Chem. Phys.* **18**, 1086 (1950).
3. Fischer EW: *Kolloid Z.* **160**, 120 (1958).
4. Mackor EL: *J. Colloid Sci.* **6**, 490 (1951); Mackor EL and van der Waals JH: *J. Colloid Sci.* **7**, 535 (1952).
5. Hesselink F Th: *J. Phys. Chem.* **73**, 3488 (1969); **75**, 65 (1971); Hesselink F Th; Vrij A and Overbeek J Th G: *J. Phys. Chem.* **76** 2094 (1971).
6. Andrews DM; Manev ED and Haydon DA: *Spec. Disc. Faraday Soc.*, **46** (1970).
7. Van Vliet T and Lyklema J: *Disc. Faraday Soc.* **65**, 5 (1978).
8. Doroszksowski A and Lambourne K: *J. Polymer Sci.* **Part C 34**, 253 (1971).
9. Barclay LM and Ottewill RH: *Spec. Disc. Faraday Soc.* **1**, 138 (1970); Barclay LM; Harrington H and Ottewill RH: *Kolloid ZZ Polymer* **350**, 655 (1972); Cairns RJR; Ottewill RH; Osmond DWJ and Wagstaff I: *J. Colloid Interface Sci.* **54**, 45 (1976).
10. Prestidge C and Tadros Th F: to be published.
11. Li-In-On FKR; Vincent B and Waite FA: *Amer. Chem. Soc. Symp. Ser.9*, 165 (1975); Cowell C; Li-In-On R and Vincent B: *J. Chem. Soc. Faraday Trans. I* **74**, 337 (1978); Vincent B; Luckham PF and Waite FA: *J. Colloid Interface Sci.* **73**, 508 (1980).
12. Asakura S and Oosawa F: *J. Chem. Phys.* **22**, 1255 (1954); *J. Polymer Sci.* **33**, 183 (1958).
13. Vrij A: *Pure and Appl. Chem.* **48**, 471 (1976).
14. Scheutijens JMHM; Fleer GJ and Vincent B: *Amer. Chem. Soc. Symp. Ser.* **240**, 245 (1984).
15. Prestidge C and Tadros Th F: to be published.
16. Goodwin JW: in Solid-Liquid Dispersions(ed); Tadros Th F; *Academic Press*, in preparation.
17. Krieger IM: *Adv. Colloid Interface Sci.* **3**, 111 (1972); Krieger IM and Dougherty H: *Trans. Soc. Rheol.* **3**, 137 (1959).
18. Heath D and Tadros Th F: *Faraday Disc. Chem. Soc.* **76**, 203 (1983).
19. Luckham PF; Vincent B and Tadros Th F: *Colloids and Surfaces* **6**, 101 (1983).
20. Gillespie T: *J. Colloid Sci.* **15**, 219 (1960).
21. Vincent B; Edwards J; Emmett S and Jones A: *Colloids and Surfaces* **18**, 261 (1986).
22. Jones A; Vincent B and Tadros Th F: to be published.

POLYMER STABILIZED LATEXES

POLYMER STABILIZED LATICES

POSITION PAPER

1. INTRODUCTION
There are few technologically (or indeed biologically) important colloidal dispersions from which polymers are absent; they may be present either as the disperse material itself and/or as additives controlling rheological, settling, optical and other properties. However, despite the practical importance of such systems there are many aspects where a greater understanding of the fundamental properties of polymer-stabilized dispersions is required. There are four main areas, in particular, where both theoretical and experimental studies should be concentrated in the immediate future. These are as follows:

(a) Stability
(b) Interparticle Forces
(c) Characterization of Adsorbed Polymer Sheaths and Depletion Layers
(d) Preparation of Model Polymer Colloids

2. Stability
It is convenient to divide colloidal systems into three classes. Firstly, 'stable' systems in which the net pair-potential remains repulsive, at all interparticle distances. Secondly, 'weakly flocculated' systems, i.e. dispersions in which a state of reversible flocculation may be attained, as a result of a shallow minimum (V_{min}) in the pair-potential, where $V_{min} > kT$. Thirdly, systems for which $V_{min} \gg kT$. In these systems the magnitude of V_{min} is sufficient to render the dispersions "irreversibly" aggregated, unless there is a sufficient energy barrier preventing such aggregation, in which case the dispersions are metastable.
For these three broad classes of colloidal dispersions, our thoughts concerning the primary objectives for future research are as follows:-
2.1. Stable Systems
It is essential that one should be able to predict the microstructure that can arise in such systems with increasing particle volume fraction since this controls the properties of the dispersion. Further experimental studies are required to determine the various structures that occur in such systems, especially those in which an order-disorder transition occurs. Ageing and gravitational effects need to be investigated.

2.2. Weakly Flocculated Systems
It is necessary to gain a better understanding of the kinetics of reversible flocculation and to investigate the effects of shear. With regard to equilibrium studies, additional statistical mechanical approaches are required which describe the phase behaviour in such systems. The importance of metastable states needs to be investigated.

2.3. Strongly Flocculated Systems
A better understanding of the manner in which floc structure evolves

is required together with quantitative descriptions for the various morphologies which can be obtained. In order to achieve this the kinetics of the flocculation process should be investigated, together with ageing effects. Also, the importance of these parameters on the bulk physical properties of flocculated dispersions needs to be determined.

In order to make progress with these objectives it will be necessary to further develop experimental techniques, such as time-averaged and quasi-elastic (light, neutron) scattering and ultramicroscopic methods. Such developments are necessary to understand better the early stages of flocculation. For the later stages, microscopy (coupled with video enhancement and image analysis) and hydrodynamic fractionation methods could be applied. For concentrated dispersions rheological methods can be applied to determine floc strength. The best methods are transient (creep) and oscillatory techniques which enable the linear elastic properties to be measured. Although theories exist which relate floc strength to the elastic properties of suspensions, further theoretical development is required.

3. Interparticle Forces
3.1. Theoretical Aspects

Future theories should provide information on both the range and the magnitude of interparticle forces. It is agreed that hard spheres are the best reference systems and that further developments which consider perturbations from the hard sphere models would be useful. A reconsideration of van der Waals forces for colloidal systems with a diffuse interfacial region (such as those with adsorbed polymer layers) is necessary, particularly where interpenetration of these diffuse layers has occurred. It is not at present clear how to account for the van der Waals forces between these layers. Should they be treated as perturbations of the particle-particle van der Waals attraction (as in the Vold approach) or included in the mixing part of the steric interaction. Indeed, can we actually separate these two forces?

3.2. Neutral Systems

Whilst current theories can cope reasonably adequately with physically adsorbed or terminally attached, flexible homopolymers, there is a need for developing understanding of adsorbed polymer layers with more complex architecture, in particular copolymers. The question of partial "burial" of adsorbed chains inside the latex particles needs to be addressed. With systems containing mixed solvents and/or electrolytes, the distribution of solvent species and/or ions near the interface needs to be determined. Furthermore, there have been very few studies reported for asymmetric systems, that is, either systems in which similar surfaces carry different polymers or systems where the surfaces (particles) are different. Recent theoretical approaches, such as that developed by Scheutjens and Fleer, should prove useful in this regard.

3.3. Charged Systems

Theories need to be developed for the pair potential in systems where neutral polymers are adsorbed on charged particles. How does the presence of adsorbed segments effect the inner and outer parts of the electrical double layer? The presence and role of charges in media of low permittivity need further analysis.

Another area where major theoretical effort is required is that of systems containing polyelectrolytes, either as stabilizers or as flocculants. Not only is their adsorption behaviour poorly understood, there is as yet no adequate theory describing interparticle pair-potentials.

3.4. Dynamics

There are almost no theories available concerning the dynamic behaviour of polymers at interfaces, apart from some thoughts by de Gennes on 'breathing modes'. Further developments here are crucial if theories of polymer adsorption rates are to be constructed. In many practical systems, the net balance between rates of polymer adsorption and the rates of particle collisions, is often the major factor determining stability.

3.5. Many-body Interactions

This is particularly relevant to concentrated dispersions, but even in dilute dispersions as flocculation proceeds, triplets and higher aggregates form. The net potential energy change associated with their formation needs to be established. With many types of interaction there are non-additive effects. For steric/depletion forces such difficulties arise where the adsorbed/depleted polymer layer is large (with respect to particle dimensions).

3.6. Experimental Techniques

Force-balance and compression methods have clearly already made a substantial contribution to our understanding of pair-potentials. However, it is felt that, where possible, such techniques should be combined with other (spectroscopic, scattering) methods which can probe the "structure" of the interfacial region, particularly as a function of surface separation, over different time scales. "Reversibility" is an important consideration here. The mica force-balance has proved exceedingly useful but there is a need for a broader range of representative surface types to be investigated. Also, it should be noted that experiments with planar or low curvature surfaces are not always directly relevant to the interactions between colloidal particles. Studies of particle/flat plate systems may lead to a better understanding of hydrodynamic effects, which are poorly understood at present. Progress in understanding dynamic effects on the pair potential and hydrodynamic effects for actual particulate dispersions may be made using acoustic techniques, although there is a need for considerable instrumental, as well as theoretical, development in this area.

4. Characterization of Adsorbed Polymer Sheaths and Depletion Layers

In order even to construct elementary theories which yield pair-potentials in colloidal dispersions containing particles with 'structured' layers, due to the presence of either adsorbed or free polymer, it is necessary to establish both the structural and dynamic properties of these layers. The key parameters which are considered to be important and methods by which they may be determined are summarized below:

4.1. Structural Parameters

These include:
- adsorbed/grafted amounts;
- segmental adsorption energies (χ_s parameters);
- segment/solvent interaction parameters (χ) as a
  function of temperature, concentration, etc.;
- the range of the structural forces (e.g. adsorbed
  layer thicknesses and depletion layer thicknesses)

and how the "thicknesses" measured by the various
experimental methods relate to one another;
- the segment density distributions both normal and
  parallel to the particle-solution interface;
- the configuration of polymer chains, especially
  any secondary structures;
- the orientation of adsorbed oligomeric and surfactant moieties and details of phase transitions
  which might occur, especially with changes in
  temperature.
- the effects of adsorbed polymers on ion distributions (including ion binding);
- the effects of ions in the double layer on the
  conformation of adsorbed polymers.

4.2. Dynamic Parameters

There is a need to establish (both translational and rotational)
diffusional rates of polymer chains and of individual segments for
physisorbed polymers. Whole molecule diffusion may well occur at low
surface coverages. Segmental diffusion rates, on the other hand, are
important in determining the rates of configurational changes for both
adsorbed and grafted polymer layers. These factors are important with
regard to polymer adsorption kinetics.

4.3. Modern Experimental Techniques

There are now a number of techniques available for determining
structural and dynamic information for adsorbed polymer layers. These
methods complement the well established, "classical" techniques and
include:
- small angle neutron scattering (both elastic and quasi
  elastic);
- fluorescence methods in which probe and quencher
  molecules are attached to the adsorbed polymer and/or
  the particle surfaces. Here one may also consider
  evanescent wave reflection techniques;
- pulsed FT NMR and ESR methods;
- PCS and electro-optical methods;
- acoustic spectroscopy;
- FTIR and laser-Raman spectroscopy.

5. Preparation of Model Polymer Colloids

Clearly, in order to gain the level of understanding demanded by the
suggestions made above it is essential that particles with well-
controlled structure are available. The properties of such model colloids
that need to be carefully controlled include particle size and particle
size distribution, the size of the stabilizing polymer chains, the
surface coverage of the stabilizer and the surface charge in systems
where it is relevant. To facilitate the synthesis of such materials it
will be necessary to develop more well-defined methods for attaching the
stabilizing polymer to the particle cores. It is equally important to
develop better methods of preparing particles with hard-cores and soft-
shells. The synthesis of microgel 'particles' possessing well-defined
properties is also worthy of further attention.

To facilitate the preparation of model colloids it is thought that
advances will have to be made in several types of polymerization
processes. At present the non-aqueous, dispersion polymerization methods

developed by ICI in the 1950's are widely accepted as the best methods available for the synthesis of polymer stabilized particles. However, these methods do have limitations. These are discussed below together with other polymerization methods which are in need of further development if they are to be of use in preparing model colloids.

5.1. Dispersion Polymerization

Here polymerization is initiated in a homogeneous solution of monomer(s), initiator and a (polymeric) stabilizer, such that the insoluble polymer produced precipitates. "Attachment" of stabilizers leads to discrete, stable particles. There is a need for dispersion polymerization to be investigated in a wide range of media, particularly polar and semi-polar solvents, so that a greater range of polymer latexes can be utilized. Such studies will require the availability of well-characterized stabilizers and stabilizer precursors, together with a much improved understanding of the mode(s) of incorporation of the stabilizing moieties into/onto the particles. In particular, greater control of "random grafting" reactions is necessary if the composition of the stabilizing moieties for the final particles is to be known unambiguously. Finally, the use of "seed" latexes prepared by emulsion polymerization in dispersion polymerization is thought worthy of attention in the future.

5.2. Polymerization of Monomer Droplets

Transparent, thermodynamically stable micro-emulsions have been known for some time. Recently, interesting work on the polymerization of these dynamic structures was reported and stable latex particles of diameters less than 50 nm were obtained. Such small particles are difficult to prepare by conventional, emulsion polymerization methods. It has been found that the composition of the starting micro-emulsion is critical in determining whether polymerization to form a latex takes place. A study of the mechanism of micro-emulsion polymerization needs to be carried out in order to determine the relationship between micro-emulsion structure and latex formation.

The polymerization of larger monomer droplets (in macro-emulsions) is known as suspension polymerization. The major problem here is polydispersity. If sufficiently stable monodisperse monomer droplets containing a suitable initiator stabilized by block or graft copolymers could be prepared, then this could prove to be a useful method to prepare larger, model polymer-stablized particles. Furthermore, with this method the possibility of chemical modification to the stabilizing moieties is much less likely than in conventional dispersion polymerization. For such studies greater availability of various types of block and graft copolymers is required.

5.3. Inverse Emulsion Polymerization

Despite the pioneering work by Vanderhoff in the 1960's and the technological importance of water soluble polymers prepared by inverse emulsion polymerization, i.e. polymerization in a non-polar medium, this method has received scant attention. Although the method is somewhat more complex than conventional emulsion polymerization, it is worthy of detailed studies since it can be used to prepare polymer stabilized colloids.

5.4. Two Step Processes

A common method of preparing polymer-stabilized latexes is to post-adsorb block or graft copolymers onto particles previously prepared by, for example, an emulsion polymerization route. The chemical attachment of polymers onto preformed particles is an area worthy of further work.

5.5. Coacervation

There exist a number of processes in which complexation or association between two chemically different polymers in a solvent bring about the formation of aggregate structures which are subsequently converted to stable polymer particles. This is achieved by the addition of monomer(s) and initiator, followed by polymerization. To date, it would seem that dispersions of broad size distribution are obtained. However, better control of such systems could lead to particles possessing unique morphological characteristics and as such the method deserves further attention.

5.6. Mechanistic Studies

The major factor restricting the wider application of the methods described above is a lack of fundamental understanding of the mechanisms of the various polymerization processes described. This situation arises from the complexity of these processes, which are multi-component and in which there are few "independent" variables. Only detailed kinetic studies of the changes in polymer molecular weight, particle size and number, monomer initiator and solvent concentration (in the different loci present), stabilizer location, etc. can lead to progress in this area. The large amount of information obtained already from similar studies of aqueous, emulsion polymerization processes, should prove a useful background framework for this task.

5.7. Non Free Radical Initiated Processes

Most particle polymerization methods utilize free radicals to initiate polymerization. While this is of obvious importance, polymeric materials that are prepared by condensation polymerization are becoming increasingly important. Materials such as polyurethanes, polyimides, polyamides, polyesters, polysulphones, aromatic polyethers and polyketones in colloidal form would find a wide range of applications. Attention deserves to be directed towards making colloidal particles of such polymers in a wide range of dispersion media.

6. SUMMARY

There is clearly much work to be carried out on polymer-stablized latex systems, both of a theoretical an experimental nature. To pick out one aspect from each of the four areas discussed which should perhaps have immediate priority for study in that area is a difficult (and highly subjective!) task. Nevertheless, we would propose the following as a guide-line:

6.1. Stability

The development of theoretical models and experimental techniques which allow the "structure" of flocculating systems to be evaluated, as a function of time.

6.2. Interparticle forces

Since many polymer-stabilized latex systems (both aqueous and non-aqueous) contain charged moeities in the interfacial region, it is vital to build theories of pair-potentials which take these into account. At the same time, experiments on well-defined systems are required.

6.3. Structure of Adsorbed and Depleted Polymer Layers at Interfaces

The most important development required here is to elucidate the way the structure of the adsorbed/depleted layer changes during interparticle encounters, over various timescales. Progress in modelling interparticle forces require this.

6.4. Preparation of Model Polymer Colloids

Control can only come from a better understanding of the complex mechanisms involved in the various polymerization processes discussed. Detailed kinetic studies are therefore required.

7. Panel

Discussion Leader:
   B. Vincent (University of Bristol, United Kingdom)

Secretary:
   T. Corner (International Paint, United Kingdom)

Participants:
   A. Ali-Khan (E.I. du Pont de Nemours and Company, Inc., USA)
   M. A. Cohen Stuart (Agricultural University, The Netherlands)
   M. D. Croucher (Xerox Research Center, Canada)
   W. E. Daniels (Air Products and Chemicals, USA)
   K. G. de Kruif (Van't Hoff Laboratory, The Netherlands)
   M. S. El-Aasser (Lehigh University, USA)
   A. Guyot (Centre National de la Recherche Scientifique Laboratoire des Materiaux Organiques, France)
   I. Piirma (The University of Akron, USA)
   P. R. Sperry (Rohm and Hass Company, USA)
   T. F. Tadros (Imperial Chemical Industries PLC, United Kingdom)

THE STABILIZATION AND CONTROLLED FLOCCULATION OF STERICALLY-STABILIZED LATICES

B. VINCENT

Department of Physical Chemistry
University of Bristol
Cantock's Close
Bristol BS 8 1TS
United Kingdom

1. INTRODUCTION: WHY STERIC STABILIZATION?

Virtually all the methods used for preparing polymer latices involve precipitation of polymer from solution during their formation. If distinct particles are to be formed, as opposed to large, ill-defined aggregates, then some form of stabilizing mechanism is required for the growing particles. In certain types of polymerization processes electrostatic stabilization alone may be sufficient. The most obvious example is aqueous emulsion polymerization used to produce fine dispersions of compact, lyophobic polymer particles (Figure 1a). Because of the long-range nature of the van der Waals (dispersion) attraction (V_A) in this case, a repulsion interaction of similar or greater range is required for stability to be achieved. Electrostatic interactions, arising from surface charge groups on the particles, achieve this requirement. If this electrostatic repulsion is removed or sufficiently reduced, such that the particles do come into primary contact (R=2a) then the force preventing further approach of the particle centers (partial coalescence) may be described as a "hard-sphere" repulsion (V_{hs}). However, because in most cases, and particularly with aqueous latices, there will be a large difference between the Hamaker constant of the particles and that of the medium, V_{min} will be of sufficient magnitude to allow irreversible aggregation (coagulation) to occur.

At the other extreme, we may consider so-called "microgel" particles, which consist of (necessarily cross-linked) lyophilic polymer chains (Figure 1c). Microgel particles have been used for many years as impact modifiers for plastics[1]. Here, because the particles are heavily swollen with solvent, the Hamaker constant difference between the particles and the medium is greatly reduced. Hence, $|V_{min}|$ is also very much reduced. Provided $|V_{min}|$ is less than some critical value (see section 7), a thermodynamically stable dispersion results.

The repulsion force which operates for R<2a with microgel particles, as opposed to compact latex particles, is now, by analogy, a "soft"-sphere repulsion. More traditionally, it is known as a "steric" interaction. Clearly some interpenetration (or compression) of microgel particles is possible during a (transient) binary collision, depending on the "softness" of this repulsion.

192

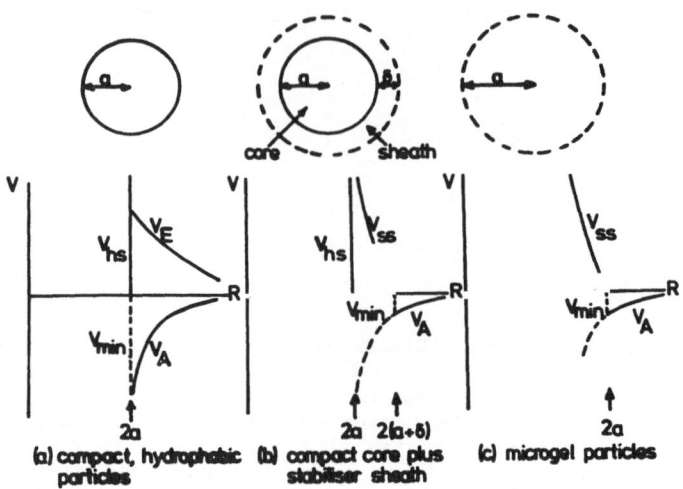

Figure 1: Types of latex particle and interaction energy (V) versus particle center-center separation (R).

Sterically-stabilized particles (Figure 1b) can be considered as being intermediate in character between (charge-stabilized) lyophobic particles (Figure 1a) and microgel lyophilic particles (Figure 1a). The heavily solvated grafted or adsorbed polymer chains confer a soft-sphere or steric repulsion at $R<2(a+\delta)$. The net effect is again to reduce $|V_{min}|$ below the critical value required for thermodynamic instability. The nature of the steric interaction is considered in more detail in Section 4. Although no electrostatic interaction is indicated in Figure 1b and, indeed, is in most cases unnecessary, under certain circumstances it is useful to introduce it by using charged polymers as the stabilizing moieties (see Section 5).

There are many disadvantages in relying on charge repulsion (V_E) *alone* (Figure 1a) for polymer latices. These may be summarized as follows:

a)   difficulties in sustaining free interfacial charge in non-polar media, or controlling such charge in polar (but non-aqueous) media, due to variation in adventitious water concentration;

b)   sensitivity to added electrolyte (particularly multivalent ions) in aqueous media;

c)   significant electroviscous effects;

d)   difficulties in maintaining stability at high particle volume fractions;

e)   sensitivity to shear;

f)   sensitivity to surface coagulation (arising from agitation);

g)   difficulties in controlling $|V_{min}|$, and hence the extent and strength of aggregation.

In principle, use of steric repulsion can overcome most of these problems.

2. THE DESIGN OF STABILIZER SHEATHS

During the early development of aqueous emulsion polymerization processes, it was found that particle stability could be enhanced by the addition of water-soluble polymers such as poly(vinyl alcohol) or polysaccharides. Since that time, a great deal of know-how has been developed in the design of polymeric stabilizers for latex systems, particularly non-aqueous latices. Work on the preparative aspects of sterically-stabilized latices is discussed more fully in the paper by Croucher in this volume.

Several criteria which must be met in stabilizer choice or design may be identified:

a) optimum coverage: if "bare patches" occur on the latex particles then aggregation may occur, either through polymer bridging (coadsorption of polymer chains on two particles during a collision), or through lateral displacement leading to pair contact (coagulation). On the other hand if the volume fraction of polymer segments in the stabilizer sheath (Figure 1b) is too high, then the effective Hamaker constant of the sheath may achieve a value significantly different from the continuous phase, leading to an increase in the van der Waals attraction (and hence V_{min}!).

b) efficient "anchoring": either strong physiosorption of the adsorbed segments (in trains) or, better still, chemical bonding to the polymer chains comprising the latex. Coagulation may result if polymer chains desorb during a binary collision.

c) a sufficiently thick barrier: δ needs to be sufficiently great so that $|V_{min}|$ (Figure 1b) is reduced below the critical value for thermodynamic instability (Section 7).

d) a good-solvent environment for the stabilizing moieties (tails or loops of segments extending from the particle surface): in a poor solvent environment the particles in effect return to being lyophobic.

Physiosorbed homopolymers do not, in general, fulfill all these conditions simultaneously. If the segments in tails (and loops) are in a good solvent environment, then the segments in trains will not be sufficiently strongly adsorbed. Indeed, if ϵ_{01} is the energy between a solvent molecule and the surface and ϵ_{02} is that between a polymer segment and the surface, then χ_s ($= \epsilon_{01}-\epsilon_{02}$) must be greater than some critical value (typically a few tenths kT), for adsorption to occur.

Cowell and Vincent[3] have shown that although homopolymer poly(ethylene oxide)(PEO), adsorbs at the polystyrene/aqueous solution interface, it is an inefficient steric stabilizer for polystyrene latex in 10^{-2} mol dm^{-3} $Ba(NO_3)_2$ solution (still a good solvent environment for PEO). For PEO of molecular weight <20,000, slow coagulation results, presumably as a result of desorption of polymer chains during particle collisions.

For this reason copolymers of various types have been developed as steric stabilizers. Indeed, so-called "poly(vinyl alcohol)" is a much more efficient stabilizer than PEO, for aqueous dispersions of hydrophobic latex particles, simply because commercial grades are, usually, actually copolymers of vinyl alcohol and vinyl acetate. The short runs of (hydrophobic) vinyl acetate segments predominantly act as the anchoring moeities (trains), while the longer runs of vinyl alcohol segments predominantly act as the stabilizing moeities (loops and tails). Most designed stabilizers for latices are based on ABA block copolymers or so-called "comb structures"[2], shown schematically in Figure 2.

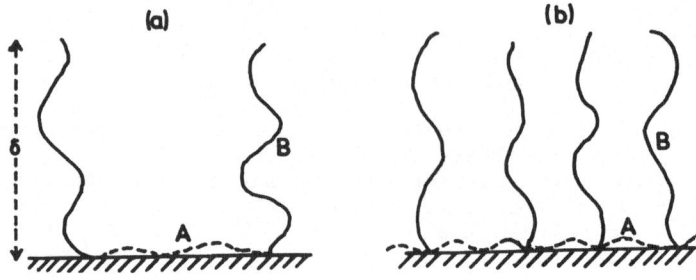

(a)　　　　　　　　　　　　　　(b)

Figure 2: Latex stabilizer types (a) ABA blocks; (b) comb structures A = anchoring moeities; B = stabilizing moeities.

The representation of the chain configurations in Figure 2 is somewhat "idealized". Depending on the polymerization method, the A chains may be "buried" in the matrix of polymer chains forming the latex particles, particularly if A is the same polymer species. It is frequently stated that the B chains simply act as terminally-anchored tails (as indeed is depicted in Figure 2). However, there is some evidence, particularly from contrast-match small-angle neutron scattering studies (see Section 3), that some of the B chains may also be "buried" in the particle matrix. Also if $\chi_s > \chi_{s,crit}$ for the B chains, then, particularly at lower coverages, there may be some "folding-back" of these chains to give a loop/train configuration (i.e. more akin to adsorbed homopolymer) rather than single tails. There is evidence for this latter effect from the work of Dobbie et al.[4], who studied the stability of aqueous polystyrene latex particle (with carboxyl surface charge groups) in the presence of AB block stabilizers (B = PEO). At low pH, and low coverages, the PEO chains adsorbed in a loop/train configuration, whereas, at high pH (ionization of carboxylic acid groups, reducing effective χ_s) and high coverages, the PEO chains adopted the more usual tail-like configuration. Clearly, the type of chain configuration in the sheath not only affects δ, but also the actual segment concentration profile. The effects of this on the steric interaction are considered in detail in Section 4.

3.　THE CHARACTERIZATION OF STABILIZER SHEATHS
In order to analyze theoretically the pair potential V(R) between two

particles, and hence be able to interpret the stability behavior of a given latex, certain characteristics of the stabilizer-plus-solvent sheath around the particles (Figure 1b) need to be established. These are identified below. Some of the experimental methods that may be used are then described.

a) Γ_2: the mass of polymer chains per unit area in the sheath (subscripts 1 for solvent and 2 for segment or polymer).
b) δ: the effective thickness of the sheath. As we shall see, different techniques lead to different forms of δ.
c) $\bar{\phi}_2$: the average volume fraction of segments in the sheath. $\bar{\phi}_2$ is obtained directly from Γ and δ

$$\bar{\phi}_2 = \frac{3\Gamma a^2}{\rho[a+\delta)^3-a^3]} \qquad (1)$$

$$\propto \frac{\Gamma}{\rho\delta} \qquad (a \gg \delta) \qquad (2)$$

where a is the core radius and ρ is the polymer density.

d) $\phi_2(z)$: the segment volume fraction distribution normal to the core/sheath interface. As we shall see in Section 4, although pragmatic theories for V(R) can be built on $\bar{\phi}_2$, more accurate theories require $\phi_2(z)$. Indeed, strictly speaking, we actually require $\phi_2(z;R)$, i.e. we need to know how the segment volume fraction profile varies with particle separation, $2a < R < 2(a+\delta)$.
e) p: the average fraction of segments of the stabilizing chains actually at the core surface in trains (or buried in the core), as opposed to those extending from the surface in tails or loops.
f) ζ: the zeta potential for charged systems. In order to characterize the electrostatic contribution $V_E(R)$, at least for $R > 2(a+\delta)$, the electrostatic potential at the periphery of the sheath is required.

The experimental techniques available for characterizing adsorbed polymers have recently been reviewed by Cosgrove, Cohen Stuart and Vincent[5].

3.1. Analytical Methods

These are used primarily to determine Γ_2. When the stabilizing polymer is post-adsorbed, i.e. after formation of the latex, then conventional adsorption isotherm methods may be used, based on the relationship:

$$\Gamma_2 = \frac{\Delta c_2^b \cdot V}{A} \qquad (3)$$

where Δc_2^b is the difference between the initial and equilibrium solution polymer concentrations, V is the total volume of the solution phase and A is the surface area of the (bare) latex particles.

In many cases, however, the stabilizing moeities become attached during the formation of the latex. There are several strategies for determining Γ_2 in this case:

i)   to use in effect equation 3, but Δc_2^b is now the difference between the polymer concentration in solution before and after formation of the latex. This assumes none of the stabilizing chains become "buried" during polymerization.

ii) to remove free stabilizer in solution, then to totally dissolve the latex particles and to analyze quantitatively the stabilizing moeities in the resulting solution.

iii) to use some spectroscopic method (see 3.2 below) to determine Γ_2 directly for the latex particles, in situ.

3.2. Spectroscopic and Scattering Methods

These are potentially the most powerful methods, particularly in combination, for determining stabilizer sheath characteristics. For example, pulsed NMR, ESR and infrared techniques have enabled p to be determined, since it is possible in principle to distinguish different segmental environments using these methods[5]. These methods, together with UV and fluorescence spectroscopy, may also be used, in certain circumstances to determine Γ_2, provided free stabilizer is first removed.

Scattering methods are potentially even more powerful. Elastic scattering methods probe structures over dimensions $\sim Q^{-1}$, where,

$$Q = \frac{4\pi}{\lambda} \sin \theta/2 \qquad (4)$$

λ is the wavelength of the light in the medium and θ is the scattering angle. Elastic (time average) light scattering is therefore of limited use since $\lambda \sim 400$ to 650 nm, although the theory of light scattering from concentric spheres (e.g. core plus shell) is established[6]. X-rays and, in particular, neutrons (λ typically 0.1 to 2nm) are much more useful. The small-angle neutron scattering of particles plus polymer sheaths has recently been described in detail by Crowley[7]. The amplitude of the scattering from a molecular system is related to its form factor. For a particle core (c) plus a sheath (s), the overall form factor F(Q) is given by,

$$F(Q) = F_c(Q) + F_s(Q) \qquad (5)$$

The intensity of scattering I(Q) is therefore given by,

$$I(Q) \sim F_s^2(Q) + F_s(Q) \cdot F_c(Q) + F_c^2(Q) \qquad (6)$$

$$= I_0(Q) + I_1(Q) \cdot \Delta\rho + I_2(Q) \cdot \Delta\rho^2 \qquad (7)$$

where $\Delta\rho$ is the difference between the neutron scattering length density of the core and the medium, I_0 is the sheath term, I_1 is a core/sheath interference term and I_2 is the core term. Both I_0 and I_1 contain information regarding the structure of the sheath, and in principle can be used to determine $\phi_2(z)$, in addition to Γ_2. Mathematically it is more straight-forward to carry out the necessary Fourier transform using the $I_1(Q)$ term, although experimentally this is less accurate. Nevertheless, this was the approach used to obtain the first reported $\phi_2(z)$ profile for a core (polystyrene latex) plus sheath (adsorbed poly(vinyl alcohol)) (PVA) system.[8] This is shown in Figure 3 in the form of a normalized segment probability distribution.

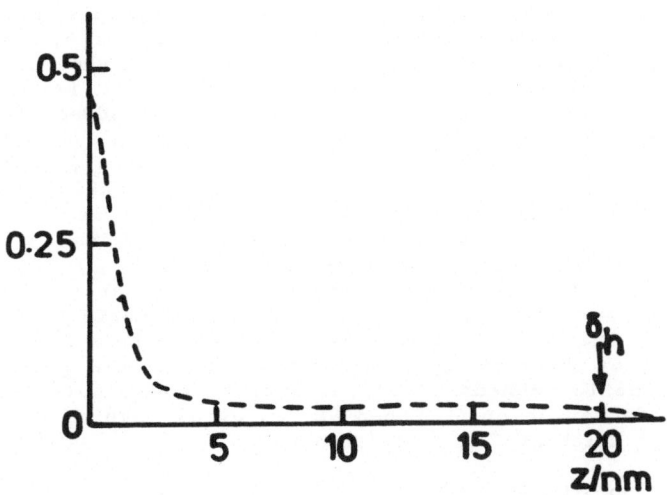

Figure 3: Plot of normalized segment probability distribution for PVA (MW 37,000) on deuterated PS latex in D_2O/H_2O.[8]

Alternatively, and more accurately, one can use the I_o term, and work under contrast-match conditions, i.e. where $\Delta\rho=0$, such that equation 7 reduces to

$$I(Q) - I_o(Q) \tag{8}$$

$$I_o(Q) - \frac{K_o}{Q^2} \left[\int_o^\delta \phi_2(z)\exp iQz \cdot dz \right]^2 \tag{9}$$

There are considerable (but not insurmountable[7]) problems in carrying out the direct Fourier transform in this case, however, to obtain $\phi_2(z)$ from $I_o(Q)$, and problems of uniqueness can be posed. An alternative procedure is to use the I_1 term to determine an approximate form of $\phi_2(z)$ and then try to fit (using a minimization procedure) the $I_o(Q)$ data by more the straight-forward reverse Fourier-transform procedure implicit in equation 9[5,9].

For $Q\delta < 1$ equation 9 can be expanded as a Taylor series leading to,

$$I_o(Q) - K_o/Q^2 \exp(-\sigma^2 Q^2)$$

Thus a plot of $\ell nQ^2 \cdot I(Q)$ vs Q^2 for the low Q data leads to K_o and $\sigma \cdot K_o$ is related to Γ_2 and σ^2 is the second moment of $\phi_2(z)$; σ is directly related to δ, e.g. for a rectangular (uniform) distribution: $\delta = \sqrt{12}\sigma$.

3.3. Hydrodynamic Methods

These lead to the hydrodynamic thickness δ_h of the adsorbed layer sheath. Several hydrodynamic techniques have been used[5] e.g. photon correlation light scattering (leads to the particle diffusion co-efficient), ultracentrifugation, electrophoresis, viscosity. Unfortunately, these methods require measurements on the core particles, as well as the core-plus-sheath particles. Thus, it is only directly applicable to latex systems where the stabilizer is added after particle formation. For those cases where the stabilizer is added during polymerization, if the stabilizer cannot be subsequently removed, then a crude, approximate thickness may be obtained by taking the electron micrograph value of the size of the dried particles as the size of the core particles (i.e. assuming that the contribution from the now collapsed stabilizer chains is negligible).

Cohen Stuart et al[10,11] have recently shown that the hydrodynamic thickness of adsorbed polymer layers on particles corresponds closely to the "real" thickness, as indicated by the $\phi_2(z)$ profile from SANS studies (see Figure 3) or obtained from compression studies (see 3.4).

Electrokinetic methods yield, of course, the ξ potential, but may also give information on δ. Cohen-Stuart et al have discussed how the electrokinetic thickness, δ_E is related to δ_h.[12]

3.4. Compression Studies

These lead directly to what may be termed the "steric" thickness, δ_s, i.e. the (time-average) distance of the farthest segment from the particle core surface. One may distinguish two-dimensional compression studies using a surface-balance technique[13] from three-dimensional compression studies using a high pressure cell device, such as that described by Cairns et al[14]. Alternatively, (slow-speed) centrifugal packing of the latex particles may be used.[15]

As with the hydrodynamic methods described in (3.3), these methods require a value for the size of the bare core particles. Nevertheless, these compression methods do enable interparticle force (F)-separation (R) curves to be determined from pressure - particle volume fraction data.[16] If theoretical values for $F(R)[F(R) = -dV(R)/dR]$ can be related to $\phi_2(z,R)$, then we can attempt to determine $\phi_2(z,R)$ using a fitting procedure. By using say the SANS technique in combination with a pressure-cell it ought to be possible to determine $\phi_2(z,R)$ directly.

So far, we have only considered the structure of the sheath. One of the key questions in stability theory, as we shall see in the next section, is to establish whether particle-particle interactions occur under equilibrium conditions or not, e.g. does $\phi_2(z,R)$ maintain the equilibrium (minimum free energy) profile? Clearly this will depend, inter alia, on the time scale of the particle-particle encounter. In a compression cell, equilibrium may be achieved. In a Brownian encounter in dilute dispersions it may not. Methods are needed therefore which allow us to examine the time scale of various polymer relaxation processes in the sheath, both for isolated particles and for interacting particles. Here, spectroscopic techniques (e.g. NMR, ESR) offer possibilities, as does quasi-elastic neutron scattering.

4. STABILITY THEORY: NEUTRAL POLYMERS

Determination of the pair-potential V(R) for a given latex system is fundamental in any interpretation of its stability behavior. For concentrated dispersions, where many-body interactions become important, one strictly requires the potential of mean force, $\phi(R)$, where

$$\phi(R) = V(R) + \psi(R) \tag{10}$$

where $\psi(R)$ is a perturbation potential. Work on this aspect of concentrated (neutral) sterically-stabilized particulated systems is rather recent.[17] Also, the contribution from $\psi(R)$ seems to be much more important in charged systems (Section 5) with long range electrostatic interactions.[18]

For the purposes of this paper, we shall restrict discussion to V(R), and therefore by implication, strictly to dilute dispersions. Mention has already been made (Section 3), of the (two-and three-dimensional) compression methods for the experimental determination of F(R), and hence V(R). Use of the more direct force-balance methods, such as those described by Ottewill[19] and Israelachvili[20], require molecularly smooth macroscopic "supports" for the adsorbed layer sheath, e.g. (deformable) silicone rubber[19] or (rigid) mica[20]. It would be an interesting challenge to develop similar smooth surfaces for other "supports" to mimic latex particles, but the elastic properties would need to be well characterized. Nevertheless, the recent work ,in particular, of Klein and Luckham[21] on the direct determination of F(R) using an apparatus based on Israelachvili's design[20], for adsorbed polymers on mica surfaces has given a great deal of insight into V(R) for this type of system with adsorbed polymers. The use of the direct grafting techniques for attachment of polymers to inorganic particle surfaces[22], would give similar information for terminally-grafted chains.

Theoretical work in this area has received more attention. A fuller discussion of this aspect of the subject appears in the paper by Cohen Stuart in this volume. A detailed review of the subject has also been given in a book by Napper.[23] Other recent reviews are by Vincent and Whittington[24] and by Buscall and Ottewill[25]. I therefore propose here only to highlight certain features.

Napper[23] has distinguished three levels of theory for the steric (soft) interaction, V_s:

a) pragmatic: these assume some form for $\phi_2(z)$, e.g. rectangular (uniform), exponential, Gaussian.

b) classical ab initio: these derive $\phi_2(z)$ but calculate separately the mixing and elastic contributions to V_s. Also, as with the pragmatic theories, it is generally assumed that $\phi_2(z)$ for each surface is independent of R [at least for $\delta < h < 2\delta$; h=R-2a].

c) self-consistent ab initio: these avoid the split into the mixing and elastic contributions, and introduce $\phi_2(z,R)$. Of course, an implicit assumption, therefore, is that the interaction follows the equilibrium path (see earlier remarks, Section 3). The first contributors in this area were Dolan and Edwards[26], but the more recent work (for flat plates) has been that of de Gennes[27], using scaling concepts, and Scheutjens and Fleer[28], using a mean-field statistical mechanical approach, in which all polymer configurations are generated on a lattice as a fraction of plate separation. (See chapter by Cohen Stuart for more details.)

Although the pragmatic theories referred to above are perhaps at best only semi-quantiative, they do give some useful insight into the important features underlying steric interactions. As an example, the equation derived by Napper[23] for the steric interaction between two spherical particles carrying polymer sheaths is given below,

$$V_s = \underbrace{\frac{kTV_2^2x^2\Gamma_2^2}{V_1}\ (1/2-\chi)\cdot S_{mix}}_{V_{s,mix}} + \underbrace{kT\Gamma_2\cdot S_{el}}_{V_{s,el}} \qquad (11)$$

where V_1, V_2 are the volume of a solvent molecule and a polymer segment respectively, x is the number of segments per chain and χ is the Flory-Huggins polymer/solvent interaction parameter. S_{mix} and S_{el} are geometric functions (of a, δ, h) which depend on the form assumed for $\phi_2(z)$, and the interaction region, i.e. $\delta < h < 2\delta$ (interpenetration only; $S_{el} = 0$) or $0 < h < \delta$ (interpenetration plus compression). Despite the rather gross assumptions on which it is based, equation 11 does illustrate clearly the dependence of V_s on Γ_2, x (and hence, MW) and χ. Most thermodynamic effects (solvent type, temperature, pressure effects) enter essentially through χ. In the more recent theory by Scheutjens and Fleer,[28] referred to above, which requires V_s to be computed numerically, the "input" parameters are χ, χ_s (net adsorption energy - see earlier), x and a lattice parameter (related to V_2 and V_1), together with the bulk solution polymer volume fraction, ϕ_2^b. Since it is an equilibrium theory, Γ_2 is now a function of ϕ_2^b and plate separation. (In the Napper approach ϕ_2^b is a fixed and independent variable.) However, the inherent power of the Scheutjens-Fleer theory emerges when it is realized that a great deal of information (in addition to V_s) may be computed, e.g. Γ_2 and the distribution of segments between trains, loops, tails and bridges, as a function of plate separation. The Scheutjens-Fleer theory was originally applied to homopolymers, but has recently been (or is currently being) extended to other classes of polymers, e.g. physisorbed block or random copolymers, and terminally-attached chains. These latter models will perhaps be more useful for applying to latex particles with in situ stabilizers. Consideration has also been/is being given to the problems of polydispersity, competitive adsorption and solvent displacement, surface non-uniformity and curvature.

An important feature in this field which has received much attention recently[23-25,28] is the effect of free (non-adsorbing) polymer on the interaction between colloidal particles. Our own interest arose originally from the observation[29-31] that polystyrene latex particles, carrying terminally-anchored PEO chains dispersed in water, flocculated weakly and reversibly over a certain concentration region of added, free PEO.

The origin of the added attractive interaction is associated with the fact that, if $\chi_s < \chi_{s,crit}$, free polymer chains, for entropic reasons, tend to be repelled from the interface, where the segment concentration is lower than in bulk. Overlap of the depletion layers on two approaching particles leads to expulsion of solvent from the overlap region into bulk solution (at a high polymer concentration); hence these solvent molecules reduce their chemical potential. Alternatively (but equivalently), one may describe the attractive interaction in terms of the osmotic pressure (Π), tending to push the two particles together. Theories for the depletion interaction have been derived by several groups.[32-36] These theories were all constructed for hard-sphere particles (Figure 1a) and, in general, seem to lead to similar predictions. For example, Scheutjens, Fleer and Vincent [36] have derived the following approximate expression for $V_{dep}(h)$

$$V_{dep}(h) = \underbrace{\left(\frac{\mu_1 - \mu_1^{\,\circ}}{v_1} \right)}_{\Pi} 2\pi a \left(\Delta - \frac{h}{2} \right)^2 \left(1 + \frac{2\Delta}{3a} + \frac{h}{6a} \right) \quad (12)$$

where Δ is the effective depletion layer thickness. $\Delta \sim r_g$ (radius of gyration of free polymer coils in solution) at low ϕ_2^b, but decreases beyond the critical free coil overlap concentration (ϕ_2^*). The result is that $|V_{dep}(0)|$, the value of V_{dep} at particle contact (h=0, R=2a), initially increases with increasing ϕ_2^b, i.e. where the osmotic pressure term dominates. However, eventually (beyond ϕ_2^*), the decrease in Δ takes over. The result is that $|V_{dep}(0)|$ passes through a maximum with increasing ϕ_2^b; this effect is illustrated in Figure 4.

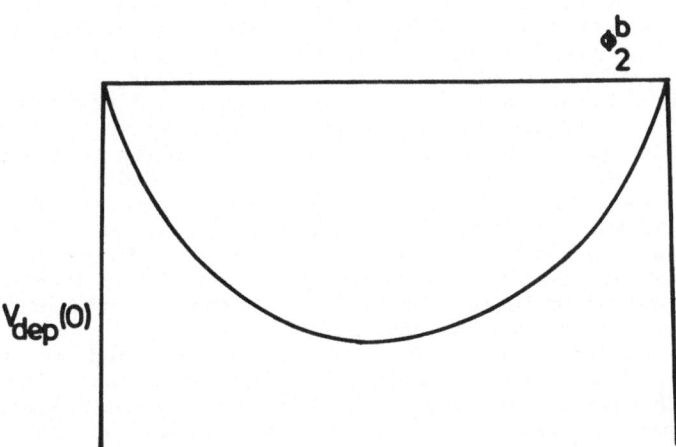

Figure 4: Schematic representation of $|V_{dep}(0)|$ as a function of ϕ_2^b for two particles.

For soft-sphere particles (Figure 1b), the theoretical analysis is less clear: the depletion effect still occurs but is weaker than in the case of hard spheres. Vincent et al[37] have recently analyzed this situation. Clearly, depletion effects are important in any practical situation which involves particles plus solvent plus free polymer. Such a situation arises for example in a drying latex paint film or surface coating. There may well be concentration regions over which flocculation may arise, giving rise to possible reduction of gloss, hiding-power capacity, etc.

The problem of stability of latex particles in polymer "melts" can be considered as a logical extension of the problem of free polymer in solution. As Figure 4 implies, hard-sphere particles should be stable in polymer melts. The same is true for soft-sphere particles, if the anchored polymer and free polymer are the same. However, (thermodynamic) instabilities may arise if the two polymers are dissimilar, although there is very little reported experimental work in this area.

As pointed out in the Introduction, there may be a significant contribution to the total pair-potential from van der Waals forces between the particle cores. Indeed these may be the dominant contribution to V_{min} (Figure 1b). Although van der Waals forces can be reasonably accurately modeled for hard-sphere particles in non-polar media[38], there are still problems, in particular, with aqueous media and also where soft-sphere particles are involved. In the latter case, strictly speaking, we again need to know $\phi_2^b(z,R)$. Estimation of van der Waals interaction in these more complex systems remains a major challenge to the theorists.

5. STABILITY THEORY: CHARGED SYSTEMS

In general, it is considerably more difficult to model the pair potential for charged systems, in comparison to neutral systems, largely because one needs to establish the charge distribution as well as the segment distribution. The problem is similar to modeling poly-electrolytes, compared to neutral polymers in solution.

Even with neutral polymers physisorbed on charged hard sphere particles, there are problems in establishing how the presence of segments in trains modify the surface charge density, the fraction of counter ions specifically adsorbed in the Stern layer, solvent dipole orientation, etc. Segments in loops and tails may affect the spatial distribution of ions in the diffuse layer. Theoretical approaches remain largely pragmatic. For example, Brooks[39] has shown that, in general, the diffuse layer around a charged particle is more extended in the presence of an adsorbed neutral polymer. An approach, frequently adopted, is simply to calculate $V_E(R)$ only for $R>2(a+\delta)$, i.e. prior to sheath overlap, and assume that the zeta potential corresponds to the electrostatic potential at the periphery of the sheath. It then has to be implicitly assumed that interaction occurs at constant potential; this may not be the case in a Brownian encounter.

Those cases where the sheath contains adsorbed or anchored polyelectrolyte has been considered by Buscall and Ottewill[25]. Dispersions with adsorbed polyelectrolytes seem to be intrinsically more stable at higher electrolyte concentrations (>5 mol dm^{-3}) than with neutral polymers. It has also been argued that the surface layer of certain, supposedly hard-sphere, charged latex particles in water may in fact be a gel-like, solvated polyelectrolyte layer. This is much more likely to be the case when the surface charge density is high, or the core latex polymer is a copolymer of a hydrophobic monomer and hydrophilic monomer (e.g. styrene and acrylic acid). It is well-known, in addition, that aqueous silica dispersions are very stable to electrolyte addition, presumably because of the gel-like nature of the surface region in water.

At low ionic strengths the ion distribution around adsorbed

polyelectrolytes is not well-defined; the electrostatic interactions are long-range (e.g. at 10^{-5} mol dm^{-3} ionic strength, the Debye length is ~ 100 nm). The adsorbed polyelectrolyte chains are greatly extended due to intra- and inter-chain repulsions. The theory of polyelectrolyte adsorption at a single interface has recently been described by van der Schee,[40] who has incorporated electrostatic free energies into the Scheutjens-Fleer theory. However, as yet, there is no theory for V(R) for particles with adsorbed polyelectrolyte, at low ionic strengths.

At high ionic strengths, greater than say 10^{-1} mol dm^{-3} the Debye length is ~ 1 nm, so that in this case the electrostatic interactions are sufficiently short-range to incorporate them into the steric interaction through the χ parameter[25]. The short-range intra and inter-electrostatic repulsions contribute negatively to χ; hence, an aqueous electrolyte-solution behaves as a much better solvent environment for a charged polymer than for an equivalent neutral polymer. Hence, the greater stabilizing power of polyelectrolytes in electrolyte solutions. Buscall[41] has studied the flocculation behavior of polystyrene latex particles, with adsorbed polyacrylate chains. The critical flocculation temperature decreases rapidly with the degree of neutralization of the poly-electrolyte chains, at a given background electrolyte concentration.

Corner and Gerrard[42] have studied the flocculation behavior of latex particles stabilized by comb-type polymers in which the backbone polymer is poly(methyl methacrylate) and the stabilizing moeities are poly(α-methylstyrene sulphonate) chains. An interesting observation was that even with low MW stabilizing chains (MW 1250; x=8), the critical electrolyte concentration for NaCl was 1.6 mol dm^{-3}; with higher chain lengths the latices were stable in saturated calcium and sodium chloride. An interesting class of stabilizers are those whose chains are essentially neutral (e.g. PEO), but have a single terminal charge group. Such systems are currently being studied by my group in Bristol.

6. THERMODYNAMIC ASPECTS OF STABILITY

Classically, the stability of colloidal dispersions has largely been considered from the kinetic viewpoint. In more recent times greater emphasis has been placed on the thermodynamic aspects. This is particularly relevent to sterically-stabilized dispersions, where in many cases flocculation is weak and reversible, and associated with the shallow minimum in the interaction (V_{min}). The analogy between weak flocculation and phase separation has been stressed[43,28-31]. Indeed, the equilibrium state observed with weakly flocculating particles (see Figure 5) is rather analogous to the vapour state/condensed state equilibrium observed with simple molecular systems.

By analogy, therefore, with (hard-sphere) molecules, one may write[30,31] for the free energy change, ΔG_f, associated with the flocculation process depicted in Figure 5,

$$\Delta G_f = \Delta G_{hs} + \Delta G_i \qquad (13)$$

where ΔG_{hs} is a configurational entropy term associated with the loss in freedom of the particles. It is essentially a function of the initial volume fraction of the particles (ϕ). ΔG_{hs} ($= -T\Delta S_{hs}$) is positive and increases as ϕ decreases; this term opposes flocculation. ΔG_i is an interaction term, and is associated with the interparticle interactions, through V_{min}. It is negative and encourages flocculation. Whether flocculation occurs, therefore, depends on a subtle balance of the ΔG_{hs}

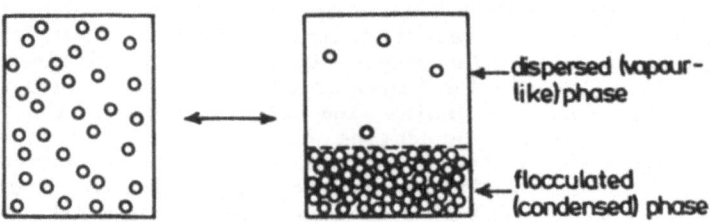

Figure 5: Flocculation equilibrium

and ΔG_i terms. These contributions may be calculated using statistical mechanical theories[44,45] and theoretical phase diagrams may be constructed[46]. To illustrate this point, Figure 6 is a schematic representation of $|V_{min}|$ versus particle volume fraction (ϕ) for a weakly interacting particulate dispersion.[46] Roman numerals indicate the number of co-existing phases. The condensed phase in this case would be a solid-like (strictly, at equilibrium, crystalline) phase. For more strongly interacting (particularly longer-range attraction) systems, the form of the V_{min} -ϕ plots are more complex, and liquid-like phases (and hence critical and triple points) can appear.[46]. However, for most latex systems stabilized by adsorbed/anchored polymer (Figure 1b), Figure 6 is probably a reasonable representation. If one starts in the stable region, say at point A, one can move into the two-phase region "vertically" by increasing $|V_{min}|$ (e.g. by a change in χ, ϕ_2^b etc.); this leads to the concept of a critical (temperature, solvency, pressure, free polymer concentration, etc.) condition for flocculation. Alternatively, one can move "horizontally" into the two-phase region; this leads to concept of a critical particle volume fraction[29-31,42] condition for the onset of flocculation.

By way of illustration, we illustrate schematically in Figure 7 the effect of added free polymer on the stability/flocculation behavior of a sterically-stabilized latex system. Consider moving in the direction of the dotted line as indicated, i.e. at fixed particle volume fraction (ϕ), but increasing polymer volume fraction (from $\phi_2^b=0$ to $\phi_2^b=1$, the polymer melt). ΔG_{hs} in equation 13 is positive and fixed (since ϕ is fixed), whereas ΔG_i (through variation in V_{min}; see Figure 4) becomes, firstly increasingly more negative, then decreases again at high ϕ_2 toward 0. Thus, there will be a range of ϕ_2 values, for which ΔG_{floc} is net negative (equation 13), such that the system is thermodynamically unstable and

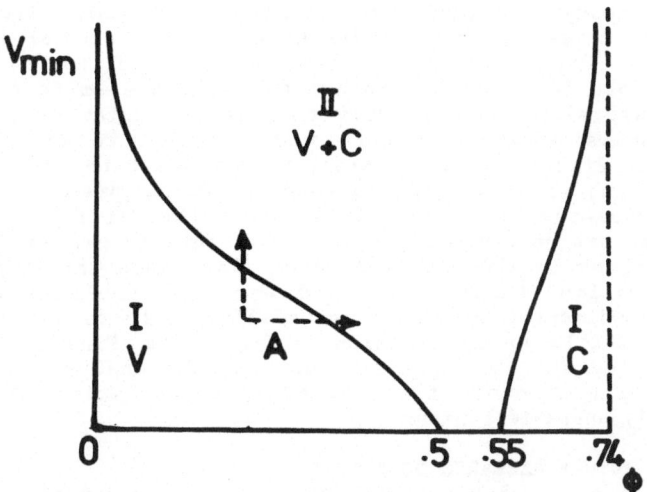

Figure 6: Schematic representation of V_{min} versus ϕ for a weakly interacting particulate dispersion V-vapour-like, C-condensed-like phase.

reversible flocculation occurs. Thus we may define one critical polymer volume fraction ($\phi_2^{\dagger}$) for the onset of flocculation, and a second one ($\phi_2^{\ddagger}$) for restabilization. The left-hand (solid) boundary line in Figure 7, therefore, represents the locus of $\phi_2^{\dagger}$ and $\phi_2^{\ddagger}$, as ϕ varies. Therefore, the depletion flocculation phenomenon is more important in concentrated latex systems than in dilute ones. Indeed, in sufficiently dilute systems, one may not observe it.

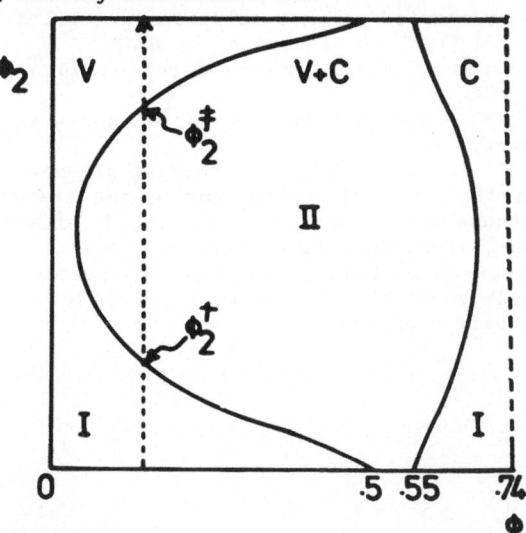

Figure 7: Stability/flocculation phase diagram for a sterically stabilized latex in the presence of free polymer. ϕ-particle volume fraction; ϕ_2-polymer volume fraction.

Although thermodynamic (and statistical thermodynamic) descriptions of stability in these weakly interacting systems are clearly very powerful, there are certain aspects which need to be watched for/accounted for. For example, metastable states seem to persist in particulate dispersions more readily than in molecular systems. Thus supersaturation is often observed[47] with respect to the left-hand boundary line in Figures 6 and 7. Also, with respect to the right-hand boundary line, equilibrium thermodynamics would predict that the condensed phase should be a crystalline phase (bcc or fcc), but often amorphous phases are observed experimentally. This is presumably due to the steric problems involved in repacking flocs into the minimum free energy configuration. Furthermore, sedimentation can mask stability effects, since sedimentation itself leads effectively to an increase in the (average) particle volume fraction in the system. Finally, to return to kinetic considerations, as yet we do not have analytical theories which successfully accounts for reversible flocculation kinetics in thermodynamically unstable regions.

7. CONCLUSIONS: FUTURE RESEARCH OBJECTIVES

Clearly, although much has been achieved in our understanding of the stabilization and flocculation of sterically-stabilized latices, there are several areas to which future research activity could be directed. With no particular order of priority (or difficulty!), some of these areas which have emerged from this review are listed below:-

a) configuration of comb, ABA block and other copolymers at latex surface (e.g. "burial"; multipoint anchoring).
b) dynamics of adsorbed polymer segments.
c) theories for $V_s(R)$ for complex copolymers.
d) to build modern concepts of polymer solution theory into theories for $V_s(R)$ (e.g. free volume effect, segment concentration effects, aqueous systems).
e) stability in concentrated polymer solutions and melts.
f) determination of $\phi_2(z,R)$ in situ, e.g. by SANS.
g) direct measurement of $F(R)$ (e.g. by force-balance techniques) and coupling to $\phi_2(z,R)$.
h) theories for $V_A(R)$ - van der Waals interactions - in the presence of adsorbed polymers.
i) theories for $V(R)$ in the presence of charged polymers.
j) statistical mechanical theories for concentrated dispersions, (particularly those which incorporate multibody effects).
k) kinetics of weak, reversible flocculation.
l) relationship between stability and other properties (e.g. rheology) for sterically-stabilized latices; effect of shear on stability/instability behavior.

REFERENCES

1. Thompson MW: in *Polymer Colloids*; (ed) Buscall R; Corner T and Stageman JF Elsevier Applied Science p.32 (1985).
2. Barett KEJ(ed): *Dispersion Polymerisation in Organic Media* Wiley, London (1975).
3. Cowell C and Vincent B: in Tadros ThF(ed); The Effect of Polymers on Dispersion Properties. Academic Press, London, p.263 (1982).
4. Dobbie JW; Evans R; Gibson DV; Smitham JB and Napper DH: *J. Colloid Interface Sci* 45, 557 (1973).
5. Cosgrove T; Cohen Stuart MA and Vincent B: *Adv. Coloid Interface Sci.*, 24 143 (1986).
6. Kerker M: The Scattering of Light and other Electromagnetic Radiation, Academic Press (1967).
7. Crowley TL: Ph.D. Thesis, Oxford (1984). See also reference 5.
8. Barnett KG; Cosgrove T; Crowley TC; Tadros ThF and Vincent B: in Tadros ThF(ed): The Effect of Polymers on Dispersion Properties: Academic Press, p.183 (1982).
9. Cosgrove T; Obey TM and Vincent B: *J. Colloid Interface Sci.* 111 409 (1986).
10. Cohen Stuart MA; Waajen FHLH; Cosgrove T; Vincent B and Crowley TL: *Macromolecules* 17 1825 (1984).
11. Cosgrove T; Vincent B; Crowley TL and Cohen Stuart MA: in Goddard ED and Vincent B(eds): Polymer Adsorption and Dispersion Stability, *Amer. Chem. Soc. Symp. Washington, DC* Ser No. 240 147 (1984).
12. Cohen Stuart MA; van den Bromgaard Th; Zourab ShM and Lyklema J: *Colloids Surfaces* 9 163 (1984).
13. Doroskowski A and Lambourne R: *J. Colloid Interface Sci.* 43 97 (1973).
14. Cairns RJR; Ottewill RH; Osmond DWJ and Wagstaff I: *J. Colloid Interface Sci.* 54 45 (1976).
15. Garvey MJ; Tadros ThF and Vincent B: *J. Colloid Interface Sci.* 55 440 (1976).
16. Evans R and Napper DH: *J. Colloid Interface Sci.* 73 43 (1978).
17. Cebula DJ; Goodwin JW; Ottewill RH; Jenkin G and Tabony J: *Colloid Polymer Sci.* 261 555.1 (1983).
18. Cebula DJ; Goodwin JW; Jeffrey GC; Ottewill RH; Parentich A and Richardson RA: *Faraday Disc. Chem. Soc* 76 37 (1983).
19. Cain FW; Ottewill RH and Smitham JB: *Faraday Disc. Chem Soc* 65 33 (1978).
20. Israelachvili JN and Adams GE: *J. Chem. Soc. Faraday Trans. I* 74 975 (1978).
21. many recent papers, see e.g.: Klein J; Almog Y and Luckham PF: in Goddard ED and Vincent R(eds) Polymer Adsorption and Dispersion Stability, *Amer. Chem. Soc.; Washington, DC,* No. 240 227 (1984).
22. Fairhurst D; Baridger K and Vincent B: *J. Colloid Interface Sci.*, 8 190 (1979).
23. Napper DH: Polymeric Stabilization of Colloidal Dispersions, Academic Press (1983).
24. Vincent B and Whittington S: in Matijevic E(ed) *Surface and Colloid Science*, Plenum Press 12 1 (1982).
25. Buscall R and Ottewill RH: in Buscall R; Corner T and Stageman JF (eds), Polymer Colloids: Elsevier Applied Science p. 141 (1985).
26. Dolan AK and Edwards SF: *Proc. Roy. Soc.* A343 427 (1975).
27. de Gennes PG: *Macromolecules* 15 492 (1982).

28. Fleer GF and Scheutjens JMHM: *Adv. Colloid Interface Sci.*, **16** 341, 361 (1982).

29. Li-In-On FKR; Vincent B and Waite FA: *Amer. Chem. Soc. Symp. Ser.* **74** 337 (1975).

30. Cowell C; Li-In-On R and Vincent B: *J. Chem. Soc. Faraday Trans. I* **74** 337 (1978).

31. Vincent B; Luckham PF and Waite FA: *J. Colloid Interface Sci.* **73** 508 (1980).

32. Urij A: *Pure App. Chem.* **48** 471 (1976).

33. Feigin RI and Napper DH: *J. Colloid Interface Sci.*, **74** 567 (1980); **75** 525 (1980).

34. Joanny JF; Liebler L and de Gennes PG: *J. Poly. Sci.* **17** 1073 (1979).

35. Sperry PR: *J. Colloid Interface Sci.* **87** 375 (1981).

36. Scheutjens JMHM; Fleer GJ and Vincent B: in Goddard ED and Vincent B(eds), Polymer Adsorption and Dispersion Stability, *Amer. Chem. Soc. Symp. Ser.* Washington, DC **240** 245 (1984).

37. Vincent B; Edwards J; Emmett S and Jones A: *Colloids Surfaces*, **18** 261 (1986).

38. Hough DB and White L: *Adv. Colloid Interface Sci.*, **14** 3 (1980).

39. Brooks DE: *J. Colloid Interface Sci.*, **43** 670, 687, 700, 714 (1973).

40. van der Schee HA: Ph.D. Thesis, Wageningen (1984); van der Schee HA and Lyklema J: *J. Phys. Chkem.* **80** (1984).

41. Buscall R: *J. Chem. Soc. Faraday Trans. I* **77** 909 (1984).

42. Corner T and Gerrard J: *Colloids Surfaces*, **5** 187 (1982).

43. Long JA; Osmond DJW and Vincent B: *J. Colloid Interface Sci.* **42** 545 (1973).

44. Snook I and van Megen W: *Adv. Colloid Interface Sci.* **21** 119 (1984).

45. Gast AO; Hall CK and Russel WB: *J. Colloid Interface Sci.* **96** 251 (1983).

46. Vincent B: *Colloids Surfaces* 1987, in press.

47. Edwards J; Everett DH; O'Sullivan T; Pangalou I and Vincent B: *J. Chem. Soc. Faraday Trans. I* **80** 2599 (1984).

CONTROL OF PARTICLE SIZE IN THE DISPERSION POLYMERIZATION OF STERICALLY
STABILIZED POLYMER COLLOIDS

Melvin D. Croucher

Xerox Research Centre
2660 Speakman Drive
Mississauga, Ontario
L5K 2L1
Canada

Mitchell A. Winnik

Department of Chemistry
University of Toronto
Toronto, Ontario
M5S 1A1
Canada

1. INTRODUCTION

   The two principal methods of preparing polymer particles in a liquid
medium are emulsion and suspension polymerization. While both preparative
techniques utilize heterogeneous reactions, the locus of the
polymerization in these two systems differs dramatically. In emulsion
polymerization the reaction proceeds within micelles in solution with the
emulsion droplet functioning as a reservoir for the monomer, while in
suspension polymerization the reaction takes place within the monomer
droplet. These procedures usually lead to small particles of $0.1 - 2.0\mu m$
diameter for emulsion polymerized particles while suspension
polymerization usually yields coarse particles that exhibit a wide
particle size distribution. Both of these particle polymerization
techniques are usually carried out in polar media such as water, although
considerable progress has been made in polymerizing water soluble
monomers in hydrocarbon fluids using the inverse emulsion technique[1].
Although conventional emulsion polymerization usually leads to submicron
size particles, major advances have been made by the Ugelstad[2] and the
Vanderhoff and El-Aasser[3] research groups in preparing monodisperse
polymer particles up to $100\mu m$ in diameter. Ugelstad et al[2] have achieved
this by utilizing a sequential swelling process followed by
polymerization of the latex while Vanderhoff and El-Aasser have achieved
the same objective by preparing the particles using a seeded emulsion
polymerization in the gravity free environment of space.

   A third particle polymerization process which has been practiced for
approximately twenty-five years is known as dispersion polymerization.
While it has not been as extensively investigated as the other
polymerization techniques, it has recently attracted considerable
interest. This method involves the polymerization of a monomer which is
initially soluble in the fluid medium. This is a major difference as
compared to emulsion and suspension polymerization in which the monomer
is insoluble in the dispersion medium. One of the requirements of the
dispersion polymerization process is that the polymer formed during the

polymerization should be insoluble in the dispersion medium. Consequently, the polymer precipitates out of solution after growing to a finite chain length to form particle nuclei. If a polymeric steric stabilizer is also present in the reaction flask it serves to stabilize the colloidal nuclei that have been formed allowing the growth of discrete particles. One of the advantages of this process is that both aqueous and hydrocarbon based liquids can be used as the dispersion medium so facilitating the polymerization of a wide range of monomers into polymer particles. To date, the majority of the studies that have been reported have used free radical polymerization to prepare disperse materials. However, dispersion polymerization is sufficiently flexible that other polymerization schemes, such as ionic polymerization, can also be used. All of the early research in dispersion polymerization was carried out in nonaqueous media such as aliphatic hydrocarbons[4]. Most of these studies produced polymer particles that were in the $0.1-2.0\mu m$ diameter range. Recent studies[5-8] have shown that it is possible to produce particles of up to at least $15\mu m$ diameter in both polar[6] and nonpolar[8] media. This had been achieved by carrying out the polymerization in mixed solvent media. By varying the composition of the mixed solvents, the temperature and the initiator concentration, it has been found possible to control both the particle size and the size distribution. In this paper we will discuss the requirements necessary to prepare polymer particles using the dispersion polymerization technique. This will be followed by a review of the current state of knowledge in this area with particular emphasis being given to control of the size of the particles that are formed.

2. THE DISPERSION POLYMERIZATION PROCESS
2.1. A Brief Historical Perspective
Dispersion polymerization, which can be looked upon as a special type of precipitation polymerization, was developed by researchers at ICI. It was devised as an analogous process to emulsion polymerization except that it was to be carried out in a hydrocarbon rather than a water medium. In emulsion polymerization the growing particles are usually protected against flocculation by ionogenic species which are embedded in the surface layer of the particle, giving rise to an ionic double layer. This provides the particle with its stabilization mechanism. Since it was believed that it was not possible to stabilize particles in hydrocarbon fluids using an ionic mechanism, an amphipathic copolymer was chosen to stabilize the growing particles against flocculation rather than an ionic stabilizer. This deductive reasoning has given rise to a new technology which is now known as dispersion polymerization. All of the early work was confined to polymerization in aliphatic hydrocarbon fluids and has been well reviewed in the classic book "Dispersion Polymerisation in Organic Media" that was written by ICI researchers. During the past five years there has been a keen interest in generalizing these results to a wide variety of dispersion media including water. Although, in principle, a wide variety of monomers can be polymerized, only a few have been investigated thoroughly. These include styrene and methyl methacrylate, and our discussion will be largely based upon these systems.

2.2. Qualitative Description of Dispersion Polymerization
The components of a typical dispersion polymerization include the dispersion medium, an amphipathic stabilizer, a monomer or comonomers and an initiator. Initially, all of the components are soluble in the dispersion medium to give a homogeneous solution. Upon initiation of the

reaction the solution remains transparent for a finite period of time after which the reaction mixture becomes cloudy as the nuclei formed act as scattering centers. Provided these nuclei are well stabilized against coagulation the particles continue to grow as discrete entities until the reaction is terminated. These reactions are usually carried out in one of two ways. The first is a "seed and feed" approach in which an initial seed stage is prepared which contains a low solids dispersion to which is added further monomer over a specific time interval. The second type of polymerization is the "one-shot" approach in which all of the monomer is added to the reaction vessel at the start of the polymerization. Since most monomers are also solvents for their polymers it is obvious that in the "one-shot" approach the "solvency" of the dispersion medium changes profoundly as the reaction proceeds. This is a feature which complicates analysis of the growth behavior of such systems. It should be pointed out that the majority of the studies reported in the literature have used the "one-shot" technique. The results reported here have all been obtained using this method.

2.3. Dispersion Medium

The reaction medium must satisfy several important criteria: It must be chosen to ensure the solubility of the amphipathic steric stabilizer and the initiator during the polymerization reaction. While the monomer must be soluble, its polymer must be insoluble in the dispersion medium if particles are to be formed. While no definitive studies have been reported it seems clear that the viscosity of the fluid should be less than ~2-3 mN.s.m^{-1} to ensure that rapid diffusion of the reactive species can take place during the polymerization. While the early studies were confined to aliphatic hydrocarbon fluids[4] more recent research has extended the range of liquids to include chlorinated hydrocarbons[8] and water based mixtures[6].

2.4. Amphipathic Steric Stabilizer

In order for the newly formed nuclei to grow as discrete entities into polymer particles, it is necessary for them to be well stabilized. Three types of polymers have been successfully used as steric stabilizers namely, block and graft copolymers, homopolymers and macromonomers.

2.4.1. Amphipathic block and graft copolymers. It has been found that preformed block and graft copolymers perform extremely effectively as steric stabilizers. Numerous molecular architectures have been used and these are shown schematically in Figure 1. In these systems one portion of the copolymer is nominally insoluble in the dispersion medium such that it will form the anchor for the stabilizer to the growing polymer particle. Overall however, the macromolecule must be soluble in the dispersion medium i.e. the dispersion medium must be a better than θ-solvent for the stabilizing polymer. One of the original stabilizers used by ICI was a "comb" stabilizer in which poly(12-hydroxy stearic acid), which is soluble in aliphatic hydrocarbons, was grafted with a nominally insoluble polymer such as poly(methyl methacrylate) to provide the anchor to the particle surface[4]. Recently, well tailored AB and ABA block copolymers[9,10] have also been used effectively to colloidally stabilize particles. Dawkins and Taylor[9] studied the dispersion polymerization of methyl methacrylate using a series of well characterized poly(styrene-b-dimethylsiloxane) stabilizers. They found that the ratio of the soluble block to the anchor block was an important variable, and that if the anchor/soluble block (ASB) ratio exceeded a critical number then stable particles could not be prepared, even though the stabilizer was soluble

in the dispersion medium. It had been suggested previously by Vincent[11] that the most efficient stabilizers have an ASB ratio within the range 0.33 to 3. The data reported by Dawkins and Taylor[9] indicate that Vincent[11] was too conservative in his estimate and ASB ratios as large as 18 have been used to polymerize particles. For ASB ratios greater than ~20 it was found that coarse, irregular particles were formed because the short soluble chains of the block copolymer were unable to colloidally stabilize the large core of the particle. A further factor which deserves attention is the molecular weight of the stabilizer. While it is conceded that a stabilizer with a molecular weight greater than ~10^4 will stabilize colloidal particles, in many cases stabilizers with molecular weights of ~10^3 have been used without any major problems. However, no definitive studies as to the optimum molecular weight of the stabilizer or to its role during the dispersion polymerization have yet been published. It is widely believed that the anchor polymer of the stabilizing molecule can be chemically very different from the polymer which constitutes the core of the particle. It should be borne in mind that most polymer-polymer pairs are incompatible[12] consequently a judicious choice for the anchor polymer would seem to be necessary for a successful dispersion polymerization.

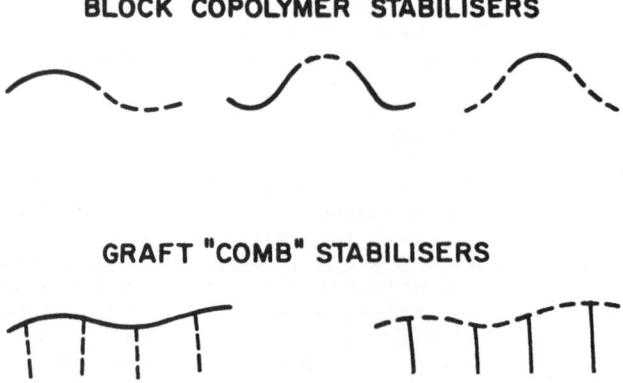

Figure 1: Shows schematically block and graft copolymers that have been used as steric stabilizers in dispersion polymerization. (——) denotes the soluble moiety, while (----) denotes the insoluble moiety.

2.4.2. Homopolymer Stabilizers. In some of the earliest experiments on the dispersion polymerization of methyl methacrylate, degraded natural rubber was used as the stabilizer and stable colloidal particles were produced when peroxides were used as initiators[4]. In more recent studies[6,13] it has been found that poly(2-ethyl hexyl methacrylate) in 2,2,4-trimethylpentane and hydroxypropylcellulose in water based mixtures also function as homopolymer steric stabilizers for growing particles in a dispersion polymerization. It is believed that in these cases hydrogen abstraction takes place which then allows grafting of the monomer in solution to take place to produce an amphipathic copolymer. Bromley has recently commented[7] that homopolymer stabilizers usually lead to "coarse" dispersions but in the work of Lok and Ober[6] monodisperse latexes were produced. From the limited information available it would appear that any homopolymer (or random copolymer) that contains labile hydrogen atoms could be used as a steric stabilizer provided the optimum initiator is used. Because of the uncertainty associated with this technique it is not to be universally recommended as a method of stabilizing growing particles.

2.4.3. Macromonomers. These are also known as stabilizer precursors and are relatively low molecular weight polymers that are terminated at one end by a reactive group. An example of this type of stabilizer is polyethylene oxide which is terminated at one end with a reactive acrylate ester group capable of undergoing polymerization[7]. The molecular weight of the polymer is critical since too low a molecular weight will not be sufficient to stabilize the particles while too high a molecular weight will lead to less efficient polymerization of the macromonomer which could lead to an unstable latex. While this concept of a stabilizer precursor is relatively new it holds the promise of providing a simpler and more flexible method of stabilizing growing particles in a dispersion polymerization.

2.5. Monomers

In principle a wide range of monomers, both water soluble and oil soluble, can be used. In practice, the limits of this polymerization process do not appear to have been widely explored, consequently the usefulness of this polymerization technique is not well established. The major requirement of the monomer is that it must be soluble in the dispersion medium but that its polymer must be insoluble.

2.6. Initiators

The most important requirement of the initiator is that it must be soluble in the dispersion medium. However, in practice the initiator is usually dissolved in the monomer and this often solubilizes the initiator in the polymerization fluid. The solubility question is important since it has been found that different initiators produce significant differences[14] in the final latex. It is also thought that partitioning of the initiator between the dispersion medium and the growing particle can take place, and that this influences the locus of the particle polymerization[15]. No definitive data have yet been accumulated but it is a subject that deserves attention.

3. THE INFLUENCE OF EXPERIMENTAL VARIABLES IN DISPERSION POLYMERIZATION

In any dispersion polymerization the variables that are usually controlled are the stabilizer concentration, the solvency of the dispersion medium, initiator concentration, temperature and stirring speed. It has been found that these variables can have a significant effect on the particle size, the molecular weight of the final polymer particles and the kinetics of the particle polymerization process. The interrelationships between these variables appear to be quite complex. We have therefore chosen to discuss these variables individually in order to delineate their effect on the properties of the particles that are formed.

3.1. Stabilizer Concentration

It has been found that the greater the concentration of the stabilizer used during the polymerization then the smaller are the particles that are produced. This is not a surprising observation since more stabilizer can stabilize a larger number of the nuclei that are formed. This will reduce the aggregation and coalescence between nuclei during the early stages of the growth process. Dawkins and Taylor[9] have studied the effect of the concentration of well tailored poly(styrene-b-dimethylsiloxane) blocks on the dispersion polymerization of methyl methacrylate in hexane. Their results are shown as curve (a) of Figure 2 and indicate that stabilizer concentration effects have a relatively modest influence on the final particle size. Similar results have been reported by Douglas, Illum and Davis[16] who investigated the aqueous polymerization of butyl-2-cyanoacrylate in the presence of dextran as stabilizer. Their results are shown plotted as curve (b) Figure 2. These workers also showed that the lower molecular weight dextrans had a smaller effect on the final particle size as a function of increasing concentration than did the higher molecular weight materials. The authors refer to their reaction as a dispersion polymerization (citing Barrett[4]) which implies that at the start of the reaction the monomer is molecularly dispersed.

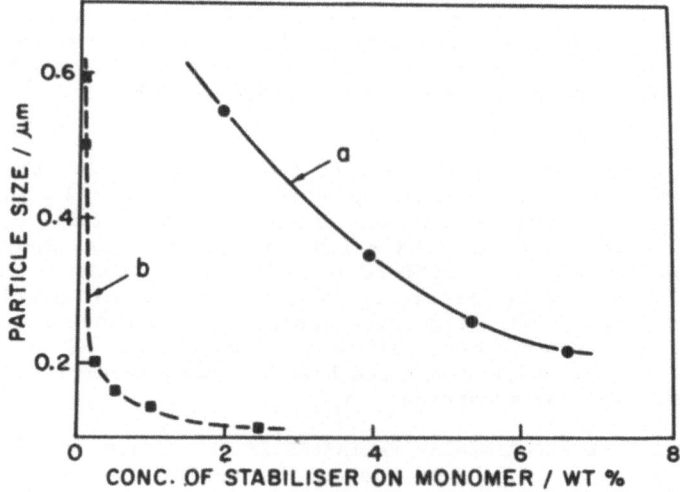

Figure 2: Shows the average particle size plotted against the stabilizer concentration. (a) refers to the reaction of methyl methacrylate in hexane using poly(styrene-b-dimethylsiloxane) as stabilizer while (b) refers to the preparation of poly-(butyl-2-cyanoacrylate) particles in water using dextran as the stabilizer.

A more interesting observation is that the degree of polydispersity of the final latex increases as the concentration of stabilizer increases[8,17]. The reason for this effect is not obvious but could be related to the fact that grafting to the stabilizer can take place during the dispersion polymerization. With more polymer chains in solution this could lengthen the time interval over which particle nucleation can take place, thereby leading to a wider particle size distribution.

3.2. Solvency of the Dispersion Medium

The majority of the polymerization studies that have been reported use the one-shot process in which all of the monomer is added at the beginning of the reaction. As the monomer is consumed the solvency of the dispersion medium decreases which makes it difficult to interpret and to model the behavior of the dispersion polymerization process. Nevertheless, some interesting observations regarding control of the particle size have been reported using this method. In a recent publication Antl et al[17] studied the effect of varying the initial concentration of monomer in the polymerization of methyl methacrylate in aliphatic hydrocarbon media using a grafted poly(12-hydroxy stearic acid) as the stabilizer. They observed that as they increased the initial monomer concentration from 35 to 50 wt% the particle size increased from $0.18\mu m$ to $2.6\mu m$. The particle size distribution for these dispersions was found to be extremely narrow. When the initial monomer concentration was varied from 8.5 to 34 wt% it was found that the dispersion formed was initially stable but subsequently flocculated before the reaction had gone to completion. These results are shown plotted in Figure 3a. Using monomer concentrations below 8.5 wt% also produced small stable particles of diameter ~80nm. Antl et al suggested[17] that at low monomer concentrations the mechanism of particle formation is different from that taking place at high monomer concentrations. Consequently, in the central portion of Figure 3a the two mechanisms are in competition with the subsequent formation of unstable latexes. The mechanistic aspect will be discussed further in section 4.2.

We have recently found[8] that it is possible to control the size of poly(methyl methacrylate) particles by carrying out the polymerization in mixed solvent media. Using a slightly unsaturated polyisobutylene as the stabilizer it was found that poly(methyl methacrylate) particles of ~1μm diameter were formed when the polymerization was carried out in 2,2,4-trimethylpentane. However, when some of the trimethylpentane was replaced by carbon tetrachloride, the particle size increased as shown in Figure 4. Fairly monodisperse particles of up to 15μm diameter have been obtained. Above ~75 wt% of carbon tetrachloride in the dispersion medium it was found to be impossible to form particles. An extremely viscous solution resulted.

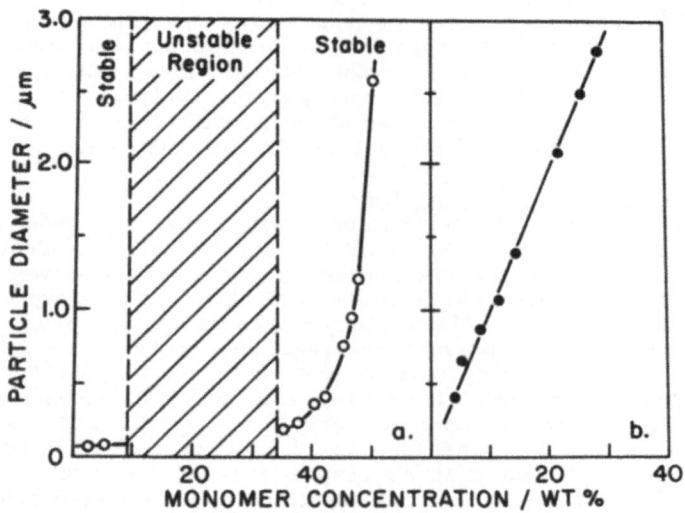

Figure 3: *Shows the final particle diameter plotted against the initial concentration of monomer in the polymerization. (a) refers to the polymerization of methyl methacrylate in an aliphatic hydrocarbon while (b) refers to the polymerization of styrene in an ethanol/water mixture.*

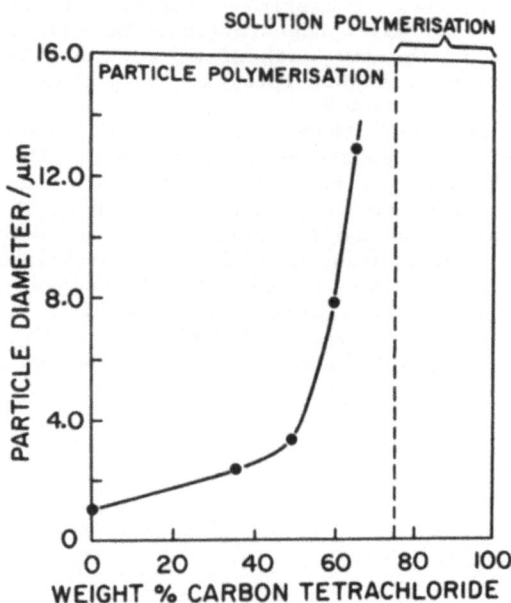

Figure 4: *Shows the particle size of poly(methyl methacrylate) latexes produced as a function of the concentration of carbon tetrachloride in the mixed solvent medium.*

The most comprehensive study to elucidate the effect of the solvency in dispersion polymerization on the final particle size is that published by Lok and Ober[6]. They studied the dispersion polymerization of styrene in alcohol based mixtures using hydroxypropylcellulose as the steric stabilizer. When particles were prepared in pure alcohols as the dispersion medium, particles of 1-2μm in diameter were obtained. Similar results were obtained when alcohol/water mixtures were used. It was also shown that the final particle size was dependent upon the initial concentration of monomer that was used in the reaction. These results are plotted in Figure 3b and parallel the results obtained in hydrocarbon media shown in Figure 3a. However, no unstable region was observed in the alcohol/water systems and the reaction proceeds normally over a wide range of monomer concentrations. When water was replaced by 2-methoxyethanol it was found that monodisperse particles of up to ~15μm diameter could be obtained. The greater the concentration of 2-methoxyethanol in the reaction mixture the larger was the final particle size. Lok and Ober[6] consider that the polarity of the dispersion medium controls the final particle size, a conclusion also reached by Almog, Reich and Levy[5]. In their study of the polymerization of styrene in methanol, ethanol, isopropanol and t-butanol Almog et al[5] were able to correlate the particle volume with the differences between the solubility parameter of the solvent monomer mixtures (δ_m) and that of the polymer (δ_p). As shown in Figure 5 the volumes of the spheres decrease as a function of ($\delta_m - \delta_p$). They argue that the increase in particle size as a function of the alcohol used can be attributed to the swelling capacity of the alcohol which is borne out by the correlation shown in Figure 5.

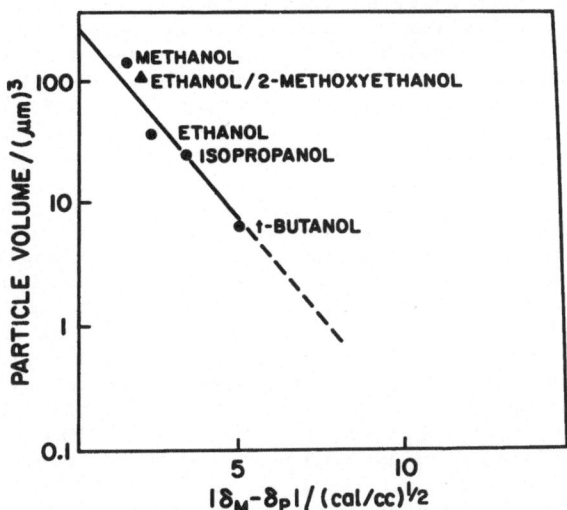

Figure 5: Shows the volume of polystyrene spheres produced in various alcohols plotted against the difference in the solubility parameter between the dispersion medium (δ_M) and the polystyrene particle (δ_p).

If we now consider the data of Lok and Ober[6] we find that for the monodisperse 3μm particles the results correlate with those shown in Figure 5. However, for larger particles the data does not fit this correlation. The reason for this is not immediately obvious, but could be related to the mechanism of particle formation which we believe to be dependent upon the solvency of the dispersion medium. Lok and Ober reported[6] that they were only able to produce monodisperse particles when the solubility parameter for the system was within the range 11.5 to 11.9. Outside of this range polydisperse samples were obtained. In order to test the validity of this assumption these authors investigated other ethanol/solvent pairs and found that large particles could also be obtained by adjusting the solvency to be within the range 11.5 to 11.9. While the results seem to demonstrate that the solvency of the dispersion medium is an important parameter it is not clear how it affects the growth process during polymerization.

3.3. Effect of Initiator Concentration and Temperature

The effect of initiator type and initiator concentration appears to have received little attention in the literature. It would be expected that the concentration of initiator will affect the number of free radicals generated. This would control the number of nuclei that are formed and consequently affect the final particle size. In our studies of the dispersion polymerization of methyl methacrylate in 2,2,4-trimethylpentane/carbon tetrachloride mixtures using azobisiso-butyronitrile as initiator we found[8] that the greater the initiator concentration, the larger the final particle size. This is shown in Figure 6a. Similar results have also been reported for the dispersion polymerization of styrene in ethanol/2-methoxyethanol mixtures using benzoyl peroxide as the initiator[6,18]. These results are plotted in Figure 6b. These authors also observed that a monodisperse latex would form only with a specific initiator concentration for a given set of reaction conditions. They also carried out the same set of reactions at five degree intervals between 55 and 80°C[18]. They found that as the temperature increased the average particle size increased as shown in Figure 7. All of the latexes produced were polydisperse in nature. When the reaction was carried out at 72°C a monodisperse latex of 7μm diameter was obtained.

In order to control and sharpen the particle size distribution for a range of temperature and initiator compositions Almog et al[5] and more recently Tseng et al[14] used low molecular weight salts as costabilizers for dispersion polymerization in alcohols. The intent seems to be to provide immediate stabilization of the nuclei that are initially formed, thus preventing their coalescence to give polydisperse particles. The low molecular weight stabilizers are presumably displaced by the less mobile polymeric stabilizer during the course of the reaction. From the results that have so far been reported, it appears that this strategy could be quite useful in preparing monodisperse particles over a wide range of polymerization conditions. Tseng et al[14] have also shown that the composition of the initiator can play an important role in controlling the polydispersity of the latex. Since the rate of decomposition of the initiator can dramatically affect the nucleation period, it can also have an effect on the polydispersity of the latex that is produced.

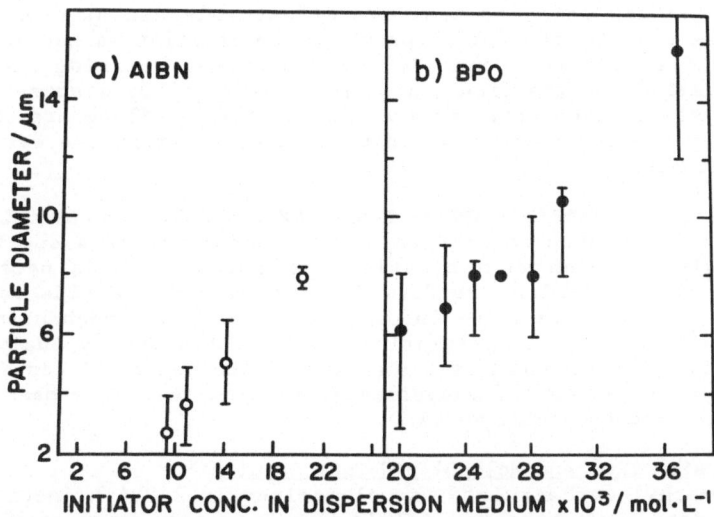

Figure 6: Shows the particle size of the latex produced plotted against the concentration of initiator used in the reaction (a) refers to the polymerization of methyl methacrylate in a carbon tetrachloride/2,2,4-trimethylpentane mixture while (b) refers to the polymerization of styrene in an ethanol/2-methoxyethanol mixture.

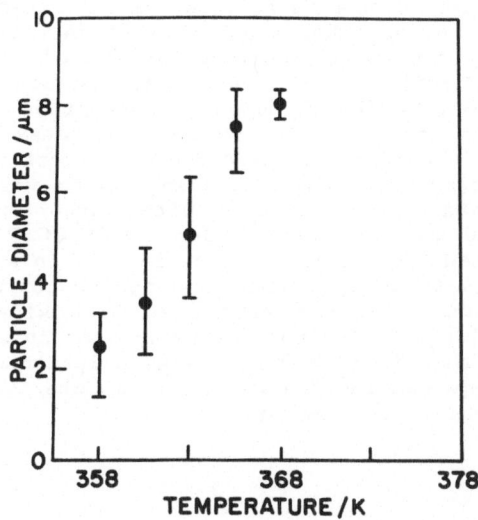

Figure 7: Shows the average particle diameter of the final latex plotted against the temperature at which the reaction was carried out.

3.4. Effect of Stirring Rate

One of the parameters that is usually carefully controlled in particle polymerization processes is the stirring rate. Numerous studies have now been conducted to test the effect this can have on particle size. It has been found[17,18] that provided the reaction is not stirred too vigorously i.e. greater than ~300 rpm, then the stirring rate has no measurable effect on the final particle size. When the system is stirred too vigorously coarse particles are often obtained. These are thought to be formed by the coalescence of unstable nuclei during the early stages of polymerization.

4. COMMENTS ON THE KINETICS AND MECHANISM OF DISPERSION POLYMERIZATION

In order to be able to prepare polymer particles of a specific size and size distribution with a variety of monomers it is necessary to understand how particles are formed and grow under a wide variety of conditions. A good deal of information about the mechanism can be obtained by studying the polymerization kinetics and by analyzing the materials that are produced as a function of the degree of conversion. In this section we review the experimental results that have been obtained in both polar and non-polar media.

4.1. Conversion Time and Molecular Weight Results

When methyl methacrylate is polymerized in 2,2,4-trimethylpentane using a slightly unsaturated polyisobutylene copolymer as the steric stabilizer and AIBN as initiator, the rate of polymerization is found to be dependent upon the initial concentration of monomer[19]. This is shown in Figure 8. The larger the initial concentration of monomer for a given set of reaction conditions, the faster the reaction proceeds and the larger is the final particle size of the latex. For instance, curve (a) of Figure 8 has double the amount of monomer as compared to curve (b). The shapes of both of these curves is indicative of a significant gel effect in these polymerizations. We have also investigated the effect of increasing the concentration of stabilizer on the kinetic behavior. While curve (b) in Figure 8 described a polymerization using 3% by weight of stabilizer compared to monomer, the reaction shown by curve (c) contained 17% by weight of stabilizer. The stabilizer concentration has only a minor effect on the kinetics of monomer consumption although it has an important effect on the particle size of the latex that is produced. Similarly, changing the stirring speed did not affect the kinetics of the polymerization process. When the dispersion medium was changed from neat 2,2,4-trimethylpentane to a mixture with carbon tetrachloride, the reaction was found to slow down[19]. Curve (d) of Figure 8 shows the kinetic behavior when the reaction is carried out in a mixture (1:2 ratio by volume) of 2,2,4-trimethylpentane and carbon tetrachloride. In this case the conversion is linear with time. The kinetic data seem to indicate that the locus of the polymerization in the mixed solvent media is different from that in pure 2,2,4-trimethylpentane. The size of the latex produced from reaction (b) was ~1μm diameter while that produced from reaction (c) was ~13μm diameter.

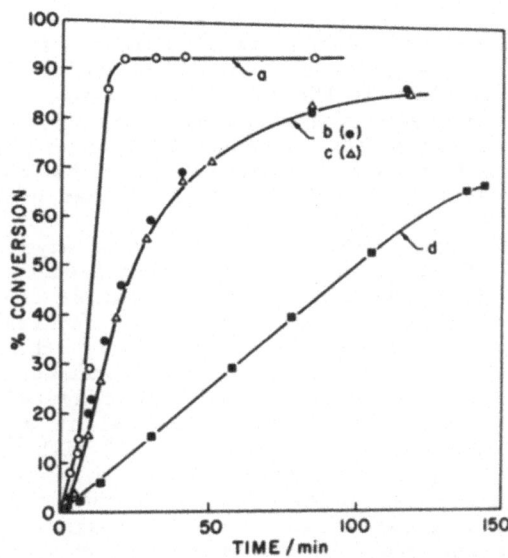

Figure 8: Shows the percentage conversion of monomer to polymer plotted against time. Curve (b) is a standard formulation for methyl methacrylate polymerization in 2,2,4-trimethylpentane. Curve (a) is the same reaction but with double the initial monomer concentration while curve (c) is the same as (b) but has four times the concentration of PIB stabilizer in the system. Curve (d) was carried out in a 2,2,4-trimethylpentane/carbon tetrachloride mixture.

An analysis of the molecular weight of the polymer particles from reaction (a) of Figure 8 revealed that the weight average molecular weight of the final polymer was 5×10^5 and the polydispersity index $(\bar{M}_w/\bar{M}_n)$ was ~4.5[8,19]. It is of interest to note that this ratio was fairly constant from about 5% to 80% conversion. For the particles formed from reaction (b) it was found that $\bar{M}_w \sim 10^5$ with $\bar{M}_w/\bar{M}_n \sim 4.5$ over the measurable conversion range. The results of the kinetic and molecular weight analyses indicate that the results are very similar to those obtained from a bulk polymerization. This suggests that once nuclei are formed, monomer diffuses into them along with initiator, and is polymerized. This interpretation is consistent with the results of Barrett et al[4] and of Antl et al[17]. In order to confirm this interpretation and its prediction that nuclei, once formed, are swollen with monomer, we measured the partition coefficient of the monomer during the reaction: i.e. the ratio of concentration of monomer inside the particles to that in the dispersion medium[19]. For the dispersion polymerization of methyl methacrylate in 2,2,4-trimethylpentane the value of the partition coefficient was found to be ~2.5 and of constant value from 20 -55% conversion. This compares well with the value of 2.6 obtained by Taylor[9] for the polymerization of methyl methacrylate in hexane. However, Barrett and Thomas[20] obtained a value of 1.0 for the dispersion polymerization of methyl methacrylate in dodecane. The reason for this difference is not understood.

When carbon tetrachloride is added to the reaction mixture (curve(d) of Figure 8) the polymerization does not exhibit a gel effect and the conversion is linear with respect to time. The final molecular weight of the particles was ~ 2×10^4. This is expected since the carbon tetrachloride functions as a chain transfer agent. A more interesting observation is that a significant amount of polymerization in this system appears to take place in the dispersion medium as observed by the increase in the viscosity of the dispersion medium. While a thorough experimental study of the carbon tetrachloride containing system is not complete, the initial results[19] suggest that the particle forming mechanism, after nucleation, is different from that postulated to take place in neat aliphatic hydrocarbon media.

In the case of the dispersion polymerization of styrene in more polar media such as ethanol/2-methoxyethanol and ethanol/water mixtures it was found that the conversion behavior displays similar characteristics to that for methyl methacrylate in hydrocarbon based liquids[6]. These results are shown in Figure 9. It can be seen that in the ethanol/water system the reaction proceeds rapidly indicating a pronounced gel effect. When the water is replaced by 2-methoxyethanol the conversion rate becomes much slower. In ethanol/water mixtures the weight average molecular weight of the particles that are formed is $\sim 10^6$ with $\overline{M}_w/\overline{M}_n \sim 2.5$, both approximately constant over the entire conversion range. The gel effect and molecular weights recorded also indicate that the polymerization takes place within the particles. For the ethanol/2-methoxyethanol reaction, $\overline{M}_w$ was of the order of 10^4 which is two orders of magnitude smaller than the molecular weight of the particles produced in the ethanol/water system. This suggests that the mechanism of particle formation in the different solvent media are somewhat different or that methoxyethanol acts as a chain transfer agent. Ober et al[15,18] suggest that once the nuclei in the ethanol/2-methoxyethanol system are formed they are swollen with monomer to a minor extent and that most of the monomer remains in the disperse phase. It was shown that the growth of the particles was a linear function of the degree of conversion which suggests that the particles must consist primarily of polymer and that little monomer is present in the particles at any given time. Evidence that polymerization is primarily occurring in solution is furnished by the molecular weight of the particles that are formed. It was found that the molecular weight was proportional to the initiator concentration used, which is typical behavior for a solution polymerization. However, it is difficult to conceive of a particle forming mechanism that is based solely upon solution polymerization. It seems more likely that polymerization is taking place within the particle and also in solution, thus, two mechanisms of growth are operative. This occurs because in changing from ethanol/water to ethanol/2-methoxyethanol the dispersion medium has become a much better solvent for polystyrene thus allowing solution polymerization to occur. At the same time the partitioning of the monomer between the solvent and the particle phase has been radically changed. As the reaction proceeds the solvency of the dispersion medium changes so that oligomers and polymers that were previously soluble become less soluble and are thermodynamically driven to become associated with the growing particle. Although the particle is not swollen to any large extent, even with the absence of a gel effect, we believe that some of the polymerization takes place within the interior of the particles as well as by capture of oligoradicals from solution. We believe that a similar growth mechanism is occurring in the dispersion polymerization of methyl methacrylate in the carbon tetrachloride/2,2,4-trimethylpentane mixture[8,19].

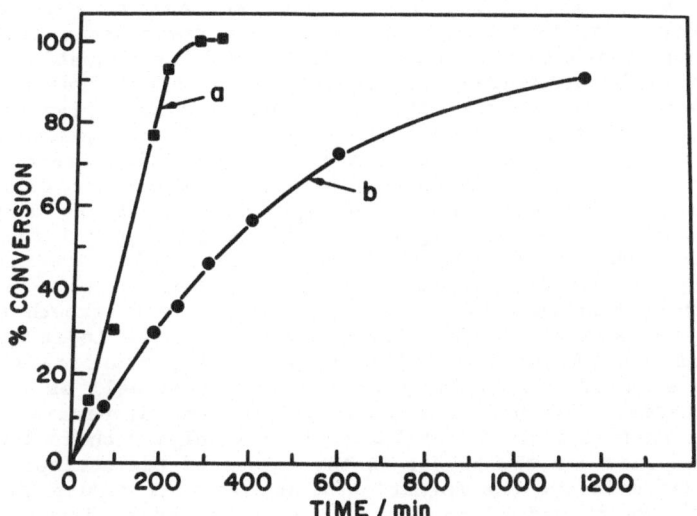

Figure 9: Shows the percentage conversion of monomer to polymer plotted against time. Curve (a) corresponds to the polymerization of styrene in an ethanol/water mixture while (b) corresponds to the reaction in an ethanol/2-methoxyethanol mixture.

4.2. The Particle Nucleation Step

In the preceding discussion we have alluded to the importance of the particle nucleation step but have not discussed it in detail. This is partially due to the fact that little quantitative work has been carried out into the earliest events that occur in such a polymerization. We feel however that it is important to understand the role of nucleation in particle formation and its specific impact on the particle size and size distribution of the final latex.

In dispersion polymerization, after initiation, the initial polymerization takes place in solution. There is usually an induction period before nuclei are formed. This is the point at which the solution becomes turbid. It is the nature of the reaction that occurs in solution during this induction period, as well as the size, structure and colloidal properties of the nuclei that are formed, that we believe to be important in determining the mechanism and kinetics of the polymerization process. This mechanism in turn determines the size and size distribution of the final latex. To date, little quantitative research has been reported and the discussion of this subject is largely speculative.

It is clear that the amphipathic steric stabilizer plays an important role in the nucleation event. For instance, when homopolymers with labile hydrogens or double bonds are present we know that grafting of the core monomers to the stabilizer takes place[6,13]. When well tailored block copolymers have been used[9] it has not been established whether or not grafting is an important first step in the nucleation process. Such copolymers will micellize in solution and the effect this has on nucleation is not understood. Concomitant with the grafting process there will also be polymerization of the core monomer to form oligomeric species. Therefore, there are two competitive reactions taking place in solution and presumably both of these species are capable of forming nuclei, either by an aggregative process or by precipitation of single chains from solution. We feel that the stabilizer must play an important role because analysis of the particles formed in nonaqueous dispersion polymerization has shown[21] that stabilizer molecules are continuously threaded throughout the bulk of the particles and can comprise more than 5% by weight of the particle. A schematic diagram of the morphology of such particles is shown in Figure 10. This morphological feature could also indicate that when initially formed, the nuclei are unstable and consequently undergo coagulation. Further experimental work is necessary to establish whether coagulation is indeed an important growth mechanism.

The experimental results discussed previously indicate that the solvency of the dispersion medium plays an important role which can affect the size of the nuclei that are formed as well as the particle growth mechanism. The better the solvency of the dispersion medium for the polymer that is formed, the larger is the polymer chain length before nuclei are formed and the bigger is the volume of the nucleus. It is also possible that delaying the formation of nuclei will reduce the number of nuclei that are formed since it is conceivable that disproportionation reactions etc. could occur over a longer time interval. Limiting the number of nuclei that are formed would allow more monomer to be polymerized per nucleus which would result in a larger particle size. We have implicitly assumed that there is a relationship between nuclei size and final particle size. It is possible that this hypothesis is incorrect and that the particle growth mechanism e.g. coagulation, controls the final particle size not the size of the original nucleus. In a paper by Antl et al[17] that was discussed in section 3.2 it was found that at low methyl methacrylate concentrations(<10wt% monomer) and at high monomer concentrations(>35wt% monomer)discrete particles were formed. At intermediate monomer concentrations stable particles could not be produced. These authors suggested that the solvency of the dispersion medium was responsible for this behavior. At large monomer concentrations they suggest that nucleation occurred by precipitation of polymer followed by adsorbtion of stabilizer chains. Growth then followed by diffusion of monomer into the nucleus with subsequent polymerization. However, at low monomer concentrations where the solvency is poor for poly(methyl methacrylate), they suggested[17] that nucleation may have occurred before the growing polymer chains were complete leading to micellization of oligomers and stabilizer. This would lead to a large number of nuclei which would give small particles as observed experimentally. In the intermediate concentration region it was suggested that both nucleation mechanisms could be competing for stabilizer with the consequent formation of an unstable system.

POLY (METHYL METHACRYLATE)
CORE

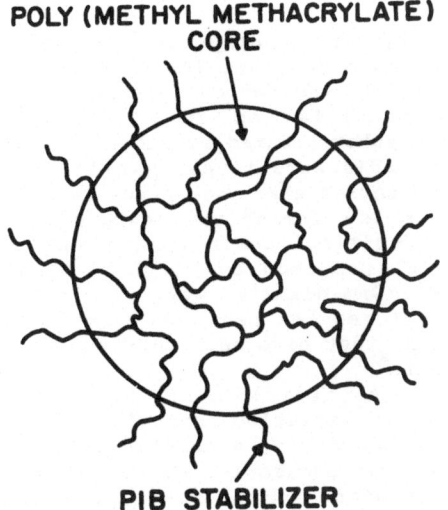

PIB STABILIZER

Figure 10: This is a schematic diagram of the morphology of a nonaqueous dispersion polymerized particle showing the stabilizer threading throughout the particle volume.

Finally, the type of initiator used has been shown [14,18] to alter both the size and size distribution of the latex that is formed. The decomposition rate of the initiator is thought to affect the nucleus size. It can be argued that if the generation rate of oligomeric radicals is much faster than the adsorption rate of the stabilizer, then larger aggregates i.e. nuclei, and hence a larger final particle size latex will be formed. Conversely, a slower initiator decomposition rate would lengthen the nucleation stage which would lead to a wider particle size distribution. While these arguments are plausible they are based on the assumption that particle nucleation occurs by aggregation of oligomeric species. However, more research is required to definitely unravel the physics and chemistry of the nucleation process.

5. CONCLUDING REMARKS

Dispersion polymerization has been known for more than twenty-five years. It was originally designed to be a process analogous to emulsion polymerization but carried out in an aliphatic hydrocarbon medium rather than in water. A steric stabilizer was used instead of an electrostatic stabilizer in the belief that electrostatic stabilization was ineffective in dielectric fluids. The process that finally evolved from the ICI group[4] is a polymerization technique that superficially bears little resemblance to emulsion polymerization[1].

Most of the early work reported on dispersion polymerization was carried out with a technological objective in mind. Since the aim was to produce surface coatings materials, these experiments were almost exclusively carried out in aliphatic hydrocarbon media. During the last five years there has been an increased interest in dispersion polymerization and an effort has been made to generalize the polymerization technique to a wide variety of monomers and to different dispersion media.

In all of the early work on dispersion polymerization the particle size reported was usually submicron. The kinetics suggested[4] that once the nucleus of the particle was formed, polymerization proceeded within the monomer swollen particles as in a bulk polymerization. This interpretation has been substantiated by other workers[17,19]. However, when more polar dispersion media were investigated[5,6], it was found that in some cases, the polymerization also took place in solution. It is thought that the oligomers that are formed are captured by existing particles before they grow to a critical chain length and are able to form new nuclei as in the model proposed by Fitch[22]. It was also found that particles in excess of $10\mu m$ diameter were also able to be formed using this technique. Similar results have now been observed in dispersion polymerization reactions that have been carried out in nonpolar[8] as well as polar dispersion media[6].

While recent results from investigations designed to probe the utility of dispersion polymerization have yielded a wealth of unexpected results much research remains to be carried out. For instance, we have almost no knowledge of the nucleation event nor do we fully understand its significance in relation to the final particle size of the latex although such a relationship is implicitly assumed in many of the published papers. Partitioning of monomer and of initiator between particle nuclei and the dispersion medium[15] appears to play a role in determining how the polymerization proceeds. As yet we do not have a firm idea as to their relative importance. Details of the kinetics of dispersion polymerization are at present limited and more data needs to be collected, perhaps on better defined systems. It would also be extremely helpful to have a straightforward theoretical model of the dispersion polymerization process since it would undoubtedly help in correlating the results and refining the ideas as to the detailed mechanism of dispersion polymerization. Finally, although the organic chemistry of dispersion polymerized particles has not been explicitly discussed in this paper, there appear to be tremendous opportunities for innovation which could lead to a variety of chemically and technologically interesting particles being synthesized.

REFERENCES

1. Vanderhoff JW: *J. Poly. Sci., Polym. Symposium*, **72**, 161 (1985).
2. Ugelstad J; Mork PC; Kaggurud KH; Ellingsen T; and Berge A: *Adv. Colloid Interface Sci.*, **13**, 101 (1980).
3. Vanderhoff JW; El-Aasser MS; Micale FJ; Sudol ED; Tseng CM; Silwanowicz A; Sheu HR and Kornfeld DM: *Polymeric Materials Science and Engineering Preprints*, **54**, 587 (1986).
4. Barrett KEJ(ed): *Dispersion Polymerisation in Organic Media*, Wiley, London (1975).
5. Almog Y; Reich S and Levy M: *Brit. Polym. J.*, **14**, 131 (1982).
6. Lok KP and Ober CK: *Can. J. Chem.*, **63**, 209 (1985).
7. Bromley CWA: *Colloids and Surfaces*, **17**, 1 (1986).
8. Williamson B; Lukas R; Winnik MA and Croucher MD: *J. Colloid Interface Sci.*, in press.
9. Dawkins JV and Taylor C: *Polymer*, **20**, 599 (1979) and Taylor G: *Ph.D. Thesis, Loughborough University, U.K.* (1977).
10. Everett DH and Stageman JF: *Colloid Polym. Sci.*, **255**, 293 (1977).
11. Vincent B: *Adv. Colloid Interface Sci.*, **4**, 193 (1974).
12. Olabisi O; Roberson LM and Shaw MT: "Polymer-Polymer Miscibility", *Academic Press, New York* (1979).
13. Egan LS; Winnik MA and Croucher MD: *J. Poly. Sci. Polym. Chem. Ed.*, **24**, 1895 (1986).
14. Tseng CM; Lu YY; El-Aasser MS and Vanderhoff JW: *Polymeric Materials Science and Engineering Preprint*, **54**, 362 (1986).
15. Ober CK; VanGrunsven F; McGrath M and Hair ML: *Colloids and Surfaces* in press.
16. Douglas SJ; Illum T and Davis SS: *J. Colloid Interface Sci.*, **103**, 154 (1985).
17. Antl L; Goodwin JW; Hill RD; Ottewill RH; Owens SM and Papworth S: *Colloids and Surfaces*, **17**, 67 (1986).
18. Ober CK and Hair ML: *J. Polym. Sci., Polym. Chem. Ed.* in press.
19. Lukas R; Winnik MA and Croucher MD: to be published.
20. Barrett KEJ and Thomas HR: *J. Polym. Sci.*, A1, **7**, 2621 (1969).
21. Pekcan O; Winnik MA and Croucher MD: *J. Polym. Sci. Polym. Lett. Ed.*, **21**, 1011 (1983).
22. Fitch RM: *ACS Symposium Series*, **165**, 1 (1981).

POLYMERS AT INTERFACES: ADSORPTION AND DISJOINING PRESSURE THEORIES

M.A. COHEN STUART

Physique de la Matière Condensée
College de France
11 Place Marcelin-Berthelot
75231 Paris Cedex 05, France

1. INTRODUCTION

Soluble polymers can produce either of two effects on colloidal dispersions: they can enhance the colloidal stability, or reduce it and promote flocculation (aggregation). Over the years a number of attempts have been made to understand these effects theoretically in terms of a particle-particle pair potential due to the presence of the polymer. A typical approach is to consider two parallel, plane surfaces separated by a slice of polymer solution, and to calculate the free energy F as a function of the separation distance H. Detailed accounts of these calculations can be found in the original literature, and a recent monograph by Napper[1] gives a good overview. The aim of this contribution is, therefore, to comment on more recent calculations, trying to elucidate important model assumptions and results.

2. POLYMERS NEAR A SINGLE WALL: ADSORPTION AND DEPLETION

The first problem to be studied is, of course, what happens if the plates are at large distances, so that no interaction occurs and $F = F_\infty$ does not depend on H. This is the case of a single wall where polymer is either attracted to or repelled by the surface so that the polymer segment density ϕ becomes a function of the distance z from the wall. Polymer adsorption is already a complicated phenomenon and we will begin to describe basic approaches and results for this problem. For more details, the reader is referred to recent reviews[2,3].

All theoretical work starts with the description of polymer chains in solution, the most familiar theory being due to Flory. The Flory theory has shortcomings, especially for the case of good solvents. To overcome these, a scaling theory was introduced, largely due to De Gennes[4]. The basic idea of the scaling theory is that the properties of the solution change rather abruptly as soon as the coils start to overlap (at $\phi = \phi^*$). For $\phi < \phi^*$ the solution is said to be dilute, and the polymer coils are highly swollen. The swelling is due to the excluded-volume effect which leads to long-range correlations between the segments up to the full coil size. In dilute solution, therefore, the correlation length, ξ, is equal to the coil radius. At concentrations larger than ϕ^*, ξ starts to decrease according to a power law: $\xi \sim (\phi/\phi^*)^{-3/4}$. This is the semi-dilute regime; it ends when ξ becomes of the order of the segment length (at $\phi = \phi^{**}$ where the concentrated regime begins). Where Flory's description is continuous over the whole range $0 < \phi < 1$, the scaling theory provides a discontinuous result with cross-overs at ϕ^* and ϕ^{**}. Napper[1] gives a good account of both theories.

For the description of polymers near an interface we have to allow for concentration gradients. Ways to do this were proposed by (among others) Edwards[5], by Cahn and Hilliard[6] and De Gennes[4,7] and, more recently, by Scheutjens and Fleer[8,9]. Since the latter method is conceptually the simplest, we shall first describe it shortly. Next, we consider approaches using the Edwards formalism, in order to elucidate differences and common points. We then discuss some results, including a scaling picture by De Gennes.

3. THE LATTICE THEORY OF SCHEUTJENS AND FLEER

In this theory, chains are inscribed on a lattice with layers numbered 1, 2, 3, ..., i, ..., M parallel to the interface. The size of a lattice site is a; segments are defined as occupying just one lattice site. The method begins by generating statistical walks of length s on this lattice, $(1 \leq s \leq N)$ and calculating the distribution of the endpoints over the layers. For example, if we start a walk in layer i (on an empty lattice), the distribution of this "monomer" is (0,...., 0, 1, 0,...., 0) i.e. only layer i contains a segment. After one random step, the distribution has changed to $(0, ..., \lambda_1, \lambda_0, \lambda_1, ..., 0)$, because each step has an a priori probability λ_1 for perpendicular steps $(i \rightarrow i \pm 1)$ and λ_0 for parallel steps $(i \rightarrow i)$, $(\lambda_1 + \lambda_0 + \lambda_1 = 1)$. After n steps one generates in this way the binomial distribution of order n. A versatile method to obtain it is to write the sets above as vectors $\overline{P}_s$.

The vector $\overline{P}_{s+1}$ is then obtained from $\overline{P}_s$ by multiplying it by a matrix $\widetilde{W}$ with diagonal elements $W_{ii} = \lambda_0$, off-diagonal elements $W_{i,i\pm1} = \lambda_1$, and all others zero.

If there were no wall, this matrix would be exactly what is called a "stochastic matrix"[10]. It is no extra effort to allow for starting probabilities in each of the M layers (i.e. a distribution of starting points) and to generate the corresponding distribution of end points: The matrix multiplication repeated s-1 times:

$$\overline{P}_s = \widetilde{W}^{s-1} . \overline{P}_1 \tag{1}$$

gives this automatically. All the vectors $\overline{P}_s$ (s ranging from one up to N) are then stored for further use in the numerical computation (see below).

The next step is to calculate the probability p(i,s;N) that a segment s in a chain of length N is found in layer i. This is done by generating two walks, one of length s end the other of length N-s+1 and to calculate the product of their end probabilities in layer i:

$$p(i,s;N) = \frac{p(i,s) . p(i,N-s+1)}{p(i,1)} \tag{2}$$

where p(i,s) stands for the component i of the vector $\overline{P}_s$. Finally, the density of segments in i, ϕ_i, is found by repeating this procedure for all possible pairs (s,N-s+1), summing this over s up to N and normalizing properly.

So far, the problem has been formulated entirely in terms of purely random walks, i.e., it was mapped on the classical diffusion problem. This is not the physics of segments in polymer chains: the path of a "walk" is filled with segments which interact with each other, with solvent and with the wall. As was shown by Scheutjens and Fleer, these interactions lead to stepweighted random walks, where each step into a layer i is weighted according to the density of segments in that layer. For each segment to be placed, one adds a factor $p(i,1) \approx p_i$ to the probability of that walk. Thus the probability $p(i,s+1)$ is found from $p(i,s)$ by:

$$p(i,s+1) = p(i,1)\{\lambda_1 p(i+1,s) + \lambda_0 p(i,s) + \lambda_1 p(i-1,s)\} \tag{3}$$

and to obtain the end point distribution of these step weighted walks one incorporates the factors p_i into $\widetilde{W}$:

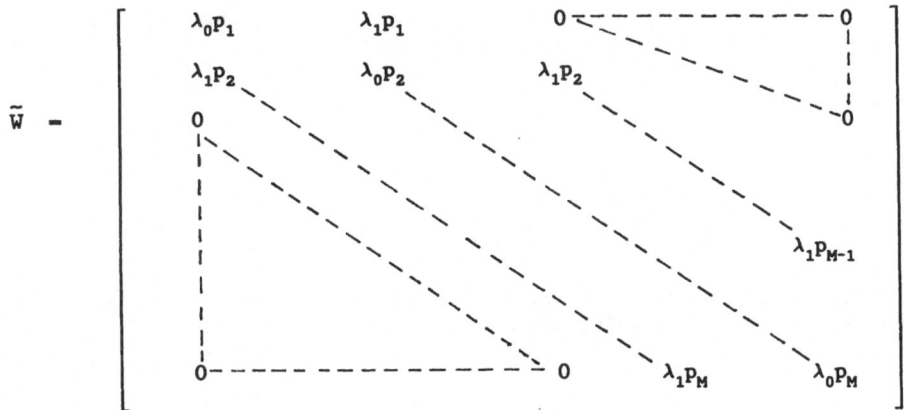

Scheutjens and Fleer demonstrated that in a Flory-Huggins model (i.e. mean field), p_i is given by:

$$P_i = \frac{1-\phi_i}{1-\phi_b} \exp\{2\chi(<\phi_i> - \phi_b) + \delta_{i,1} \chi_s\} \tag{4}$$

where ϕ_b is the bulk concentration, χ the Flory Huggins interaction parameter and $<\phi_i> = \lambda_1\phi_{i-1}+\lambda_0\phi_i+\lambda_1\phi_{i+1}$, i.e. an average volume fraction around a site in i. For an adsorbing wall, an adsorption energy $-kT\chi_s$ is added through an extra weighting factor $\exp \chi_s$ for layer 1 only. Equation 4 states that the weighting factors are obtained from the initial density profile, and from equation (1,2,3) we see that they can then be used to generate a new density profile. Repeating (by iteration) this procedure one finally obtains the equilibrium ('self-consistent') profile.

An advantage of this dicrete approach is that it allows, in principle, for straightforward extension to more complicated systems such as, e.g., copolymers (more than one expression for p_i), polydisperse polymers, solvent and/or polymer mixtures, terminally attached chains etc. One can do more: by calculating an analogous set of vectors P_s^f for chains which do not touch the surface (this amounts to suppressing all steps into layer i-1, i.e. $p_1=0$, and setting $p(1,1)$ equal to zero) one can split up the total end point distribution into entirely free (f), and with at least one segment adsorbed (a) walks. By making suitable combinations when evaluating the products in equation 2 (adsorbed-free-adsorbed = loop, adsorbed-free = tail, etc.) one can extract any conformational detail.

4. CONTINUUM THEORIES

Let us now consider the continuum approach. The main ideas are the same as for the lattice theory described above, but certain approximations are made. First, relatively long walks are only considered. This enables to expand the distribution functions (the vecors $\overline{P}_s$ in a Taylor series up to second order. For purely random walks, this leads to the well-known Gaussian distribution. To see the effect on the step weighted random walk we expand the recursion formula (3):

$$p(i,s) - p(i,1) \left\{ p(i,s) - \frac{\delta p(i,s)}{\delta s} + \lambda_1 \frac{\delta^2 p(i,s)}{\delta i^2} \right\} \qquad (5)$$

which, by rearrangement, leads to

$$\lambda_1 \frac{\delta^2 p(i,s)}{\delta i^2} - \frac{1-p(i,1)}{p(i,1)} p(i,s) - \frac{\delta p(i,s)}{\delta s} \qquad (6)$$

Defining a continuous variable $z-a.i$, and taking $\lambda_1 - 1/6$ for a completely isotropic (cubic) lattice, we can rewrite this

$$\frac{a^2 \delta^2}{6 \delta z^2} P_s(z) - \frac{1-p_1(z)}{p_1(z)} P_s(z) - \frac{\delta P_s(z)}{\delta s} \qquad (7)$$

Since (6) holds for any component $p(i,s)$ of $\overline{P}_s$, we have replaced $p(i,s)$ by the whole set of $p(i,s)$ values, i.e. the function $P_s(z)$. Equation 7 is the equation used by Edwards and others. They use the notation $G_s(z/\overline{z})$ for the distribution of end points, given a starting point distribution $\overline{z}$, and a walk of s steps. Note that for $p(z,1) - 1$, (i.e. in the bulk solution) we recover the well-known diffusion equation; the second term plays the role of a potential energy, i.e. the loss/gain in free energy when a segment is placed at z.

The problem can now be redefined in terms of solving equation (7) self consistently. In principle, the same procedure can be applied by choosing equation 4 for the potential energy and calculating the function $P_s(z)$ for all starting points z. Applying the composition law (equation 2) and integrating over all s, N-s one obtains the segment density distribution. A calculation of this kind was carried out by Dolan and Edwards[5] by numerical methods.

However, it has been proposed that suitable approximations may lead to analytical solutions[11,12,13,14]). First of all, for not too high segment densities, the mean field potential $[1-p(z,1)]/[p(z,1)]$ can be expanded to $(1-2\chi)\phi(z)$. This is probably acceptable for the semidilute part of the layer where only binary interactions between segments are important. Further progress was suggested by the formal equivalence of equation 7 with the time-dependent Schroedinger equation of quantum mechanics (P_s being the wave function and s (imaginary)time). The solutions of this equation can always be expressed as a series of eigenfunctions Ψ_k:

$$P_s(z) - \sum_k a_k \Psi_k e^{-\epsilon_k s} \qquad (8)$$

where the spectrum of eigenvalues ϵ_k is determined largely by the shape of the potential. In many problems in quantum mechanics this spectrum is discrete, i.e. the levels are well separated provided the time is long ($s \to \infty$). The minimum of free energy than corresponds to the system being in the 'ground state', i.e. we can neglect all eigenfunctions except $\Psi_o - \Psi$ which adequately describes the distribution of end points of all (long) walks. According to equation (2) the segment density is then proportional to Ψ^2. Inserting this into equation 7 gives a self consistent equation which can be solved analytically:

$$\frac{a^2}{6} \frac{\delta^2 \Psi}{\delta z^2} - (1-2\chi) \; \Psi^2 \; \Psi - \epsilon_o \Psi \qquad (9)$$

However, the (exact) numerical results of Scheutjens and Fleer do not confirm the validity of this approximation. In a recent paper[15] it is shown that two eigenfunctions, both with eigenvalues very close to unity, give results which agree much better with the full numerical computations. The occurrence of long dangling tails in saturated adsorption layers (interesting for their pronounced hydrodynamic and steric effects[16]) is intimately related to this double eigenfunction structure. An analytical solution of equation 7 taking this into account is still lacking.

4.1. Results for adsorbing homopolymers

All theories find that attractive and repulsive regimes are separated sharply by a non-zero, slightly attractive, value (kT χ_{sc}) of the segmental adsorption energy kT χ_s. This is because the elastic term is always repulsive and needs to be offset by a weak attraction for adsorption to occur. Adsorption increases with increasing χ_s but levels off already at $\chi_s \approx 2$-3. Lattice theory predicts that adsorption increases with molecular weight, but in good solvents a plateau is reached, in agreement with the conclusion from continuum theory in reference 17.

The continuum approach gives for the shape of the segment density profile:

(positive) adsorption:

$$\phi(z) - \phi_b \; coth^2 \; [z/\xi_E + c_a] \qquad (10)$$

depletion:

$$\phi(z) - \phi_b \; tanh^2 \; [z/\xi_E + c_r] \qquad (11)$$

where ϕ_b is the polymer concentration in the bulk solution, c_a and c_r are constants related to the segment-wall interaction (χ_s) and ξ_E is the 'mean field correlation length' defined by Edwards:

$$\xi_E - R_{coil} \quad \text{(dilute)} \qquad (12a)$$

$$\xi_E - a \; \{3(1/2-\chi)\phi_b\}^{-1/2} \quad \text{(semidilute)} \qquad (12b)$$

Note that for z/ξ_z large, both profiles relax exponentially to ϕ_b; for large deviations from the bulk density (close to the wall) one finds a quadratic behaviour; e.g. from equation 10:

$$\phi(z) \approx \phi_b(z/\xi_z + c_a)^{-2} \; \alpha \; z^{-2} \qquad (13)$$

It may be interesting to note that the relation between the bulk (correlation) length ξ_z and the bulk density, equation 12b, is of the same type as the relation between the density $\phi(z)$ in the adsorbed layer and the distance z: $\phi(z)^{-1/2} \; \alpha \; z$. This seems to be a general property of adsorbed polymer layers and has been referred to as 'self-similarity'.

In theta solvents the correlation length goes as ϕ^{-1}(1,4). We therefore conjecture here that for that case the segment density profile goes as z^{-1}, and by integration over z up to $z_{max} \sim N^{1/2}$ we conclude that the adsorbed mass should increase logarithmically with molecular weight. This agrees surprisingly well with the numerical data of reference 8.

The mean-field approach was criticized by several physicists, among which de Gennes[4]. The fact that polymers in good solvents are self-avoiding up to a length scale ξ invalidates the calculations for densities between ϕ^* and ϕ^{**}. De Gennes[16] constructed a scaling theory by making use of the property of self-similarity discussed above. Since the presence of the wall fixes the length scale (= the distance to the wall), the concentration as a function of distance is thereby also fixed. This leads to:

$$\phi(z) - \phi_b \left[\frac{a}{z + 4/3 \; D} \right]^{4/3} \qquad (14)$$

where a is the segment size and D is given by the segment-surface interaction. [namely by $a/D \propto \chi_s^{3/2}$; $kT\chi_s$ is the net adsorption energy per segment.] For large distances, the osmotic effects become small and the profile relaxes exponentially. The treatment disregards the occurrence of long tails, but does not disprove them. Polydisperse polymers are interesting because subtle entropic effects lead to preferential adsorption: from dilute solution long chains adsorb preferentially, whereas from very concentrated solutions the preference reverses. Here, we have, so far, only results from the lattice theory[18,19,20].

In conclusion, mean-field continuum theories yield simple analytical results for the segment density profiles, i.e. equations 10,11, (for good solvents), but they are not corroborated by the full calculation, their validity is restricted, and no conformational details can be extracted. For flexible chains in good solvent, we have a scaling result (equation 14). For more complicated (and practically relevant!) cases such as polydisperse polymers and copolymers, there is as yet no alternative to the lattice theory.

5. CHAINS BETWEEN TWO WALLS/DISJOINING PRESSURE

Most of the earlier mean-field calculations have been reviewed by Napper[1]. We therefore limit ourselves to recent calculations by Scheutjens and Fleer[21], de Gennes[22], and Klein and Pincus[23]. It should be said immediately that we do not know much about the dynamical situation when two polymer-covered particles meet in a Brownian encounter. Two relatively well-defined cases could be envisaged:

(I). Full equilibrium between the polymer in the gap and an external solution, throughout the approach of the two walls, i.e. the thermodynamical potential of both solvent and polymer remain at their bulk values, even if this calls for chains desorbing and diffusing out of the gap.

(II). Equilibrium for the solvent alone, but for the polymer a constant amount of attached chains Γ_p between the surfaces which could be taken to be equal to the value at large separation, where no interaction occurs. ('restricted equilibrium').

In practice, one expects that case I is only of interest for walls which repel the polymer. For attractive walls, even if these approach each other slowly, it is doubtful whether case I is relevant under experimental conditions. Theoretically, all recent calculations agree that at full equilibrium two walls with adsorbing polymer always attract each other. The situation of lowest free energy is, here, one where most polymer is desorbed during the approach, while the remaining polymer is confined between the walls in a completely two-dimensional conformation, and interacts with both walls. This minimizes the energy (many favourable segment-wall contacts) and, since $\chi_s > \chi_{sc}$, the elastic term in the free energy is of no consequence.

It appears that if there is at the same time desorption and bridging of the chains between both walls, the energy of the system is constant, so that the attractions must be due to an entropy increase; one might call this 'entropic attraction'.

We know that, in practice, adsorbed polymer may well give rise to repulsion. Here, case II is hopefully the relevant case, at least for slow approaches. We may think of concentrated dispersions, where particles are in contact for extended periods. For dilute dispersions, even case II may be irrelevant, and we may have to take a dynamic point of view.

6. ATTRACTION BETWEEN POLYMER AND WALL/RESTRICTED EQUILIBRIUM CASE

De Gennes[22] studied long chains (long enough to have a vanishing molecular weight effect in adsorption) in good solvents, concentrating on the case of saturated adsorption. From his comparison of mean-field and scaling calculations it appears that the disjoining pressure is the outcome of a subtle balance of forces: up to the interactions between segments in pairs, the force in mean-field was zero, whereas the one found by a scaling approach was clearly repulsive.

The force was found to vary as H^{-3} at the larger distances and as $H^{-9/4}$ at smaller separations. Pincus and Klein[23] turned their attention to poor solvent conditions. Expanding a free energy of mixing of the Flory-Huggins type up to the third order term, they found an attractive region in the force-distance curve, due to osmotic effects.

Scheutjens and Fleer[21] find, for saturated layers, the same type of behaviour: repulsion in good solvents (and for long chains) and attraction in poor solvents. However they have also studied the more interesting effect of unsaturated layers, by varying the coverage Γ_p. Their conclusions are summarized in Figures 1,2, and shed much light on the questions of dispersion stability. Figure 1 shows that for coverage below saturation, there is a minimum in the potential/distance curve, i.e. an attractive region, even if the solvent is good. Closer inspection of the conformations reveals that the attraction at large H is due to the

236

formation of bridges, just as in the full equilibrium case. At shorter distance, osmotic forces start to build up a repulsion so that the potential rises steeply with decreasing distance.

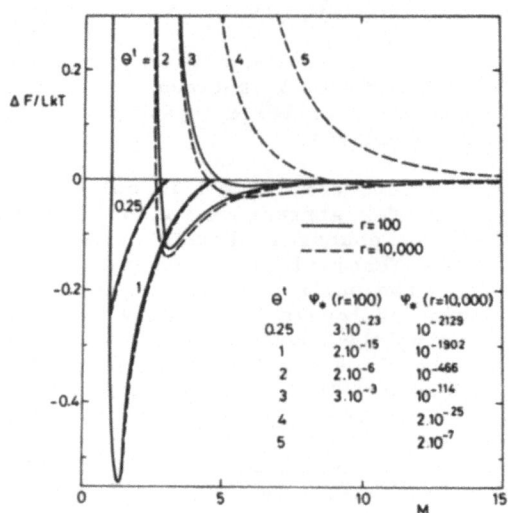

Figure 1: Interaction curves (free energy per lattice site) at different (fixed) amounts of polymer Γ_p between the surfaces, for two different chain lengths. Theta solvent. The probabilities φ_b to have free chains (bulk concentration) corresponding to each adsorbed amount Γ_p are given. Comparing the two curves labeled '3' we see that stability for the shorter chain requires $\phi_b - 3.10^{-3}$, whereas for the longer chain $\phi_b - 10^{-114}$ is sufficient, i.e. essentially no free polymer (from reference 21).

In Figure 2, the depth of this minimum ΔF_{min} is plotted as a function of coverage. Clearly, the deepest minimum occurs around an adsorbed amount $\Gamma_p - 2x0.4$ monolayers. Shallower minima occur for larger adsorbed amounts, in good solvents up to $\Gamma_p - 2x0.7$. Under practical circumstances, this means that short chains, which are unable to sufficiently saturate the surfaces, cannot give stability. In theta solvents, stability calls for $\Gamma_p > 2x3$ and in poor solvents, stability disappears altogether, even for very thick layers. These results agree with what has been found in references 22 and 23; in fact, these latter calculations represent the limit of very large adsorbed amount.

On the basis of Figure 2, we may formulate some general remarks on stability in the presence of adsorbing polymers. Homopolymers, by virtue of the fact that each of their segments can in principle adsorb, have a strong tendency towards bridging. They can, however, stabilize but will only do so at high molecular weights and with saturated adsorption layers. At lower coverages, we have attraction provided the chains will find time to cross the gap. All this agrees with common practice, and qualitatively, with data of Klein et al[24]. Quantitatively, the forces calculated by Scheutjens et al seem to be smaller than experimentally

found, although the calculated shape of the force profile agrees fairly
well with the data. The results also allow to understand the dramatic
effects of coverage found in early flocculation studies.

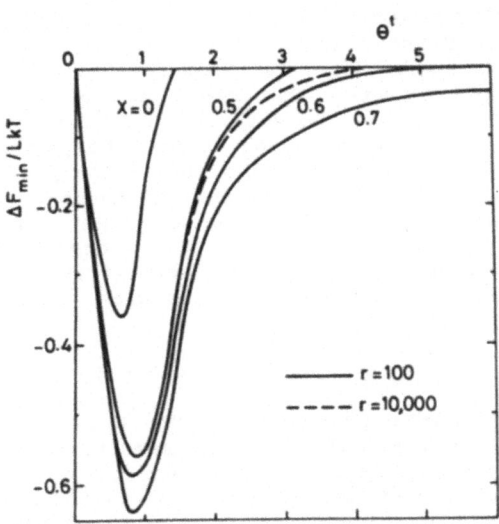

*Figure 2: Depth of the interaction minimum, ΔF_{min}, as a function of
coverage, for various solvents: good solvent (χ-0), theta solvent
(χ-0.5), poor solvent (χ-0.6, 0.7) (a minimum equal to zero means no
minimum in the interaction curve). It can be seen that a stability
requirement can be formulated in terms of coverage. The deep minimum
corresponds to extensive bridging; the plateau developing for poor
solvents (χ-0.7) corresponds to osmotic effects. From Scheutjens et al.,
reference 21.*

For copolymers, the situation is quite different. Copolymers are
expected to have one type of segment preferentially adsorbed[3], leaving
the other type predominantly in loops and tails. They may, however, be
able to bridge and induce flocculation, because potentially adsorbing
segments do still occur in loops and tails. A representative case is
perhaps partially hydrolyzed poly(vinylacetate)[25].

Block copolymers of the A-B type are, as known, powerful
stabilizers, because they have the least tendency to bridge. Recent
measurements by Tirrell et al.[26] confirm this picture. The stability of
systems which are unable to bridge is to a large extent governed by
osmotic effects. It is therefore logical that such systems tend to
aggregate close to the θ-temperature of the stabilizing moiety.
Calculations to quantify the adsorption and disjoining pressure of
copolymers are now in progress.

As to θ-temperatures, it should be said that these may manifest
themselves either as an upper critical solution temperature (UCST) or a
LCST, many systems have both. Phase behaviour of polymers as a function
of temperature is usually rationalized in terms of the two opposing
tendencies of free volume dissimilarity and contact energy dissimilarity,

the former being important at high temperature and the latter at low temperature. As a result, χ passes through a minimum. Aqueous systems often deserve special attention because of effects of salts.

7. REPULSION BETWEEN POLYMER AND WALL: FULL EQUILIBRIUM

Although, at first sight, the repulsive case might seem uninteresting, it is in fact very relevant for colloidal stability. As is now well established, the walls that repel the polymer will attract each other if the solution between the walls is in contact with a bulk polymer solution. As the walls approach each other, the polymer molecules will leave the gap and the average concentration in the gap will start to drop strongly at about $H \approx 2\delta$, where δ is the thickness of the depletion zone (which is of the order of ξ). The result is an osmotic underpressure between the plates, which shows up as attraction. The resulting flocculation has been called 'depletion flocculation'.

The attractive potential ΔF (per unit area) is (for flat walls) of the form[27,28]:

$$\Delta F = \Pi_b(\phi_b) [H-2\delta] , H < \delta \approx \xi_b \tag{15}$$

where Π_b is the osmotic pressure of the bulk solution and ξ_b the correlation length of that solution. As several versions exist for the dependencies $\Pi_b(\phi_b)$ and $\xi_b (\phi_b,n)$ (e.g., mean-field, scaling) so does one find various results for ΔF. In order to obtain an expression for spherical particles, equation 12 is usually integrated over overlapping spherical shells (so-called Deryagin integration); this introduces a quadratic dependence on distance into the potential[27,28].

Usually, rather concentrated solutions are needed to produce a sufficiently deep minimum for aggregation. Since Π_b increases with concentration, and ξ_b decreases, the result is that ΔF itself may pass through a minimum as a function of concentration. The fact that the potential well is relatively shallow leads to a complicated phase behaviour of the dispersion. Gas-like, liquid-like and solid-like systems may all occur at appropriate potentials and densities. The full description therefore calls for a rather sophisticated theory of dense particle systems[29].

An extra complication is the effect of a 'steric' layer of adsorbed or chemically attached polymer on the particle wall. Here, one expects the 'depletion' effect to develop less strongly, and to depend on the density of the steric layer. Explicit calculations have been made for a simple model[30].

8. CONCLUSION

Recent theoretical work has provided much insight into polymer effects on disjoining pressure. However, the reader should be aware that in all these calculations some well-defined equilibrium was considered. Clearly, there is a need for more understanding of the dynamics, in order to appreciate the role of relaxations. The situation is similar to the one for charge-stabilized dispersions, where, also, the dynamic aspects are beginning to receive more attention[31]. It should be said that, recently, some experimental[32] and theoretical[33] studies began to appear. It is our hope that these valuable efforts will be followed by many more.

9. ACKNOWLEDGEMENT

Discussions with W.B. Russel and B. Vincent at the Workshop were most helpful in preparing the final draft of this contribution. Some expressions in this review,, in the section dealing with comparison between lattice and continuum theories, were taken from preliminary notes by H. Ploehn, working with W.B. Russel; I should give due credit to him. I am also indebted to Gerard Fleer for a critical reading of the manuscript, which prompted me to rewrite large parts of it.

REFERENCES

1. Napper DH: Polymeric Stabilization of Colloidal Dispersions, *Academic Press*, NY (1983).
2. Fleer GJ; Lyklema J: in Adsorption from Solution at the Solid/Liquid Interface, Parfitt GD and Rochester CH(eds); *Academic Press*, NY; p.153 (1983).
3. Cohen Stuart MA; Cosgrove B; and Vincent B: *Adv. Colloid Interface Sci.* 24, 143-239 (1986).
4. De Gennes PG: Scaling Concepts in Polymer Physics, *Cornell University Press*, Ithaca, NY (1979).
5. Dolan AK and Edwards SF: *Proc. Roy. Soc.*, London A343 427 (1975).
6. Cahn JW; and Hilliard JE: *J. Chem. Phys.* 28 58 (1958).
7. De Gennes PG: *Rep. Prog. Phys.* 32 187 (1969).
8. Scheutjens JMHM and Fleer GJ: *J. Phys. Chem.* 83 1619 (1979).
9. Scheutjens JMHM and Fleer GJ: *J. Phys. Chem.* 84 178 (1980).
10. (See e.g., Perrin CL: Mathematics for Chemists, Woley NY (1970).
11. Richmond P and Lal M: *Chem. Phys. Lett.* 24 594 (1974).
12. Joanny JF; Leibler L and De Gennes PG: *J. Poly. Sci. Pol. Phys. Ed.* 17 1073 (1979).
13. Ober R; Paz L; Taupin C and Pincus P: *Macromolecules* 16 50 (1983).
14. Jones IS and Richmond P: *J. Chem. Soc., Faradday Trans.* 2 73 1062 (1977).
15. Scheutjens JMHM; Fleer GJ and Cohen Stuart MA: *Colloids and Surfaces* 21 285 (1986).
16. Cohen Stuart MA; Waajen FHWH; Cosgrove T; Vincent B and Crowley T: *Macromolecules* 17 1825 (1984).
17. De Gennes PG: *Macromolecules* 14 1637 (1981).
18. Roefs SPFM and Scheutjens JMHM: *Macromolecules* (in preparation).
19. Scheutjens JMHM and Fleer GJ: in The Effect of Polymers on Dispersion Stability, Tadros Th. F(ed): Academic Press, London p.145 (1982).
20. Cohen Stuart MA; Scheutjens JMHM and Fleer GJ: *J. Poly. Sci. Pol. Phys. Ed.* 18 559-573 (1980).
21. Scheutjens JMHM and Fleer GJ: *Macromolecules* 18 1882 (1985).
22. De Gennes PG: *Macromolecules* 15 492 (1982).
23. Klein J and Pincus P: *Macromolecules* 15 1129 (1982).
24. Klein J and Luckham: *Nature* (London) 308 836 (1984).
25. Fleer GJ and Lyklema J: *J. Colloid Interface Sci.* 55 2208 (1976).
26. Madriiannou G; Patel S; Granick S and Tirrell M: *J. Am. Chem. Soc.* 108 2869 (1986).
27. Fleer GJ; Scheutjens JMHM and Vincent B: in Polymer Adsorption and Dispersion Stability; Goddard ED and Vincent B(ed): ACS Symposium Series 240 p.245 (1984).
28. Joanny JF; Leibler L and De Gennes PG: *J. Pol. Sci. Pol. Phys. Ed.* 17 1073 (1979).
29. Gast AP; Hall CK and Russel WB: *J. Colloid Interface Sci.* 96 251 (1983).
30. Rao IV; Ruckenstein E: *J. Colloid Interface Sci.* 108 389 (1985).
31. Overbeek J.Th: *Adv. Colloid Interface Sci.* 16 17 (1982).
32. Pefferkorn E; Carroy A and Varoqui R: *Macromolecules* 18 2252 (1985).
33. De Gennes PG: *Compt. Rend. Ac. Sci.* (Paris) 301 1399 (1985).

NEW TECHNIQUES IN CHARACTERIZATION OF POLYMER COLLOIDS

NEW TECHNIQUES IN CHARACTERIZATION OF POLYMER COLLOIDS

POSITION PAPER

1. INTRODUCTION

After a polymer colloid has been prepared it is important to characterize as completely as possible the properties of the particles formed. Usually, this means using more than one technique and requirements may well differ according to the ultimate use of the particles, whether it is, for example, basic academic research, industrial production or biomedical applications. Thus the characterization required will vary widely from user to user. However, the following properties are clearly important:-

Particle size
Particle size distribution
Chemical nature of surface groups
Concentration of surface groups, i.e. surface charge density
Location of surface groups, uniform or patchy
Presence of residual monomer
Morphology of the particles, e.g. "core-shell" or "currant-bun"
Composition of co-polymer latices
Concentration of adsorbed emulsifier
Concentration of adsorbed polymeric stabilizer
Thickness of steric stabilizing layer
Reaction of the particle to environment, e.g. swelling of poly-
    electrolyte latices or on uptake of organic solvent
Structure of concentrated latex dispersions
Nature of the interactions between the particles
Surface potential of the particles in dilute and concentrated
    dispersions
Particle surface morphology, rough or smooth

It is also important that the characterization techniques chosen should be capable of giving reproducible results in different laboratories. Consequently, there is a need to test the methods which have been developed widely and then to describe carefully those which are reproducible. Methods by which some of the information required could be obtained have been lacking hitherto. However, at the present time a number of new techniques are being developed which will substantially increase the information which can be obtained.

2. PARTICLE SIZE DISTRIBUTIONS
2.1. Scattering Methods

Three types of radiation are currently used in scattering methods:-

LIGHT: the visible region is most common, covering the wavelength region approximately 400-650 nm. Monochromatic coherent laser sources are now readily available.

X-RAYS: available in laboratory generators to give wave lengths of ca
            0.08 to 0.15 nm; the development of rotating anode generators
            promises an increase in flux of about an order of magnitude.
            Moreover, synchrotron radiation sources are becoming widely
            available worldwide and offer possibilities of longer
            wavelengths.

NEUTRONS: provide radiation in the wavelength range 0.2 to 2.0nm. An
            increasing number of neutron facilities, either with constant
            flux or pulsed sources are becoming available.

2.2. Small Angle Neutron Scattering, SANS

SANS provides a means of investigating many problems within the
Polymer Colloid field. These include the determination of average
particle size and particle size distributions. Neutron beams are
particularly useful for the small particle size region, e.g. for particle
diameters up to $0.5\mu m$. Since the curve of intensity against scattering
vector ranges over several orders of magnitude for a small angular range
the method gives high sensitivity in this region compared with time
average light scattering. Particle size distributions can be obtained
from curve fitting or by inversion of the basic data if a form is assumed
for the distribution. Currently maximum entropy methods are being
developed which should allow determination of a size distribution without
assuming a form for it.

Because neutrons are scattered by the nuclei of atoms, of size ca
10^{-12} cm, the scattering centres are well separated and hence multiple
scattering effects are less pronounced than with X-rays or light.
Consequently, very concentrated dispersions of polymer colloids can be
examined and information about the structure of such systems elucidated.
This provides a means of linking microscopic structure to the bulk
behaviour of materials, e.g. their rheology. Future developments will
include the examination of systems under sheared conditions and during
the application of an electric field. This, in principle, leads to a
method for characterizing non-spherical particles.

2.3. Light Scattering

The basic principle of time average light scattering is very similar
to that described above for SANS. It becomes a better method for particle
diameters above ca $0.3\mu m$ because the angular scattering curves show a
variation of intensity with angle; computer procedures can then be used
to invert the data in order to obtain a particle size distribution.
Whereas conventional time average light scattering is directly related to
the mass of material, photon correlation spectroscopy examines the
intensity fluctuations produced by the Brownian motion of the particles.
The quantity obtained is the translational diffusion coefficient which is
directly related to the hydrodynamic radius of the particles. The latter
appears as an exponential term in the experimentally measured correlation
function. Thus for a polydisperse system the correlation function is the
sum of many exponential terms and hence the determination of
polydispersity is a difficult mathematical problem. Nevertheless several
methods have been proposed for the determination of a particle size
distribution and with the greater computer power now available a more
sophisticated analysis can be attempted.

Recent developments have been made with the use of crossed laser
beams and two correlators. This leads to:-

(a) determination of the rotary diffusion coefficient for non-spherical particles and hence to information on particle shape.
(b) a modulated signal arising from the particles traversing a spatial interference pattern from which particle size can be inferred.
(c) under near refractive index matching conditions a means of examining concentrated systems under minimal multiple scattering conditions.

2.4. Other Methods

The determination of particle size distributions of polymer colloidal particles in the submicron range still needs considerable improvement despite the development of techniques such as HDC (Hydrodynamic Chromatography) and SFFF (Sedimentation Field Flow Fractionation). In the case of HDC efforts should be directed towards better resolution. This might be achieved by designing columns that resemble the theoretical model for HDC packed columns, viz bundles of parallel monosized channels.

Recent work has shown that Capillary Chromatography, which has been used for particles greater than one micron, might be applied to submicron particles by adsorbing a high molecular weight polymer to the inner side of the capillary and/or to the particles to be analyzed.

SFFF may give problems in systems where, apart from a particle size distribution, there is a distribution in polymer composition and therefore in density over the various particle sizes. On the other hand, if the form of the particle size distribution is known, it might be possible to determine the compositional distribution by SFFF.

3. THE NATURE OF SURFACE GROUPS

3.1. Chemical Derivatization

If the surface groupings can be transformed chemically then it is possible to use fluorescent labelling. For example, a fluorescent energy donor is attached to a small fraction of the sites and an energy acceptor (non-fluorescent) to all the other sites. The fluorescent decay profile of the donor then provides a measure of the distribution of quenchers about the donor site. Similar experiments have been carried out by binding dyes to vesicle surfaces in aqueous media.

3.2. NMR Methods

No publications have yet appeared on the application of solid state NMR to dried preparations of polymer colloids. However, high resolution methods may provide chemical information on polymer colloid surfaces provided that the experiment can be tailored to excite or probe only the surface layer. This technique might then be used, for example, to establish the relationship between the number of functional groups at the surface of the particles and the total number of end groups measured by solution methods.

The dynamics of polymer chains at the surface of a latex particle are likely to be different from those in the interior. The differences are temperature dependent, i.e. above a critical temperature the relaxation times of both converge. Peak width and relaxation time experiments may be used to probe both the nature of surface groups and their motions. It may therefore be possible to determine the depth of the interphase region, if it exists.

The increasing availability of 300 MHz and higher NMR facilities along with the continuing development of sophisticated pulse sequence program packages has added a new probe for the molecular understanding of the colloid interface. Probe molecules such as Xe are detected by the different chemical shift within the hydrocarbon tails of surfactants as opposed to the signal from Xe dissolved under pressure. Other species, such as cobalt acetylacetonate show a different signal and associate differently with micellar structures.

3.3. Surface Techniques

Up to the present the types of charged groupings on the surface of a polymer latex particle have largely been inferred from the polymerization chemistry and then confirmed by the use of potentiometric or conductometric titration. The presence of nonionizable groups is often neglected because of the difficulty of detecting them directly. However, in the case of hydroxyl groups oxidation to carboxyl groups is often employed.

Provided that the dispersion medium can be removed from the latex without introducing atmospheric contamination e.g. by controlled evaporation in an evacuated pre-chamber it is possible to probe the surface chemistry of the polymer particles using a range of UHV surface analytical techniques. To date these techniques have not been widely used for polymer colloids but recent advances in the instrumentation should make their contribution to this area highly significant.

Most promising among these techniques is secondary ion mass spectrometry (SIMS) which has good surface sensitivity (5-10 A sampling depth). The new generation time-of-flight SIMS instruments which can minimize ion beam damage of polymers will be most appropriate for polymer colloid work. Using this technique procedures will need to be developed for surface group characterization. Analysis by mass spectrometry of the fragments coming from polymer surfaces being irradiated by X-rays (X-ray induced Ion Mass Spectrometry) also needs further investigation.

Whilst SIMS (Secondary Ion Mass Spectrometry) can provide qualitative information on the nature of surface groups it is by no means certain at present whether the technique can be made quantitative and development of SNMS (Secondary Neutral Mass Spectrometry) appears to be more promising in this respect. Quantitative surface chemical information can be obtained as well from X-ray photoelectron spectroscopy (XPS) but surface sensitivity (50-100 A) is not as good as SIMS. However selective derivatization of surface groups with molecules containing heavy atom labels with high sensitivity in XPS could be used to probe the outermost surface layer in which the surface groups reside, thereby providing an estimate of the complete range of surface groups that are present. While a number of derivatization reactions have already been investigated new techniques need to be developed before this can be used routinely for polymer latex studies.

Among other techniques that offer possibilities of obtaining information about the interphase between a polymer particle surface and a liquid phase SERS (Surface Enhanced Raman Scattering) appears promising. Currently investigations should be encouraged to proceed in two directions:-

(a) probing the surface by depositing a metal colloid on it to provide the SERS effect.
(b) depositing the polymer colloid on a metal electrode.

In addition to increasing the general understanding of the SERS technique there are possibilities of developing work on catalysis and effects at electrode surfaces.

4. RESIDUAL MONOMER

The measurement of residual monomer in polymer colloids is important so that the exact physical and chemical nature of the polymer colloid can be described. In addition the industrial application of polymer colloids requires for legislative reasons, that the residual monomer be below specified levels. The specifications for the levels of residual monomer are continually decreasing and hence more sensitive chromatographic techniques will need to be developed. Coupling of chromatographs with FTIR or mass spectrometers allows identification of the monomer as well as quantitative measurement.

5. MORPHOLOGY
5.1. Electron Microscopy

Various possibilities are available for the examination of latex particle morphology and these include:-
(a) negative staining by materials containing heavy metal atoms, e.g. phosphotungstic acid
(b) positive staining by the use of volatile heavy metal oxides such as osmium or ruthenium tetroxides
(c) use of a combination of (a) and (b) together
(d) sectioning. After freeze-drying, particles can be embedded in resin and then sectioned with an ultra-microtome to give thin sections. Although the carbon hydrogen and oxygen content of many polymers is very similar it nevertheless appears possible to distinguish between some polymers on exposure to the electron beam. For example, in 2-phase polymer particles containing poly(styrene) and poly(methyl methacrylate), poly(styrene) appears dark whereas poly(methyl methacrylate) appears light. This type of contrast does not appear to be directly related to the electron density of the polymer; it seems more likely to be connected with the susceptibility of the polymer to electron beam damage. Additional contrast in the sections can also be obtained by selective staining using the methods listed above.
(e) selective dissolution. In some cases one of the polymers in a copolymer particle can be dissolved preferentially by suitable choice of a solvent
(f) freeze fracture and plasma etching would appear to offer possibilities for exploitation
(g) video enhanced image analysis, now widely used in optical microscopy could well be extended to electron microscopy to improve contrast.

5.2. Fluorescence Techniques

Particle morphology can be studied by fluorescence quenching techniques if individual components of the preparation can be labelled by covalent attachment of an appropriate dye (F). Gross morphology (core-shell, interpenetrating network, currant-bun,...) can often be inferred through sorption studies. In this case one measures the time necessary for a sorbent, which is also a fluorescence quencher (Q), to decrease particle fluorescence intensity. Similar methods can be used to examine local morphology, especially the extent of interphase formation or the sharpness of the interface between phases within the particle. Potentially more powerful are double labelling experiments in which one phase is labelled with F; and the other, with Q. These measurements are closely related to energy transfer experiments on polymer blends where they have been shown to be very powerful.

The sorption experiments are closely related to swelling experiments studied by SANS and light scattering. In principle, fluorescence quenching measurements should allow one to observe sorption into individual phases within the particle and to determine partitioning of sorbents between these phases.

The glass transition temperature of the individual phases in a multicomponent particle will normally be different and it has been shown that the fluorescence of the "twisting dyes" is sensitive to Tg. These dyes might be used as probes of Tg within the individual phases.

5.3. Small Angle Neutron Scattering (SANS)

A substantial asset of the SANS technique is the variation of the coherent scattering length density between various polymers; moreover, this can be enhanced by deuteration of either whole molecules or parts of molecules. The variation of contrast between the particle and the dispersion medium by the use of mixtures of hydrogenated and deuterated solvents, e.g. H_2O and D_2O, means that conditions can be chosen so that one part of the particle can be matched out and the scattering from the unmatched part examined. Hence, this leads to a means of examining the internal structure of particles. In the case of adsorbed layers of surfactant or polymers grafted to the surface the core particle can be matched and the adsorbed layer examined.

5.4. Solid State NMR Experiments

Recent pioneering work has shown that using multiple quantum coherence techniques in zero field, the extent of domain structure in heterogeneous solids can be determined. In the future this type of experiment might well be extended to study: 1) microdomain structure in copolymer or multipolymer particles, and 2) microdomain structure in homopolymer colloidal particles in order to further elucidate their morphogenesis.

5.5. Light Scattering

The availability of high-speed computers has changed the inversion of the light scattering data on concentric shell structures from a research problem to a diagnostic tool that may be used to provide precise information on particles of core-shell morphology. Improvements in apparatus design and inversion algorithms will allow attention to be directed to the variables controlling the syntheses required for new submicron colloidal morphologies.

6. CHEMICAL CHARACTERIZATION OF LATICES

It is often necessary to determine the chemical composition of latex particles prepared from more than one monomer after isolation and purification of the copolymer particles; the fact that all the monomer may not end up in the particles is easily overlooked. In addition, phase separation may occur within the particle in which case it is desirable to determine the relative fractions of homopolymer and graft copolymer.

The gross composition of the particle can be inferred from combustion analysis (C, O, H, N) and possibly confirmed by nmr measurements. If the particle can be dissolved it is important to determine both the molecular weight distribution and the compositional heterogeneity. The latter can be determined by adsorption chromatography. Moreover, thin layer chromatograplhy has recently become a quantitative technique through the introduction of automatic scanning and flame-ionization detection. HPLC techniques (silica gel or reversed phase columns) are even more versatile. There are still some detection problems here when one uses gradient elution techniques.

Determining molecular weight distributions is particularly difficult for heterogeneous copolymer samples, especially if composition changes with molecular weight. When gel permeation chromatography is used in conjunction with low angle light scattering detection, one must first fractionate the sample by adsorption chromatography to obtain fractions of similar chemical composition. The data from these individual samples, analyzed by GPC, can then be interpreted to yield molecular weight distributions.

7. CHARACTERIZATION OF LATICES BY ADSORPTION STUDIES
7.1. Surfactants

Adsorption of surfactants onto latex surfaces offers a possible method of assessing the polarity of the polymer at the surface. Currently, the relationship of the area occupied per molecule at saturation adsorption is empirical for homopolymer latices. It would be of interest to see also whether such methods could be developed for the characterization of co-polymer latices.

A point of significant interest is the adsorption of surfactants at very low concentrations, that is, in the Henry's law region where there should be a linear relationship between the amount adsorbed and the concentration of the adsorbate. This, so far, is an experimentally difficult region and attention should be given to assaying accurately very low concentrations of surfactant or weakly adsorbing materials such as hydrotropes. Furthermore, enthalpies of adsorption, either from calorimetric measurements or from the temperature dependence of adsorption, should give further insight into the energetics of latex surfaces.

A third area of interest is that of competitive adsorption when mixed surfactants are used either for emulsion polymerization or subsequently to confer colloid stability on the particles.

7.2. Polymeric Stabilizers

Fluorescence methods can be used to determine the concentration of adsorbed polymeric stabilizer by mixing a stabilizer containing a small fraction of covalently bound fluorescence energy donor, D, with stabilizer containing a similar amount of chemically linked acceptor, A. The particles are prepared using this mixture. The stabilizer dynamics can be suppressed by increasing the solvent viscosity, for example, by adding mineral oil or glycerine to the continuous phase.

The fluorescence decay profile of D can then be measured and fitted with a distribution function of D-A separations. In turn these can be interpreted in terms of the concentration of segments of stabilizing polymer.

The type of experiment described above can also be repeated for a system in which a small fraction of the stabilier contains both D and A groups, randomly distributed, leaving the remaining stabilizer unlabelled. If diffusion is slower than energy transfer, the experiment yields information on the conformation of the polymer chains. Recent theoretical studies also indicate that it is possible to obtain the radius of gyration of the stabilizer from this type of experiment.

In low viscosity media, both conformation and dynamics contribute to energy transfer processes. Information on dynamics can be obtained by comparing results in viscous and very fluid media. Similar experiments are possible using excimer formation from pyrene groups attached to the polymer.

A complication in these and all labelling experiments with

stabilizer chains is that if, during particle synthesis, a significant amount of stabilizer becomes buried in the particle interior then signals from both buried and surface polymer are detected in the experiment.

8. SURFACE POTENTIAL
8.1. Electrophoresis
Recent instrumentation offers the possibility of simultaneous measurements of the electrophoretic mobility and the hydrodynamic radius. Such measurements coupled with automated dielectric relaxation equipment should allow us to treat these rapid, _in-situ_, electrical signals as probes to investigate the effects of the adsorption of surfactants, polymers and polyelectrolytes.

8.2. Dielectric Spectroscopy
The properties of the diffuse part of the electrical double layer can be probed by this powerful technique. Concentrated and opaque systems are especially amenable to study, i.e. those which are difficult to study by more conventional methods. Total surface charge, dimensions of the particle, the double layer and the fraction of counterions bound in the Stern layer can be determined. The technique should be especially valuable in looking at specific ion effects and the rheological properties of concentrated systems. Although sophisticated equipment is now available for acquiring and manipulating data, the "wet" techniques associated with this kind of work demand great care and attention to cleanliness.

8.3. Acoustophoresis
The physical principle of this effect is that the acoustic wave causes the medium to oscillate whilst the particle remains essentially stationary. Thus the counter-ions are separated from the particle at the "plane of shear". The attractive advantages are the non-invasive, rapid, and _in-situ_ capability of measuring zeta-potentials. The method appears to be applicable to concentrated systems.

9. CROSS-LINKING FREQUENCY
It is often desirable to determine the degree of cross-linking in latex particles. Such particles are insoluble but swell in solvents which can dissolve the uncrosslinked polymer. The extent will depend upon the solvent, temperature and distance between cross-links.

The frequency of cross-links can also be calculated from the well-established theory of rubber elasticity applied to measurements obtained on a film formed from a latex. This forms the basis of a technique known as Dynamic Mechanical Spectroscopy.

Another technique which can be potentially developed to determine the extent of cross-linking is thin-layer chromatography.

10. TECHNIQUES NEEDING FURTHER DEVELOPMENT
10.1. Surface Energy of Latex Particles
A very direct need exists to develop methods for determining the surface energy of latex particles, both for homopolymers present in liquid media and for the interface between polymer particles in other polymer matrices.

The most probable approach would appear to be to base the method on the determination of contact angle between the particle surface and a suitable fluid (e.g. air), either using a microscope technique and large latex particles or assemblies of latex particles at an interface, for example, spread on the surface of a Langmuir trough.

10.2. Small Angle X-Ray Scattering

This technique may well prove to be useful for the study of latex surfaces by the use of heavy metal derivatives bound to the surface to enhance the contrast.

10.3. Shelf-Life of Latices

Determination of the shelf-life of polymer colloid systems, especially colloidally stable materials is a slow process. Well-defined accelerated ageing studies would be of considerable interest and this topic needs to be developed.

11. Panel

Discussion Leader:
R. H. Ottewill (University of Bristol, United Kingdom)

Secretary:
D. G. Rance (Imperial Chemical Industries PLC, United Kingdom)

Participants:
M. H. Andrus, Jr. (3M, USA)
F. Candau (Centre National de la Recherche Scientifique CNRS, France)
W. A. B. Donners (DSM Research, The Netherlands)
R. M. Fitch (S. C. Johnson & Son, Inc., USA)
A. Klein (Lehigh University, USA)
B. Kronberg (Ytkemiska Institutet, Sweden)
R. L. Rowell (University of Massachusetts, USA)
L. Tsaur (S. C. Johnson & Son, Inc., USA)
M. A. Winnik (University of Toronto, Canada)

CHARACTERIZATION OF POLYMER COLLOIDS

R.H. OTTEWILL

University of Bristol
Bristol BS8 1TS
England

1. INTRODUCTION

Once a polymer latex has been prepared a number of features need to be determined. Depending on the experimentalist involved requirements will differ and each will have his own list. However, a number of features of high priority can be listed as follows:

Particle size, mean or modal.
Particle size distribution.
Nature of surface groups.
Concentration of surface groups, i.e. surface charge density.
Location of surface groups, uniform or patchy.
Presence of residual monomer.
Morphology of the particles, e.g. "core-shell" or "currant-bun".
Composition of co-polymer latices.
Concentration of adsorbed emulsifier.
Concentration and conformation of adsorbed polymeric stabilizer.
Reaction of the particle to environment, e.g. swelling of polyelectrolyte latices or on uptake of organic solvent.
Thickness of steric stabilizing layer.
Structure of concentrated latex dispersions.
Nature of the interactions between the particles.
Surface potential of the particles in dilute and concentrated dispersions.
Nature of the particle surface, rough or smooth.

The list is a long one and no one technique can be used to determine all the parameters needed. However, in this section we will endeavour to cover many of the newer techniques which are being applied to the characterization of Polymer Colloids.

For this lecture I will concentrate on examining the use of scattering techniques to characterize latices remembering that the three main types of radiation currently applicable for use in scattering studies are, light, X-rays and neutrons.

The first of these has been in use for a long time since the theory of light scattering was first investigated by Lord Rayleigh in 1871[1] and then subsequently developed in the early part of the 20th Century by Mie[2] and Debye[3]. The experimental investigation of scattered light was greatly enhanced in the 1940's by the development of the photomultiplier and then in the 1960's by the invention of the laser as a coherent light source. The latter feature led to the emergence of quasi-elastic light scattering (photon correlation spectroscopy) as a powerful method for the investigation of the diffusional motion of particles[4].

The study of crystals by X-rays was also well-developed by ca. 1930[5] but although the theory of small angle X-ray scattering was nicely presented by Guinier and Fournet in 1955[6] it has not become a widely used experimental technique. However, the development of multidetectors for X-ray instruments and the production of X-ray beams from Synchroton Radiation Sources may well drastically change this scene in the next few years[7].

The neutron was discovered by Chadwick[8] in 1932 and although it was shown by Mitchell and Powers[9] in 1936 that neutrons could be diffracted, it was not until the 1970's with the building of high flux reactors[10] that neutron scattering began to develop.

In Table 1 a summary is given of the range of wavelengths available from these various sources of radiation.

TABLE 1.

Radiation	Radiation Sources	Wavelength Å
Light		ca. 4000 - 6500
X-rays	Generators	0.4 - 1.8
	Synchroton Radiation	1 - ≥ 5
Neutrons	Thermal	ca. 1.0 - 5.0
	Cold	ca. 5.0 - 20.0

The use of the light scattering technique to study Polymer Colloids has been reviewed by Rowell[11] and in this article I will mainly give attention to the use of neutron scattering methods for the characterization of latex systems.

2. BASIC SCATTERING CONCEPTS

The neutron is a fundamental particle of mass, $m = 1.675 \times 10^{-27}$ kg, a spin of 1/2 and a magnetic moment of -1.913 nuclear magnetons; it has zero electrical charge. Its velocity is directly related to its energy or wavelength by de Broglie's relation, so that for a neutron travelling with velocity v the wavelength is given by:

$$\lambda = h/mv$$

with h = Planck's constant.

The wavelength λ is a scalar quantity but the beam has a definite direction and velocity and hence can be represented by a vector $\underline{v}$; whence we can write for the momentum of a single neutron:

$$m\underline{v} = [h/(2\pi)][(2\pi)/\lambda] = \hbar\ \underline{k}$$

where $\underline{k}$ is termed the wave vector.

If the direction of propagation of the incident neutron beam travelling with a velocity $\underline{v}_o$ is represented by the $\underline{k}_o$ and that of the scattered neutron beam travelling with velocity $\underline{v}$ by $\underline{k}$ and the scattered beam makes an angle θ with respect to the incident beam then $\underline{Q} = \underline{k}_o - \underline{k}$. Hence,

$$Q^2 = k^2 + k_o^2 - 2\ k\ k_o\ \cos\theta$$

and

$$\hbar\ Q = m\ \underline{v} - m\ \underline{v}_o$$

The change of energy in the scattering process is given by:

$$\Delta E = m\underline{v}^2/2 - m\underline{v}_o^2/2$$

The velocity of a neutron corresponding to $\lambda = 10$ Å is 100 m s^{-1} (360 km h^{-1} = 225 miles per hour) so that the velocity can be measured (time of flight) before and after scattering and the energy change determined. Hence a wide range of inelastic phenomena can be examined (ΔE finite). For elastic scattering $\Delta E = 0$ and $k = k_o$ since $\underline{k}$ changes in direction but not in magnitude and the magnitude of Q becomes:

$$Q = |\underline{Q}| = 2\ k_o\ \sin(\theta/2)$$

$$= (4\pi/\lambda)\ \sin(\theta/2)$$

Q is a general quantity called the scattering vector. For a light beam it is written as $4\pi n_o \sin(\theta/2)/\lambda$ where n_o = refractive index of the medium; it is often given, in this case, the symbol K.

Scattering from a single atom can be represented schematically as shown in Figure 1. A planar wave, $\psi(z) = \exp(i\ \underline{k}_o\ z)$ is incident upon the particle and is then scattered by the particles in all directions in the form of a spherical wave, which can be represented by $\psi(r) = b_o \exp(i\ \underline{k}\ \underline{r})/r$. The scattered energy is distributed over the surface of a sphere of surface area $4\pi r^2$, the minimum value of which is $4\pi b_o^2$. The quantity b_o is known as the neutron scattering length and since neutrons are scattered by the nuclei of atoms, b_o is of the order of 10^{-12} cm for all nuclei.

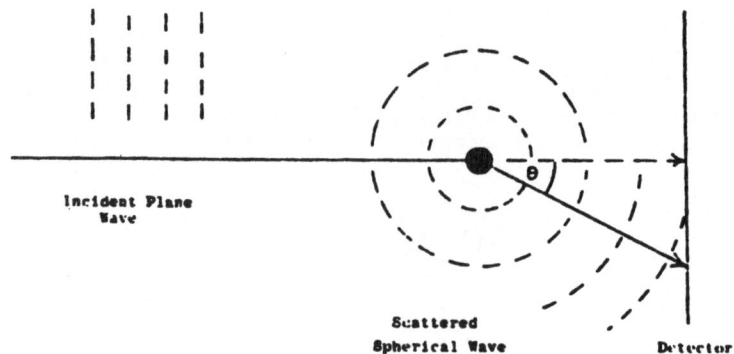

Figure 1: Schematic illustration of scattering of a neutron beam by a single atom.

The scattering cross-section, σ_s, which is the outgoing flux of neutrons divided by the incident flux is given by:

$$\sigma_s = 4\pi b_o^2$$

Contributions to σ_s arise from both coherent and incoherent scattering events. The latter arise as a consequence of random scattering events since not all nuclei of a given element are identical either because of different isotopes or different nuclear spin states. Incoherent scattering is isotropic and produces a background which needs to be subtracted from the measured total scattering in order to determine the intensity of the coherent scattering. Hence, we may write:

$$\sigma_s = \sigma_{inc} + \sigma_{coh}$$

with

$$\sigma_{coh} = 4\pi b_{coh}^2 = 4\pi b^2$$

The value of b, the COHERENT SCATTERING LENGTH depends on the isotope and hence does not scale with atomic number. The values are also independent of wavelength and scattering angle. A number of values are listed in Table 2 and a comprehensive list is given in Bacon[12].

TABLE 2. Coherent Scattering Length for Various Atoms.

Atom	Nucleus	b - Coherent Scattering Length/10^{-12} cm
Hydrogen, H.	^{1}H	- 0.374
Deuterium, D.	^{2}H	0.667
Carbon, C.	^{12}C	0.665
Oxygen, O.	^{16}O	0.580
Fluorine, F.	^{19}F	0.560
Chlorine, Cl	^{35}CL	1.180
Sulphur, S	^{32}S	0.280

In the case of a polymeric material we need to define the coherent scattering length density corresponding to the assembly of atoms which make up the molecule or particle compound of molecules. This is given by:

$$\rho = \Sigma \ (b_i / V)$$

Hence for polystyrene taking $[C_6H_5\text{-}CH\text{=}CH_2]_n$, as the segmental unit we have 8 x 0.665 x 10^{-12} cm for the 8 carbon atoms and 8 x -0.374 x 10^{-12} for the 8 hydrogen atoms giving $\Sigma b_i = 2.326$ x 10^{-12} cm. The molecular volume of the monomer unit, V = 104.1/(6.02 x 10^{23} x 1.054) cm^3 = 1.64 x 10^{-22} cm^3, taking 1.054 g cm^{-3} as the density of polystyrene. This gives the value of ρ_p for polystyrene as 1.42 x 10^{10} cm^{-2}. The values for a number of materials are given in Table 3.

TABLE 3. Neutron Scattering Length Densities for Various Molecules

Material	Formula	Coherent Neutron Scattering Length, ρ /10^{10} cm^{-2}
Water	H_2O	- 0.56
Deuterium oxide	D_2O	6.40
h_{14}-Hexane	C_6H_{14}	- 0.58
h_{26}-Dodecane	$C_{12}H_{26}$	- 0.46
d_{18}-Octane	C_8D_{18}	6.43
d_{26}-Dodecane	$C_{12}D_{26}$	6.71
h_8-Toluene	C_7H_8	0.94
d_8-Toluene	C_7D_8	5.63
Polystyrene	$[C_8H_8]_n$	1.42
d-Polystyrene	$[C_8D_8]_n$	6.47
Polymethylmethacrylate	$[C_5H_8O_2]_n$	1.07
d-Polymethylmethacrylate	$[C_5D_8O_2]_n$	7.03
Polyacrylonitrile	$[C_3H_3N]_n$	2.27

2.1. Scattering from a Homogeneous Spherical Particle

For a single homogeneous spherical particle of radius R the intensity of neutrons scattered at a particular value of Q is given by:

$$I(Q) = A(\rho_p - \rho_m)^2 \; V_p^2 \; P(Q) \tag{1}$$

where A = an instrumental constant, ρ_p = the coherent neutron scattering length density of the particle, ρ_m = the coherent neutron scattering length density of the medium and V_p = the volume of the particle, i.e. $4\pi R^3/3$. The particle form factor is given by:

$$P(Q) = \left[\frac{3(\sin QR - QR \cos QR)}{Q^3 R^3} \right]^2 \tag{2}$$

Provided the system is sufficiently dilute that particle-particle interactions can be neglected then for a number concentration of N_p particles per cm^3, equation (1) can be rewritten, assuming single scattering, as

$$I(Q) = A(\rho_p - \rho_m)^2 \; N_p \; V_p^2 \; P(Q) \tag{3}$$

and the volume fraction $\phi = N_p V_p$.

The form of the curves for completely monodisperse spherical particles with R = 250 Å and R = 1000 Å are shown in Figure 2.

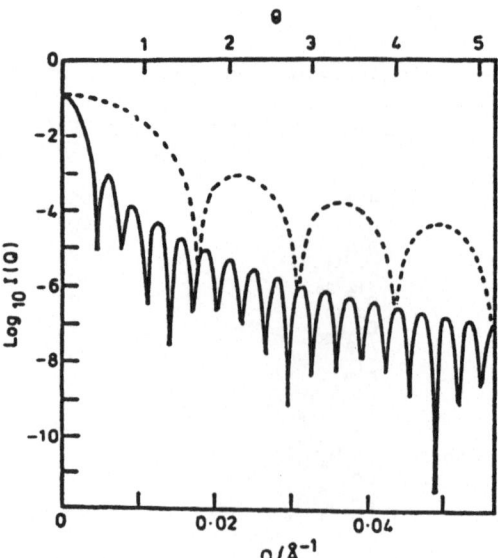

Figure 2: $Log_{10}|I(Q)|$ against Q for spherical particles calculated using equation (3) with---R = 25nm; —— R = 100nm, and λ = 10 Å.

As can be seen, a number of maxima and minima in intensity occur in the scattering curves and these increase in number as the particle size increases. With increase in particle size the first minimum moves closer to the ordinate and for large particles is so close to the axis that observations become experimentally inaccessible. This is quite an advantage since dust particles, the bane of light scattering, in general, do not interfere seriously with neutron scattering results. The results shown in Figure 2 are for a completely monodisperse system and resolution of such high quality is difficult to achieve in practice owing to particle size polydispersity and other effects (see later). It should be noted that the ordinate is plotted as $\log_{10} I(Q)$ so that for the larger particles the fall off in scattered intensity between the first and second peaks is several orders of magnitude. The abscissa at the top of the figure is scaled in terms of the scattering angle θ to emphasize the small angular range over which the scattering features can be observed.

2.2. Scattering at Zero Angle

At zero scattering angle, i.e. $Q = 0$. the particle form factor $P(Q)$ becomes unity and hence,

$$I(0) = A(\rho_p - \rho_m)^2 \phi V_p \qquad (4)$$

An immediate use of this relationship is to determine the neutron scattering length of particles by the contrast variation method. The neutron scattering length density of the medium, ρ_m, can be varied from that of pure water, -0.56×10^{10} cm^{-2}, to that of pure D$_2$O, 6.40×10^{10} cm^{-2}, by the use of D$_2$O:H$_2$O mixtures. Rewriting the equation in the form

$$\pm \sqrt{I(0)} = (\rho_p - \rho_m) \, [\phi A V_p]^{1/2} \qquad (5)$$

we see immediately that for

$$\rho_p = \rho_m$$

then $\sqrt{I(0)} = 0$ independent of volume fraction. Hence, extrapolating $I(Q)$ against Q data to zero Q enables a plot to be obtained of $\sqrt{I(0)}$ against ρ_m, of the form shown in Figure 3 for experiments carried out on h-polystyrene and d-polystyrene. The figures obtained by experiment are in good agreement with those obtained by calculation and listed in Table 3.

260

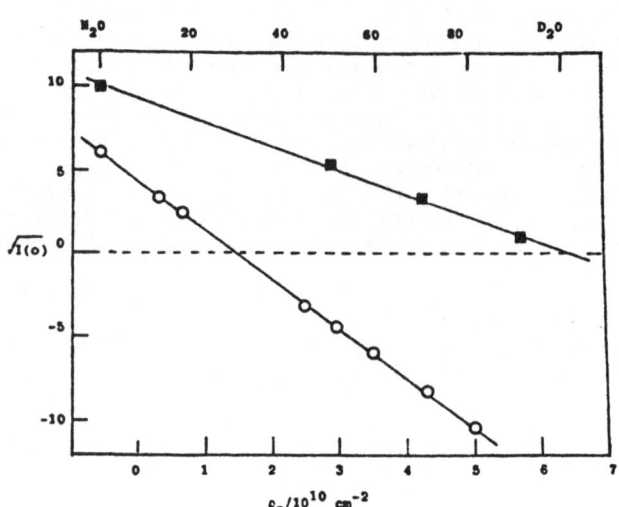

Figure 3: $\sqrt{I(0)}$ *against scattering length density of dispersion medium,* ρ_m.
—O—, *polystyrene latex;* —■—, *polydeuterostyrene latex*

From equation (5) we find also that the slope of the curve is given by

$$d\sqrt{I(0)}/d\rho_m = \pm \left[\phi A V_p\right]^{1/2} \tag{6}$$

and the intercept by

$$\rho_p \left[\phi A V_p\right]^{1/2}$$

whence the ratio of the intercept to the slope is given by

$$\text{Intercept/Slope} = \rho_p \tag{7}$$

Moreover, if the instrument constant is known and the volume fraction of the dispersion is determined, then $V_p^{1/2}$ can be obtained from the slope and hence the radius of the particles deduced.

2.3. Scattering at Small QR

For values of QR<1 the terms in P(Q) can be expanded and we obtain as a limiting law, often called the Guinier equation,

$$I(Q) = I(0) \exp \left(-Q^2 R_g^2/3\right) \tag{8}$$

where R_g — the radius of gyration of the particles. For spherical particles,

$$R_g^2 - 3R^2/5 \tag{9}$$

3. EXPERIMENTAL RESULTS ON HOMOPOLYMER LATEX PARTICLES

The experimental scattering spectra obtained on three monodisperse polystyrene latices, containing particles of radius 178 Å, 660 Å and 1010 Å are shown in Figure 4. These spectra show clearly the high quality of the data which can be obtained and also demonstrate how as the particle size increases the observed features move to low Q values.

It also becomes evident from the experimental results shown in Figure 4 that the deep minimum shown in the theoretical calculations (Figure 2) are not quite as clearly observable in the experimental curves. The reasons for this are:

polydispersity of particle size

polychromaticity of the incident beam; typically $\Delta\lambda/\lambda \approx 5$ to 10% as determined by the velocity selector

finite size of the detector elements

angular divergence of the incident beam.

For many colloidal dispersions, including polymer colloids, it has been shown that the particle size distribution can be simulated by a logarithmic distribution function[13,14] of the form,

$$p(R) - \frac{\exp(-[\ln R - \ln R_m]^2/2\sigma_o^2)}{(2\pi)^{1/2} \sigma_o R_m \exp(\sigma_o^2/2)} \tag{10}$$

where R_m — the modal radius of the particles and σ_o a parameter which gives a measure of the width of the distribution. For a narrow distribution the shape is Gaussian and the standard deviation, σ is given by

$$\sigma \approx \sigma_o R_m \tag{11}$$

Once the instrumental factors have been determined computer fitting procedures can be utilized to match the experimental data and hence σ and the distribution of particle radii determined. In Figure 5 a comparison is made of particle size histograms obtained from transmission electron microscopy and by use of the above distribution to fit small angle neutron scattering results.

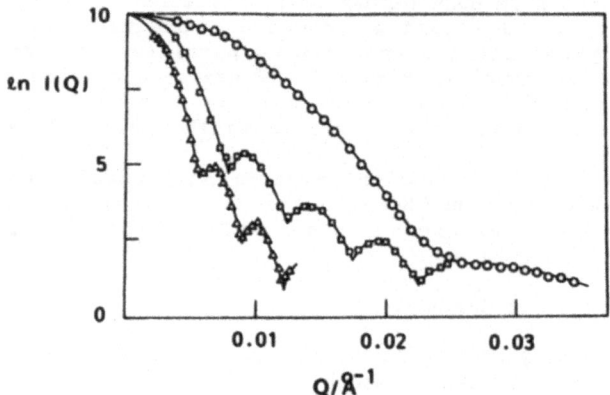

Figure 4: *ℓn I(Q) against Q for polystyrene latex particles of various diameter. Ordinate - arbitrarily scaled to show lateral displacement with increase in size:* O, *diameter 346 Å;* □ *diameter 1320 Å;* △ *2020 Å.*

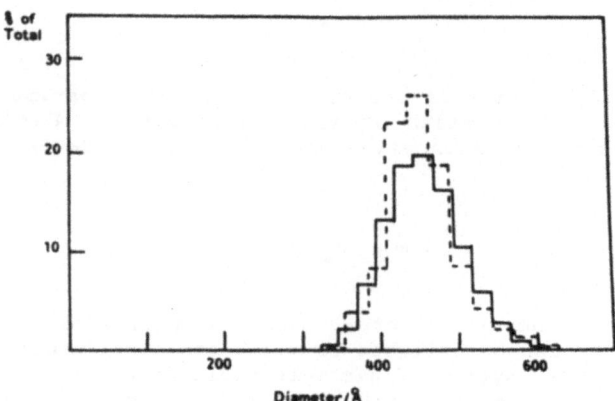

Figure 5: *Particle size distribution.———, small angle neutron scattering; ---, electron microscopy.*

4. DETERMINATION OF PARTICLE COMPOSITION

The contrast variation method described in the previous section can also be used as a means of determining the composition of a copolymer particle. An example is given in Figure 6 for a latex prepared from d_8-styrene and acrylonitrile. The intercept at $\rho_m - 4.94$ x 10^{10} cm^{-2} is determined by the composition of the particle, assuming volume additivity, such that,

$$\bar{\rho} - [V_{dPSL}\,\rho_{dPSL} + V_{PAN}\rho_{PAN}]/V_T \qquad (12)$$

or

$$\bar{\rho} - \alpha\,\rho_{dPSL} + (1 - \alpha)\,\rho_{PAN}$$

where V_T = total volume of the particle, determined from the scattering curve, V_{dPSL} and V_{PAN} = volumes per cent occupied by d-polystyrene and polyacrylonitrile. This gave the composition as 63.6% d-polystyrene and 36.4% polyacrylonitrile.

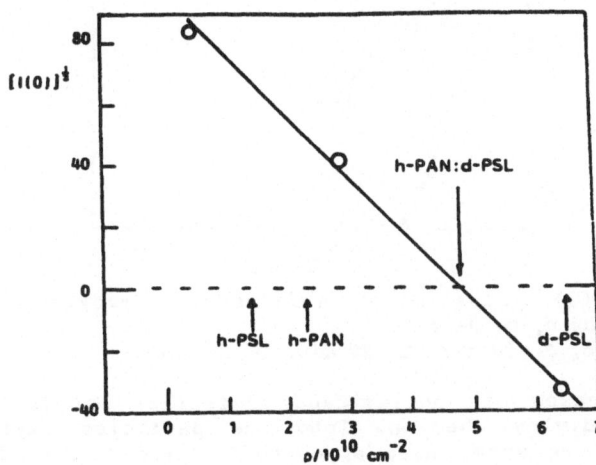

Figure 6: $[I(0)]^{1/2}$ *against coherent scattering length of medium for a polystyrene-polyacrylonitrile copolymer latex.*

5. PARTICLE SWELLING

The large variation in scattering length between H and D atoms can also be used as a means of examining the take up of an organic phase by particles. This effect is illustrated in Figure 7 where a polystyrene latex, diameter 1485 Å, was initially dispersed in an aqueous medium composed of 30% D_2O and 70% H_2O; this gives ρ_m - 1.50 x 10^{10} cm^{-2} close to the contrast match point of polystyrene (1.41 x 10^{10} cm^{-2}). d_8-toluene was then added to the system and the scattering followed at various time intervals[15]. The uptake of the deuterated solvent increases (ρ_m - ρ_p) and also the size of the particle so that I(O) increases and the peak maxima move to lower Q. The curves, however, all exhibit a form consistent with the homogeneous sphere scattering function and a good correspondence is obtained between the experimental results and the simulated fits for uniform spherical particles. The diffusion coefficients obtained for d_8-toulene in the latex particles were of the order of 6 x 10^{-12} cm^2s^{-1} suggesting that the diffusion of toluene was not the rate determining step but rather the actual diffusion of polymer molecules, i.e. a diffusional breathing motion.

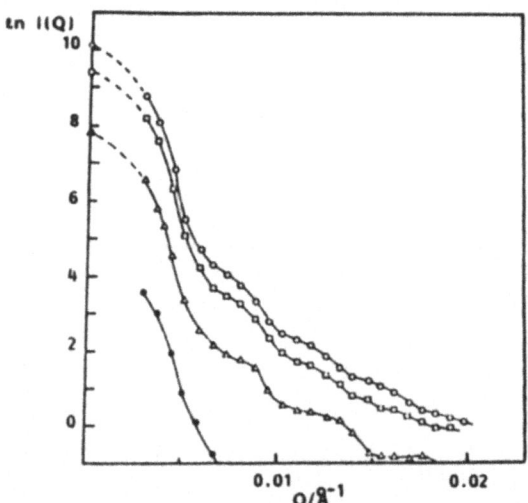

Figure 7: ℓn I(Q) against Q for swelling of polystyrene at various times after adding d_8-toluene.
•, zero time; △, 10 min; □, 20 min; ○, 90 min.

In the case of polystyrene particles cross-linked by incorporating divinyl benzene into the particles fitting by a homogeneous sphere model also appeared to give a good fit to the results. The initial swelling ratios, however, were slower and the swelling was restricted. The apparent diffusion coefficients obtained were also an order of magnitude smaller than those obtained from a non cross-linked latex particle.

The mode of preparation of the latex was also found to influence the rate of swelling of polystyrene particles by d_8-styrene monomer[16]. The effects observed are illustrated in Figure 8. The latex was prepared by an emulsifier free method[17] and it is interesting to note that, after treatment by dialysis alone, this material showed a distinct reduction in particle size after the initial swelling process. A possible explanation for this was the release of low molecular weight material from the particle.

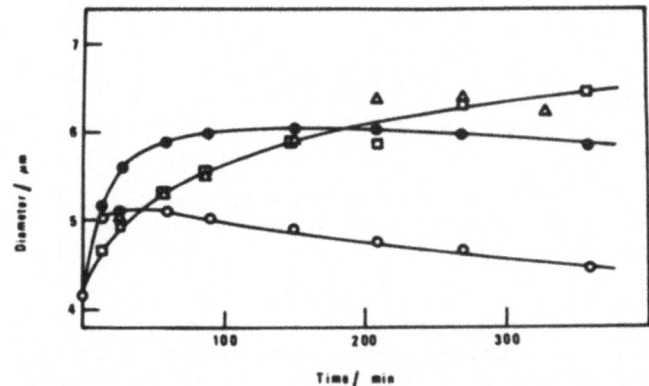

Figure 8: *Radius of latex particles as a function of time after the addition of deuterostyrene:○, latex I -- dialyzed; ●, latex II -- steam-stripped; △, latex III -- ion-exchanged; □, latex IV -- steam-stripped and ion-exchanged.*

6. CONCENTRIC SPHERES

The case of a spherical particle of radius R_1 and a neutron scattering length density of ρ_p has been extended to that of the same particle with a concentric layer of neutron scattering length density, ρ_A, such that the overall diameter of the particle becomes R_2 as illustrated in Figure 9.

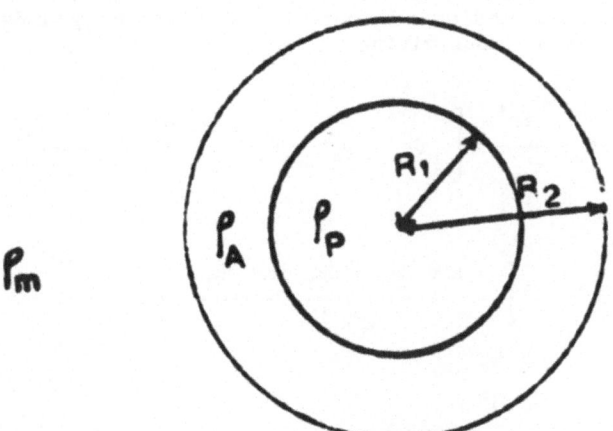

Figure 9: *Spherical particle with a neutron scattering length density of ρ_1 and radius R_1, with an attached layer of neutron scattering length density, ρ_A, overall radius R_2, in a medium of neutron scattering length density, ρ_m.*

The intensity of scattering for this situation[14,18] is given by,

$$
\begin{aligned}
I(Q) = A\, N_p \frac{16\pi^2}{9} \Bigg[& (\rho_A - \rho_m) \left\{ 3R_2{}^3 \left(\frac{\sin QR_2 - QR_2 \cos QR_2}{Q^3 R_3{}^3} \right) \right. \\
& \left. - 3R_1{}^3 \left(\frac{\sin QR_1 - QR_1 \cos QR_1}{Q^3 R_1{}^3} \right) \right\} \\
& + (\rho_p - \rho_m)\, 3R_1{}^3 \left(\frac{\sin QR_1 - QR_1 \cos QR_1}{Q^3 R_1{}^3} \right) \Bigg]^2
\end{aligned}
\tag{13}
$$

It follows from this equation that if the scattering length density of the medium is made equivalent to that of the attached layer, i.e. $\rho_A = \rho_m$, we obtain

$$
I(Q) = A\, N_p \frac{16\pi^2}{9} \left[(\rho_p - \rho_m)\, 3R_1{}^3 \left(\frac{\sin QR_1 - QR_1 \cos QR_1}{Q^3 R_1{}^3} \right) \right]^2
\tag{14}
$$

which is identical with equation (3). Under these conditions, therefore, the size of the core particle can be obtained, that is, R_1. However, if ρ_p is known, and it frequently is, then by selecting a suitable dispersion medium of deutero and hydro-compounds it can be arranged that $\rho_m = \rho_p$, thus giving

$$
\begin{aligned}
I(Q) = A\, N_p \frac{16\pi^2}{9} \Bigg[& (\rho_A - \rho_m)\, 3R_2{}^3 \left(\frac{\sin QR_2 - QR_2 \cdot \cos QR_2}{Q^3 R_2{}^3} \right) \\
& - 3R_1{}^3 \left(\frac{\sin QR_1 - QR_1 \cos QR_1}{Q^3 R_1{}^3} \right) \Bigg]^2
\end{aligned}
\tag{15}
$$

and hence the scattering is that of the attached layer, that is the system scatters as a shell of scattering length density ρ_A and thickness $t = R_2 - R_1$. The possible scattering situations are illustrated in Figure 10.

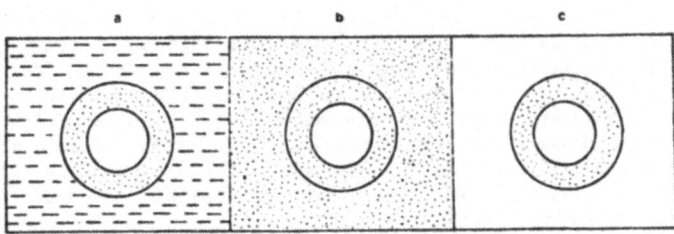

Figure 10:   Schematic illustration of contrast matching conditions

a) $\rho_m \neq \rho_A \neq \rho_p$
b) $\rho_m = \rho_A$
c) $\rho_m = \rho_p$

An example of this approach can be provided from data obtained on a poly-deuteromethyl methacrylate particle stabilized by a chemically grafted layer of poly-12-hydroxystearic acid.   The latter layer provides the steric stabilization necessary to maintain the particles as discrete entities in a hydrocarbon environment.

Figure 11 shows curves of I(Q) against Q obtained in h_{14}-hexane, in order to contrast match the poly-12-hydroxystearic acid.   A good fit to the experimental data was obtained using an R_1 value of 18 nm[19].

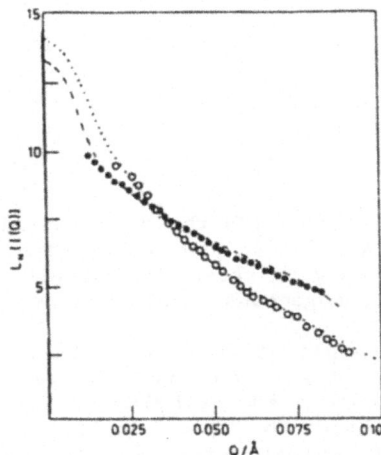

Figure 11:   SANS results obtained on dPMMA latex ($\phi = 0.1$): •, in d_{18}-octane; O, in h_{14}-hexane. ——, calculated curve for dPMMA cores of R = 18 nm with t = 6nm in d_{18}-octane.   - - - -, calculated curve for dPMMA cores with R = 18 nm, t = 0 in h_{14}-hexane

Another series of experiments was carried out using d_{18}-octane as the dispersion medium, a close match for the scattering length density of the poly-deuteromethyl methacrylate core of the particle, thus giving the scattering from a concentric shell in the manner described by equation (15). The scattering data in this case were fitted with $R_1 = 18$ nm and $R_2 = 24$ nm, thus giving the apparent thickness of the adsorbed layer, t, as 6 nm. A fully extended poly-12-hydroxystearic acid chain is considered to have a length of ca. 9 nm, so that the estimated thickness of the shell from small angle neutron scattering is slightly less than this. On the other hand[19] interaction studies with this system showed the hard sphere radius to be 28 nm, suggesting an adsorbed layer thickness of 10 nm. The difference is almost certainly a consequence of the penetration of solvent into the attached layer of poly-12-hydroxystearic acid.

7. PARTICLE MORPHOLOGY

In order to obtain information about the structure of the particle, a useful function is the radial density distribution, P(r), which is obtained by Fourier transformation of the scattering data in the form I(Q) against Q. It is defined by,

$$P(r) = \frac{1}{2\pi^2} \int_0^\infty Q r\, I(Q)\, \sin(Qr)\, dQ \tag{16}$$

Figure 12a shows the form of the graph of P(r) against r obtained for a polystyrene latex particle prepared by emulsion polymerization and subsequently-treated by mixed-bed ion-exchange resin[20].

For a homogeneous sphere P(r) is given by the expression[6,21],

$$P(r) = \text{Const} \left[\frac{r^2}{R} - \frac{3r^3}{4R^2} + \frac{r^5}{16R^4} \right] \tag{17}$$

and for comparison in Figure 12a the curve is given calculated from this expression using $R = 165$ Å. As can be seen the experimental and calculated curves are in good agreement suggesting that in this case the latex particle is homogeneous within the detectable limits of the experimental procedure.

An interesting example of the use of this approach is given in the work of O'Reilley et al[22] who prepared latices by the emulsion polymerisation of deuterated methylmethacrylate and deuterated styrene. This resulted in latex particles with a core-shell morphology. The $\rho(r)$ against r function for the shell is given in Figure 12b.

This is clearly a technique which can be exploited to obtain information about particle morphology but it must be emphazised that very good experimental data is required before Fourier transformation can be undertaken.

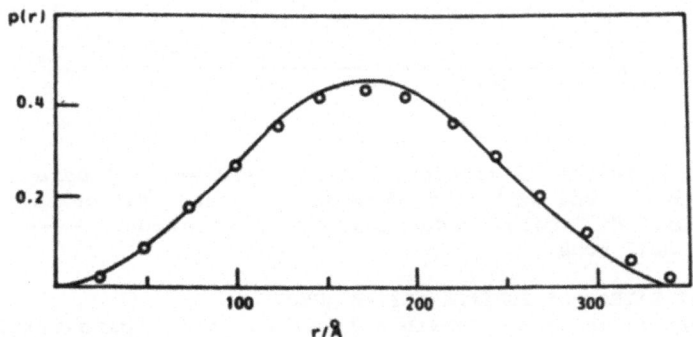

Figure 12a: P(r) against r for a polystyrene latex particle diameter 320 Å.

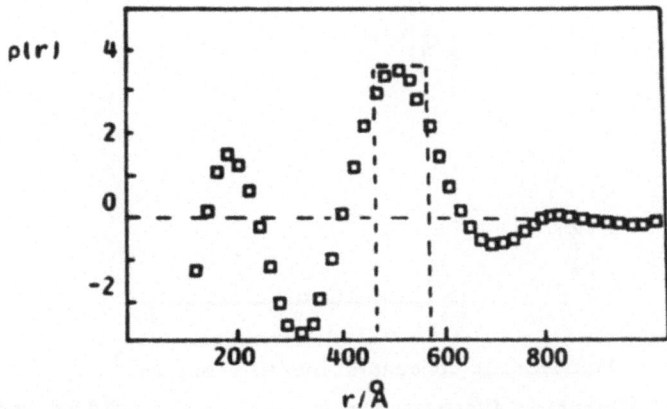

Figure 12b: ρ(r) against r for a shell of perdeuterated PMMA on a core of deuteropolystyrene.

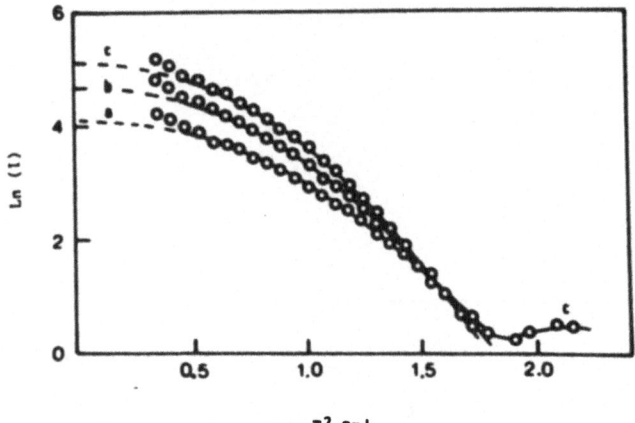

$Q/10^{-2}\ \mathring{A}^{-1}$

Figure 13: Examples of scattering data obtained. a) bare latex; b) latex in 4 x 10^{-3} mol dm^{-3} total dodecanoate concentration; c) latex in 1.2 x 10^{-2} mol dm^{-3} total dodecanoate concentration. ————, fits of concentric shell model.

8. ADSORBED LAYERS OF SURFACE ACTIVE AGENT

In many cases it is possible to start with a bare particle and then to adsorb molecules on to the surface. For example, the adsorption of d_{23}-dodecanoic acid onto polystyrene latex particles has been investigated in some detail[23]. Initially the latex particles were examined in a series of H_2O - D_2O mixtures and then re-examined in the same mixtures in the presence of d_{23}-dodecanoic acid. The scattering curves obtained are shown in Figure 13 and the adsorption isotherm obtained in Figure 14a.

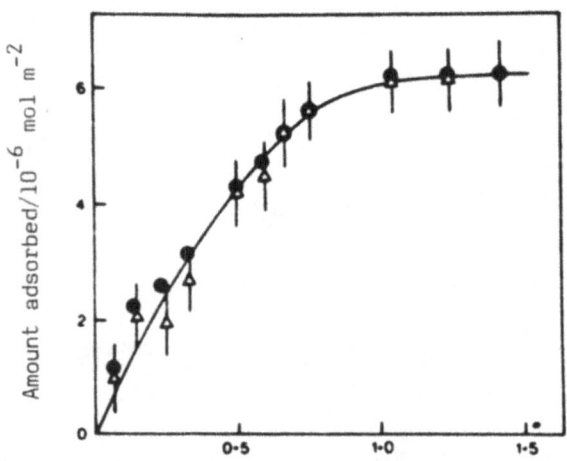

Equilibrium concentration/10^{-2} mol dm^{-3}

Figure 14a: Adsorption isotherm for d_{23}-dodecanoic acid on polystyrene latex at pH 8.1 and 22°C with a total electrolyte concentration of 2.2 x 10^{-2} mol dm^{-3}. △, calculated by fitting a spherical shell model plus error bars; •, points obtained from Guinier analysis.

Figure 14b shows the variation of the limiting area per molecule as a function of pH and suggests that at pH 7.0 and lower pH values the fatty acid forms an adsorbed layer on the polystyrene particle surface in which the fatty acid molecules are close-packed and vertically oriented. The experimental values appear to be approaching the cross-sectional area of a monobasic acid molecule of 20.2 Å².

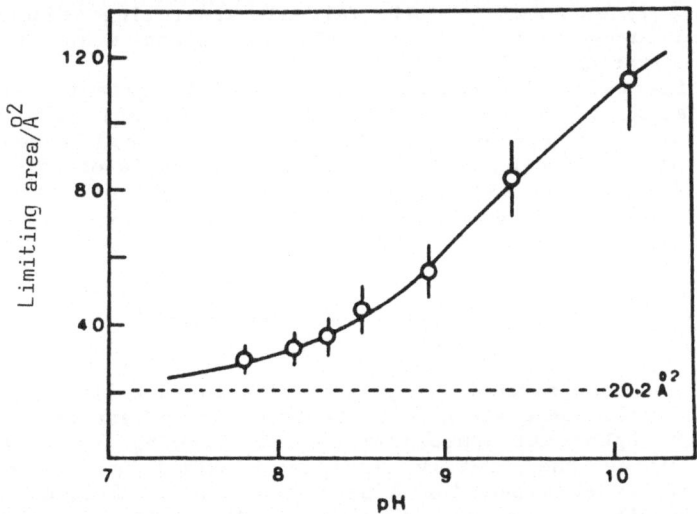

Figure 14b: *Variation of limiting area per molecule with pH. Total* d_{23}-*dodecanoate concentration,* 10^{-2} *mol* dm^{-3} *in a total electrolyte concentration of* 2×10^{-2} *mol* dm^{-3}.

9. CONCENTRATED LATICES

The majority of latices used commercially are in concentrated form and hence it is important to develop techniques to study these; in addition it is a subject of considerable fundamental importance. It has been demonstrated over the past few years that scattering techniques provide one of the few methods of studying concentrated systems where the particles are in constant interaction with each other. Consequently an ordering of the system occurs which is dependent on the number concentration and the strength of the repulsive interactions. The spatial correlations thus produced lead to interparticle interference effects which can be expressed as a structure factor, $S(Q)$, given by,

$$S(Q) = 1 + \frac{4\pi N_p}{Q} \int_0^\infty [g(r) - 1] \, r \cdot \sin Qr \cdot dr \qquad (18)$$

where $g(r)$ = the pair correlation function and r = the centre to centre interparticle separation. For interacting homogeneous spherical latex particles therefore the scattered intensity is given by,

$$I(Q) = A(\rho_p - \rho_m)^2 \, V_p \, \phi \, P(Q) \cdot S(Q) \qquad (19)$$

Thus S(Q) becomes an experimentally accessible quantity and a distinct advantage can be obtained from using neutrons to determine it[21]. Firstly, since the Q range available on small angle instruments is approximately from 10^{-2} to 0.2 Å^{-1} in reciprocal space, the distances in real space sampled under these conditions range from ca. 6000 Å to 30 Å; a range unattainable by light scattering. Secondly, since neutrons are scattered only by atomic nuclei, multiple scattering effects are much less pronounced since the nuclei are dilute even at high molecular concentrations.

An experimentally determined curve of S(Q) against Q for a polystyrene latex, of volume fraction 0.08, in 10^{-2} mol dm^{-3} sodium chloride solution is shown in Figure 15. Such curves can be modelled theoretically by introducing a pair-potential for the interaction[24,25]. For example, for aqueous latices of reasonably small particle size where the interaction is dominated by electrostatic repulsion, the pair potential is given by,

$$V(r) - 4 \pi \, \varepsilon_r \varepsilon_o \, R^2 \Psi_s^2 \, \exp(2 \, \kappa R) \cdot \exp(-\kappa r)/r \qquad (20)$$

where Ψ_s — the surface potential, ε_r — relative permittivity, ε_o — the permittivity of free space and κ — the Debye-Huckel parameter. By the use of the mean spherical approximation and allowing for many-body interactions using the Orstein-Zernicke approach a theoretical expression can be obtained for S(Q) which can be fitted to the experimental results. An example is shown in Figure 15. For this fit only the surface potential, Ψ_s, was allowed to float and the best fit was obtained with $|\Psi_s|$ — 52 mV. This value can be compared with that obtained from electrophoretic mobility measurements, using photon correlation spectroscopy, on dilute systems (ϕ — 10^{-4}), as a zeta-potential, viz., -54 ± 5 mV. This evidence strongly indicates that interactions in concentrated dispersions are dominated by the interactions of diffuse electrical double layers. It must also be concluded that the examination in this way allows a determination of the surface potential of particles in concentrated dispersions.

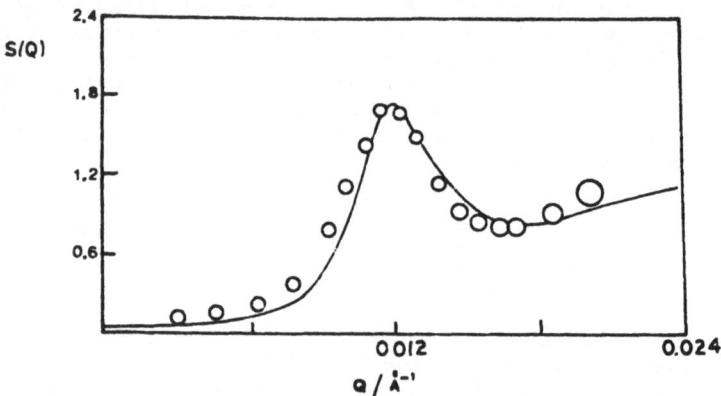

Figure 15: Plot of S(Q) against Q for a polystyrene latex at a volume fraction, ϕ — 0.08, in 10^{-3} mol dm^{-3} sodium chloride solution. O, experimental points; ———, curve calculated using MSA theory with $|\Psi_s|$ — 52 mV

The quantity g(r) given in equation (18) is also an important one since it gives the probability of finding a particle at a distance r from a reference particle. It can be written as

$$g(r) = \frac{N_p(r)}{N_p} \tag{21}$$

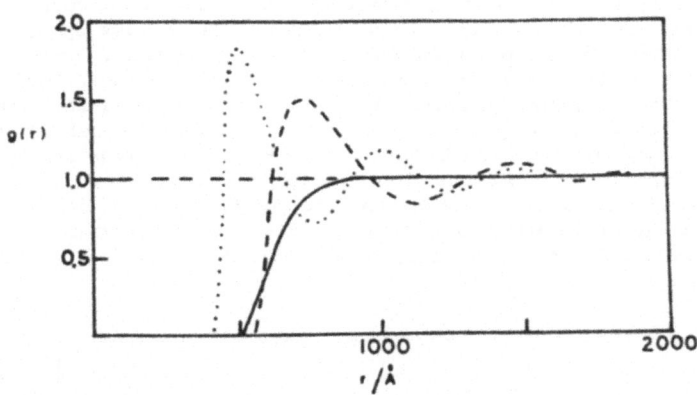

Figure 16: Plot of g(r) against r for polystyrene latices in 10^{-4} mol dm^{-3} sodium chloride solution at various volume fractions, ———, $\phi = 0.01$; – – –, $\phi = 0.04$; ····, $\phi = 0.13$

where N_p is the average number concentration, a MACROSCOPIC quantity, and $N_p(r)$ is the radial number density, a MICROSCOPIC quantity. The quantity $4 \pi r^2 N_p(r)$ is known as the radial distribution function. Once experimental data is available in the form of S(Q) against Q then g(r) against r can be obtained since it is given by the Fourier transform of equation (18) as,

$$g(r) = 1 + \frac{1}{2\pi^2 r N_p} \int_0^\infty [S(Q) - 1] Q \sin Qr \cdot dQ \tag{22}$$

An example of the results obtained is given in Figure 16 for three volume fractions $\phi = 0.01$, $\phi = 0.04$ and $\phi = 0.13$ in 10^{-4} mol dm^{-3} sodium chloride solutions. With the most dilute system the curve shows clearly an excluded volume effect at small r values and then rises to unity as r becomes greater than 500 Å. The form of the curve indicates that the particles are randomly distributed but interact weakly on approaching to distances of the order of 1000 Å and very strongly at distances of the order of 500 Å. At the higher volume fractions strong interactions occur and "structure" is produced, which becomes more pronounced as the volume fraction increases; this is evidenced by the increase in height of the peaks and the increase in depth of the minima. Ultimately with continued increase in volume fraction a solid-like lattice is produced.

274

This technique can also be applied to non-aqueous latex dispersions and in recent work[26] systems of polymethylmethacrylate with a grafted layer of poly-12-hydroxy stearic acid (PHS) have been examined at volume fractions up to 0.42. It was found with these systems that the extension of the steric stabilizing layer, the PHS, was dependent on volume fraction and became more compressed as the volume fraction increased. The curves of g(r) against r for several volume fractions are shown in Figure 17. From this work it was concluded that at low volume fractions, when there is a large distance on average between the particles and only occasional contacts, the PHS chains are fully extended and interaction only occurs between the most extended parts of the chains. With increase in volume fraction and consequently less free space in the system, it appears that the PHS chains compact either as a consequence of a change in solvency of the PHS layer or the greater frequency of collisions between the particles causes the solvent to be squeezed out from between the chains so that the surface layer becomes much more compact. This compaction can be seen from Figure 17 in that the distance of close approach between the molecules decreases with increase in volume fraction. Desorption of the stabilizing molecules in this system cannot occur since the stabilizing moieties are chemically bonded to the surface.

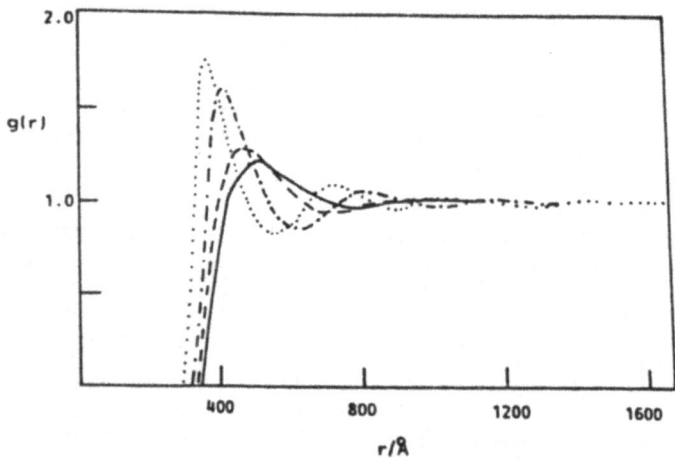

Figure 17: g(r) against r for sterically stabilized polymethyl methacrylate latices in dodecane at various volume fractions: ——, ϕ = 0.23; ----, ϕ = 0.28; —·—, ϕ = 0.36; ····, ϕ = 0.42.

REFERENCES

1. Lord Rayleigh, *Phil. Mag.*, 1871, 41, 107, 274
2. Mie G: *Ann. Physik*, 1908, 25, 377.
3. Debye P: *Ann. Physik*, 1915, 46, 809.
4. Berne BJ and Pecora R: *Dynamic Light Scattering*, John Wiley and Sons, Inc., New York, 1975.
5. Bragg WH and Bragg WL: *The Crystalline State*, Bell, London, 1933.
6. Guinier A and Fournet G: *Small Angle Scattering of X-Rays*, John Wiley and Sons Inc., New York, 1955.
7. Synchroton Radiation Research, *Annual Report*, Daresbury Laboratory, 1985.
8. Chadwick G: *Nature*, 1932, 129, 132; Proc. Roy. Soc., 1932, A136, 692.
9. Mitchell DP and Powers PN: *Phys. Rev.*,, 1936, 50, 486.
10. Neutron Research Facilities at the ILL High Flux Reactor, Institut Laue-Langevin, Grenoble, France, 1983.
11. Rowell RL and Kidnie KM: *Science and Technology of Polymer Colloids*, NATO ASI Series E67, 1983, vol.II, 264.
12. Bacon GE: *Neutron Scattering in Chemistry*, Butterworths, London, 1977.
13. Espenscheid WF; Kerker M and Matijevic E: *J. Phys. Chem.*, 1964, 68, 3093.
14. Markovic I and Ottewill RH: *Colloid and Polymer Sci.*, 1986, 264, 65.
15. Goodwin JW and Ottewill RH: to be published.
16. Goodwin JW; Ottewill RH; Harris NM and Tabony J: *J.Colloid Interface Sci.*, 1980, 78, 253.
17. Goodwin JW; Hearn J; Ho CC and Ottewill RH: *Colloid and Polymer Sci.*, 1974, 252, 464.
18. Markovic I; Ottewill RH; Cebula DJ; Field I and Marsh JF: *Colloid and Polymer Sci.*, 1984, 262, 648.
19. Cebula DJ; Goodwin JW; Ottewill RH; Jenkin G and Tabony J: *Colloid and Polymer Sci.*, 1983, 261, 555.
20. Ottewill RH and Richardson RA: *Colloid and Polymer Sci.*, 1982, 260, 708.
21. Kratky O: *Nova Acta Leopoldina* NF55, Nr.256, 1983.
22. O'Reilly JM; Melpolder SM; Fisher LW; Wignall GD and Ramakrishnan VR: private communication.
23. Harris NM; Ottewill RH and White JW: in *Adsorption from Solution*, Academic Press, London, 1983, 139.
24. Ottewill RH: Ber. Bunsenges. *Phys. Chem.*, 1985, 89, 517.
25. Cebula DJ; Goodwin JW; Jeffery GC; Ottewill RH; Parentich A and Richardson RA: *Faraday Discuss. Chem. Soc.*, 1983, 76, 37.
26. Markovic I; Ottewill RH: Tadros Th F and Underwood SM: Langmuir, 2, 625, 1986.

THE CHARACTERIZATION OF POLYMER COLLOIDS BY FLUORESCENCE QUENCHING TECHNIQUES

Mitchell A. Winnik

Department of Chemistry
University of Toronto
Toronto, Ontario
M5S 1A1
Canada

and

Melvin D. Croucher

Xerox Research Centre
2660 Speakman Drive
Mississauga, Ontario
L5K 2L1
Canada

1. INTRODUCTION

Analytical techniques based upon fluorescence and phosphorescence spectroscopy have seen numerous applications in the biological area[1], in the study of micelles and bilayer systems[2], and in certain areas of polymer science[3]. It is only recently, however, that these luminescence techniques have been applied to the study of colloidal dispersions. It is clear from the work already published that many kinds of information can be obtained in a straightforward manner from these kinds of experiments which would be difficult to obtain with other techniques. In addition, some kinds of information which these techniques can provide are simply not accessible at this time by other methods.

Since luminescence techniques are new to the colloid field, and the number of publications is still relatively small, one cannot yet pinpoint the full scope or limitations of these methods. This is an interesting situation for scientists working in the colloid field. The techniques themselves are well documented[4]. Good equipment is commercially available and reasonable in price. While there are the normal pitfalls for anyone working for the first time with a new technique, progress, for the moment, is limited only by the imagination of the scientists who choose to work with these techniques.

The subject of this paper is limited to luminescence techniques applied to polymer colloids. Within this subject, emphasis will be focussed on fluorescence quenching experiments as applied to non-aqueous dispersions. It is important to point out that there have been some interesting applications of luminescence techniques to study aqueous dispersions formed from polymerized microemulsions[5], and colloidal inorganic dispersions, particularly clays. The reader is referred to the recent work of J.K. Thomas[6a], and of H. Van Damme[6b].

A quenching process is defined as an interaction of an excited chromophore (B^*) with a ground state species (Q) which leads to a decrease in the emission intensity from B^* and increase in its decay rate. This is a phenomenological definition, and encompasses all quenching mechanisms, including electron transfer, energy transfer, excimer and exciplex formation, and stimulated singlet-triplet interconversion[7]. Each of these processes is normally characterized by a rate constant which, for immobile B^* and Q, is a function of the distance between them. In thinking about these processes, it is useful to define a characteristic distance R_o where the quenching rate equals the decay rate of unquenched B^*. For some processes such as energy transfer by the dipole coupling (Forster) mechanism[7], R_o can be calculated theoretically from the spectroscopic properties of B and Q. For other processes such as energy transfer by the exchange (Dexter) mechanism[8a] and electron transfer[8b], R_o can be determined experimentally. Various quenching processes and their range of R_o values are summarized in Table 1. These techniques are particularly useful for examining the transition zone between discrete polymer phases. From a strategic point of view, one uses the recipe for synthesis of a material to introduce a derivative of B covalently into one of the discrete phases. One now has a number of alternatives, for example in the study of polymer blends[9], one can introduce a derivative of Q into the other phase. Communication between B^* and Q can occur only at the interface between the phases.

TABLE 1. Bimolecular Excited State Quenching Processes.

Interaction mechanism	Effective distance[a,b] (in Å)
1. Energy transfer by dipole coupling	10 to 100
electron exchange	4 to 15
reabsorption	as far as emission reaches
2. Electron transfer	4 to 25
3. Exciplex formation	4 to 15
4. Excimer formation	ca. 4
5. Non-emissive self-quenching	4 to 15
6. Heavy atom effect	ca. 4
7. Chemical bond formation	ca. 2 to 4

[a]The minimum interaction distance is arbitrarily taken to be 4 Å except where new chemical bonds are formed.
[b]Each pair of chromophores, for each interaction mechanism, has its own characteristic distance R_o at which the interaction rate equals the decay rate of the unquenched excited state.

Alternatively, one can choose a quencher that has a different solubility or diffusivity in the different phases. One might call this the mobile quencher approach. We have made use, for example, of the pronounced difference in oxygen sorption in hydrocarbon-like rubbery long chain methacrylate polymers compared to that in glassy poly(vinyl acetate) (PVAc)[10]. By labelling the latter with a fluorescent dye, and exposing the system to air or oxygen, quenching measurements gave important information on the interface between the two polymers.

These kinds of experiments provide information about polymer morphology. In the case of sterically stabilized dispersions of polymer colloids, which are composed of two or more normally incompatible polymers, these techniques allow one to determine the location of the various polymer materials within the particle. As in the case mentioned above, a particularly useful kind of information which one can obtain concerns the sharpness or diffuseness of the interface between the different polymer phases.

Sorption processes can be studied under a variety of circumstances. If the sorbent itself is an effective quencher (e.g. ketones, CCl_4, CS_2), uptake of the sorbent leads to a decrease in luminescence intensity[11]. Alternatively, the sorbent may change the solubility or diffusivity of oxygen in the system. Changes in oxygen quenching provide an indirect measure of changes occurring within the particle itself.

More recently we have learned how to use fluorescence quenching techniques to study the collapse of the steric barrier chains during particle flocculation[12]. This kind of experiment is quite straightforward. If the stabilizer contains both donor B and acceptor A chromophores for energy transfer, collapse of these chains brings these groups closer together. The decrease in polymer coil dimensions is accompanied by a large increase in the efficiency of energy transfer. One can observe the initial stages of flocculation. In addition, one can study the later stages of flocculation to gain information about the conformation and dynamics of the steric barrier within the floc.

In the sections that follow, three particular types of experiments are described that provide information about polymer colloid systems. The first employs oxygen quenching to elucidate the structure of the interface in a dispersion of PVAc stabilized by poly(2-ethylhexyl methacrylate) (PEHMA). The second involves energy transfer experiments which provide information on the morphology of a sterically stabilized poly(methyl methacrylate) (PMMA) particle. The third employs energy transfer experiments to study molecular details of solvent-induced flocculation in a sterically stabilized PVAc particle system.

2. INTERPHASE STRUCTURE IN PVAc PARTICLES

The study of labelled particles begins with their synthesis. It is normally most appropriate, for sterically stabilized polymer particles, to prepare independently materials labelled in the particle core and in the stabilizer layer. These reactions are described in Schemes I and II for the example of PVAc particles stabilized by PEHMA. Central to the design of the experiments is the choice of phenanthrene (Phe) as the fluorescent dye to be incorporated into each phase. Phe fluorescence is relatively insensitive to its environment, and it has little tendency to form excimers. As a consequence, its fluorescence decay I(t) normally has a simple exponential form.

280

SCHEME I

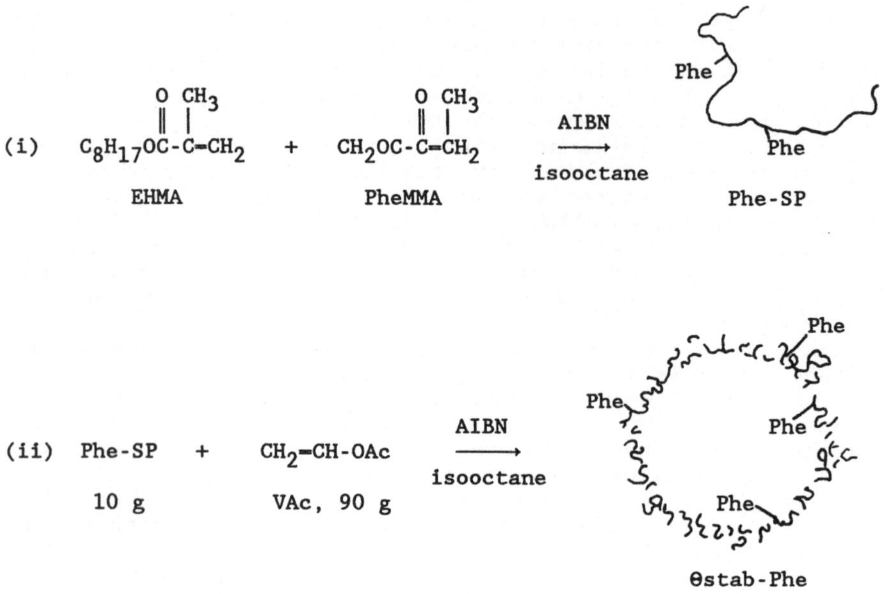

(i) $C_8H_{17}\overset{\overset{O}{\|}}{O}C-\overset{\overset{CH_3}{|}}{C}=CH_2$ + $CH_2\overset{\overset{O}{\|}}{O}C-\overset{\overset{CH_3}{|}}{C}=CH_2$ $\xrightarrow[\text{isooctane}]{\text{AIBN}}$

       EHMA   PheMMA   Phe-SP

(ii)   Phe-SP   + $CH_2=CH-OAc$ $\xrightarrow[\text{isooctane}]{\text{AIBN}}$

       10 g   VAc, 90 g   θstab-Phe

SCHEME II

(i)   EHMA   +   AIBN $\xrightarrow[\text{isooctane}]{}$ PEHMA

(ii)   PEHMA +   VAc   +   PheMMA $\xrightarrow[\text{isooctane}]{}$ θcor-Phe

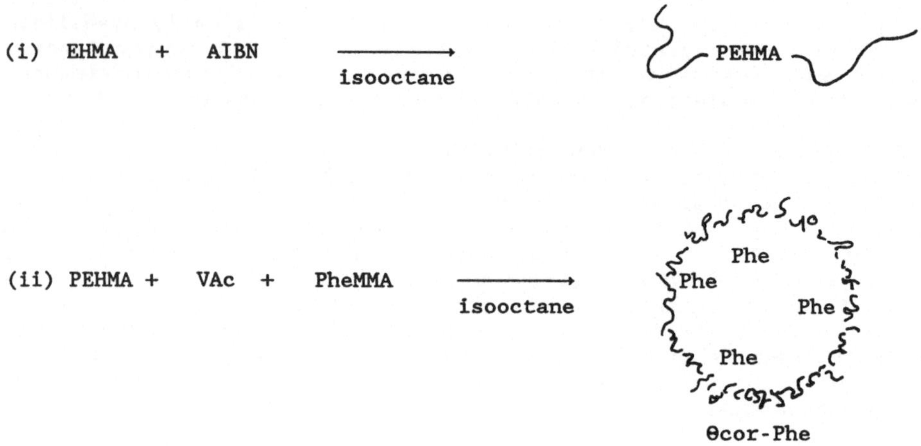

This property contributes two important features to the experiment: First, the fluorescence decay time τ^o (in the absence of quencher) is a well-defined quantity, facilitating interpretation of steady state fluorescence experiments. Second, in complex systems, non-exponential decay traces may be observed.

These can be interpreted in terms of a non-uniform distribution of quenchers in the system. In principle, this distribution function can be extracted from the fluorescence decay data.

In the synthesis of the stabilizer-labelled particles described in Scheme I, there is a good match of the reactivity ratios of the EHMA and PheMMA monomers. Under the conditions of reaction[1], the amount of PheMMA is small enough (1%) that one can assume that it is randomly distributed in the polymer. In the synthesis described in Scheme II, there is a serious mismatch of reactivity ratios between PheMMA and VAc that favors formation of PheMMA blocks. We were still able to overcome the reactivity ratio difficulty by using only a tiny trace (100 ppm) of PheMMA in the reaction mixture. Particle diameters here were ca. 300 nm [13].

Information about the particles is obtained by comparing four systems[10]. First, in order to establish a benchmark, one looks at a small molecule, here 9-phenanthrylmethyl pivalate, 1, as a model for the chromophore in the polymer. Second, one examines the copolymer Phe-SP (the labelled prestabilizer) to establish the behavior of the polymer-bound dye. These form the basis for interpreting experiments on the two particle systems labelled in the stabilizer (Θstab-Phe) and in the core (Θcor-Phe).

In a well-behaved system, the fluorescence intensities (I) and lifetimes (τ) follow the Stern-Volmer equation 1, where the superscript o implies the value in the absence of added quencher. The derivation of this equation presumes uniform concentration of quencher in solution, and a unique second order rate constant k_q to describe the quenching process.

$$\frac{I^o}{I} = \frac{\tau^o}{\tau} = 1 + k_q \tau^o (Q) \qquad (1)$$

Exposure of solutions of the model compound 1 and the copolymer Phe-SP to various oxygen concentrations causes both I and τ to decrease. The data in Figure 1 follow the Stern-Volmer relationship. Oxygen quenching is diffusion controlled. The decrease in slope for the copolymer is due to its smaller diffusion coefficient compared to 1. When the stabilizer-labelled particles are exposed to oxygen, the data also follow equation 1. This result is somewhat unexpected, since I(t) here, in the presence of oxygen, deviates from an exponential form, implying a distribution of oxygen solubilities and mobilities in the stabilizer layer of the particle. The data recover their simple form when one calculates mean lifetimes $\langle \tau \rangle$ from the I(t) measurements and plots them according to equation 1.

One can conclude from the particle data in Figure 1 that the entire region containing the stabilizer polymer is highly permeable to oxygen. Oxygen quenching is almost equally effective for the particle-bound stabilizer as for the free prestabilizer in solution. These data are too simple to provide deeper insights into the structure of the particle.

A very different situation is found for the core-labelled particles. By DSC the dried particles behave exactly like PVAc powder, with an identical glass transition temperature (T_g). From what is known about oxygen solubility and oxygen diffusion in PVAc, one would expect little or no detectable fluorescence quenching of phenanthrene embedded in a pure PVAc matrix. Its fluorescence lifetime is too short (44 ns).

Exposure to oxygen of the core-labelled particles, dispersed in cyclohexane, caused a pronounced decrease in both I and $<\tau>$. The observation of strong quenching is surprising. The curvature in the data plotted according to equation 1 provides more information, Figure 2a, and suggests that not all of the Phe groups in the system are readily quenched. Biologists have observed a similar phenomenon in globular proteins, where tryptophan groups inside the protein are protected against fluorescence quenching by oxygen, whereas those on the surface are exposed to quencher (4). Such systems follow equation 2, where f_a represents the fraction of fluorescent groups able to be quenched.

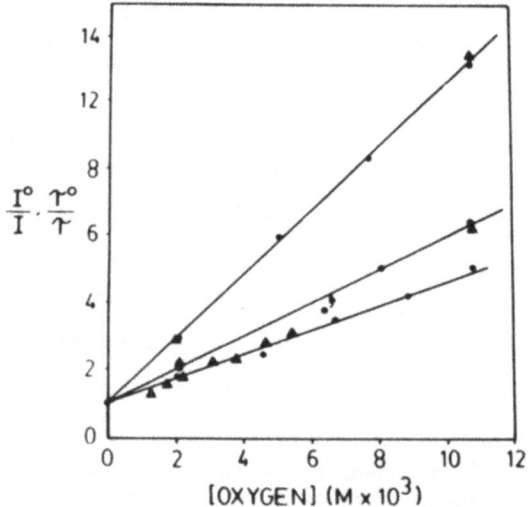

Figure 1. A plot of I°/I (●) and τ°/τ (▲) vs oxygen concentration in the solvent (cyclohexane) for: top line, 9-phenanthrylmethyl pivalate (1); middle line, Phe-SP; bottom line, PVAc particles Θstab-Phe. For the particles, $I(t)$ is non-exponential in the presence of O_2. Here the y-axis represents $\tau^{\circ}/<\tau>$.

$$\frac{I}{I^{\circ} - I} = \frac{1}{f_a} + \frac{1}{f_a k_q \tau^{\circ}(Q)} \qquad (2)$$

When our data are replotted in Figure 2b according to equation 2, we obtain a straight line with an intercept of ca. 2. Hence 50% of the Phe groups in these particles are protected against oxygen quenching. Those that are exposed to oxygen are quenched with great effectiveness.

How can these results be explained? One begins by assuming that the Phe serves as a marker for the location of the PVAc polymer, which comprises ca. 93% of the dry volume of the particle. Our observations indicate that half of the PVAc has been transformed into a phase with high oxygen permeability. This new phase must be an interphase containing significant amounts of PEHMA. The presence of stabilizer polymer in the interphase would promote penetration of cyclohexane into these regions of the particle.

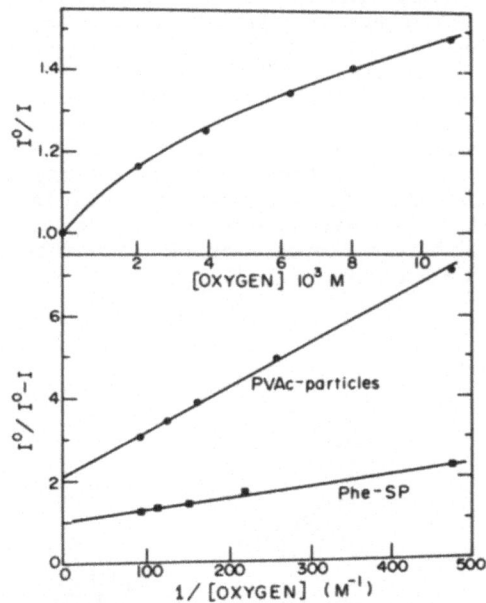

Figure 2. Top: a plot of I^o/I vs. solvent concentration of O_2 for θcor-Phe. Bottom: a plot of $I^o/(I^o - I)$ vs. $1/(O_2)$ for the same data. The lower line is a replot of the data for Phe-SP from Figure 1. Note that the y-intercept is 1.0.

Three models for particle morphology are presented in Figure 3. All three models incorporate a surface covering of PEHMA to provide colloidal stability. The first model is the classical core-shell model. It cannot explain the observation that half of the PVAc has been transformed into a phase of high oxygen permeability. This model can be modified to include an interphase surrounding a core of PVAc. Alternatively, one can imagine a more interesting structure of the interpenetrating network (IPN) type[14], in which small microphases of essentially pure PVAc are surrounded by an interphase containing PEHMA. Both models are consistent with the DSC and the oxygen fluorescence quenching data. To distinguish between them, one needs a probe which can "see" across the phase boundary to sense the dimensions of the PVAc domains.

3. MICROPHASE STRUCTURE OF PMMA PARTICLES

Energy transfer by the dipole coupling mechanism can occur over distances as large as 80Å to 100Å, depending upon the choice of B and A[7]. In this way an energy acceptor (A) in the mobile phase can interact with a B^* buried within the glassy phase. Such experiments provide information complementary to those obtained by oxygen quenching. The important considerations are the R_o value characterizing a particular B/A pair and the observation [1,4] that energy transfer can be detected experimentally for B/A separations up to about $2R_o$.

We have prepared PMMA particles (2μm diameter) sterically stabilized by polyisobutylene (PIB, 9 mol%), which have been labelled with naphthalene (N) groups in the PMMA component[15]. Energy transfer experiments were carried out in which the energy acceptor was anthracene (An), which was simply dissolved in the continuous phase (cylohexane). For this pair of chromophores, the R_o value is 23Å.

The most important result of these experiments is that most if not all of the N groups in the particle could be quenched by energy transfer to An. Since energy transfer can occur over at most ca. 45Å, and it would take years for the An to diffuse into the glassy PMMA domains, this result indicates that the PMMA domains are sufficiently small that most of the interior points are within 45Å of a phase boundary. This result[15] is consistent only with a microphase or IPN structure as suggested by the third model shown in Figure 3.

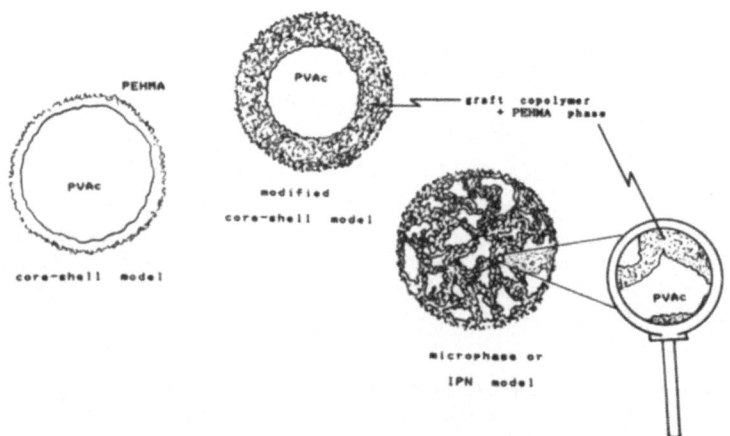

Figure 3. Three models for the internal structure of sterically stabilized polymer particles. The modified core-shell and microphase models are drawn to emphasize the large interphase content found the PVAc particles described in the text.

In order to confirm this structure, we developed a novel approach based upon X-ray scattering[16]. Tetraphenyl lead is soluble in PIB but can diffuse only slowly into PMMA below its T_g. The particles were treated with a dilute solution of Ph_4Pb in cyclohexane, and the solvent was removed under vacuum. Here Ph_4Pb serves as a marker for the PIB phase.

The X-ray scattering experiments on this material confirmed an IPN-type structure, with a Pb-containing PIB phase penetrating throughout the particle adjacent to a PMMA phase in which Pb was absent.

4. FLOCCULATION OF PVAc PARTICLES

The classical conception of flocculation for sterically stabilized particle dispersions is that changes in temperature or solvent composition neutralize the normally repulsive interaction between steric barrier chains on separate particles, permitting aggregates to form[17]. If the solvent becomes a poor solvent for the steric barrier, there will be a net attractive interaction between these polymers. Flocculation will be preceded or accompanied by conformational changes within this polymer.

In order to examine this process, we prepared PVAc particles very similar to those described above, except that the PEHMA steric barrier contained a statistical distribution of small amounts of N (0.7 mol%) and An (2.0 mol%) groups. A control material contained a similar amount of only the N substituents. We then measured the extent of energy transfer from N* to An as the solvent composition was changed from pure cyclohexane to mixtures of cyclohexane and ethanol. The particle dispersions were indefinitely stable in solvents containing up to 67% ethanol by volume. We refer to the particles as metastable at 70% ethanol since flocculation occurred (apparently irreversibly) over a period of three weeks. Above 72% ethanol, visible flocs appeared in a matter of seconds to minutes.

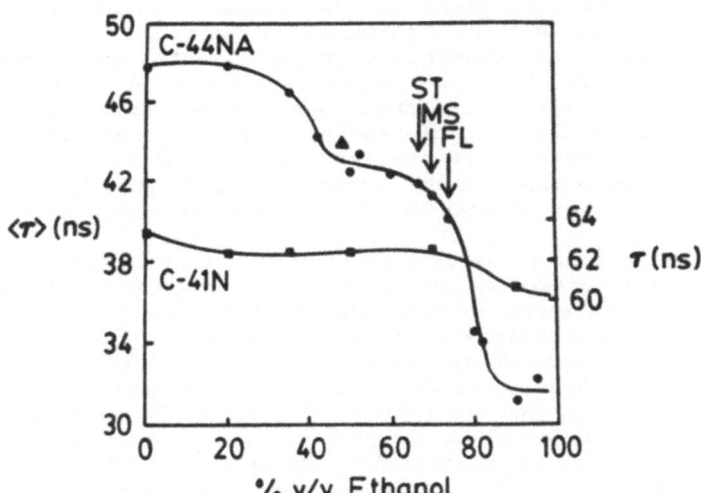

Figure 4. Plots of τ and <τ> vs. solvent composition for two colloidal PVAc dispersions. The sample C-41N contains 1.5 mol% N groups in the PEHMA chain. I(t) here is exponential (τ values, right hand axis). C-44NA contains 0.7 mol% N and 2.0 mol% An groups in the PEHMA chain (<τ> values, left hand axis). The arrows labelled ST, MS, and FL refer, respectively, to stable, metastable, and rapidly flocculated dispersions.

In these experiments, the fluorescence decay profiles of N^* were measured. Because the decay forms were non-exponential, we calculated mean decay times. These are plotted as a function of solvent composition in Figure 4. The $<\tau>$ values of the sample (C-44N in Figure 4) containing only naphthalene groups are relatively unaffected by changes in solvent or by the flocculation process. When flocculation is very fast, we actually had to tilt the cell to make measurements on the floc itself.

In the sample (C-44NA) containing both N and An groups, energy transfer is apparent even in cyclohexane. Here sensitized An* emission can be observed in the fluorescence spectrum, and the $<\tau>$ value of N^* is reduced to 48 ns from its value of 63 ns in the C-41N sample.

Two regions of change are evident in the energy transfer experiment. There is a small decrease in $<\tau>$ in the solvent composition range of 40% to 50% ethanol, and a more dramatic change in the range corresponding to particle flocculation. These indicate a substantial increase in energy transfer in the more polar media.

Both static and dynamic processes contribute to the rate of energy transfer, since local motions within the stabilizer may be fast enough to allow N^* - An distances to change during the lifetime of N^*. At elevated ethanol concentrations, particularly at the flocculation point, static factors, we believe, make the dominant contribution to the increase in energy transfer. This could result from collapse of the stabilizer chains in solvents of unfavorable composition, which would lead to a decrease in N/An separations. When these experiments are repeated on a mixture of particles, one labelled in the stabilizer with N, the other with An, no energy transfer is observed! As little as 1% energy transfer could be detected. These experiments lead to the remarkable conclusion that solvent-induced flocculation of NAD particles is accompanied by contraction of the dimensions of the polymer chains in the steric barrier. No interparticle chain interpenetration occurs.

In this kind of experiment, one would like to have a material in which only the stabilizer on the surface contained the fluorescent labels. Here we know that large amounts of the stabilizer are incorporated within the particle and form an interphase with the PVAc. Consequently, the changes in energy transfer that we observe contain contributions from changes in the conformation and dynamics of the steric barrier and also from changes in the stabilizer incorporated into the interphase.

The importance of these experiments is that it demonstrates for the first time that fluorescence quenching experiments can be applied to the study of the flocculation process. They provide, for example, the first molecular-level observations on solvent-induced NAD flocculation, with the striking result that no interpenetration of the steric barrier chains occurs between adjacent particles. There is much more information here than we have reported, and these kinds of experiments can be applied in a rather straight-forward way to the process of film formation. It seems, therefore, that if one is interested in the molecular details of particle-particle interactions, fluorescence quenching studies of labelled particles can provide a wealth of new information.

5. SUMMARY

Fluorescence quenching techniques provide a variety of different experiments which can be used to study structure and transport phenomena across phase boundaries in polymer colloid systems. Detailed studies of two non-aqueous dispersions indicate that the particle morphology is not of the core-shell type, but rather a more intricate interpentrating network of core and stabilizer polymers is generated during particle formation. Fluorescence quenching experiments also offer great promise as a means of studying the molecular details of stabilizer conformation and dynamics in intact dispersions as well as during particle flocculation.

REFERENCES

1. (a) Chen RF and Edelhoch H(Eds.): *Biochemical Fluorescence*, Marcel Dekker, New York (1975).
    (b) Schlessinger J and Elson EL: *Methods Expt. Phys.* 30, 182 (1982).
2. (a) Kunitake T; Tawaki S and Nakashima N: *Bull. Chem. Soc.* Japan, 56, 3235 (1983).
    (b) Lianos P; Zana R: *J. Phys. Chem.*, 84, 3339 (1980).
    (c) Turro NJ; Baretz BH; and Kuo P-L: *Macromolecules* 17, 321 (1984).
3. (a) Winnik MA(ed): *Photophysical and Photochemical Tools in Polymer Science*, D. Reidel, Dordrecht, Netherlands, (1987).
    (b) Guillet JE: *Polymer Photochemistry and Photophysics*, Cambridge University Press, Cambridge, UK, (1985).
4. Lakowicz R: *Principles of Fluorescence Spectroscopy*, Plenum Press, New York, (1983).
5. Atik S; Thomas JK: *J. Am. Chem. Soc.*, 105, 4515 (1983); and earlier papers.
6. (a) Della Guardia RA; Thomas JK: *J. Phys. Chem.*,87, 3550 (1983).
    (b) Habti A; Keravis D; Levitz P and van Damme H: *JCS Far. Trans 2*, 80, 67 (1984).
7. Birks JB: *Photophysics of Aromatic Molecules*, Wiley-Interscience, New York (1971).
8. (a) Inokuti M and Hirayama F: *J. Chem. Phys.*, 43, 1978 (1965).
    (b) Miller JR; Beitz JV and Huddleston RK: *J. Am. Chem. Soc.*, 106, 5057 (1984).
9. Morawetz H: *Ann. N.Y. Acad. Sci.*, 366, 404 (1981).
10. Egan LS; Winnik MA and Croucher MD: *Polym. Eng. Sci.*, 26, 15 (1986).
11. Winnik MA; Egan LS; Owens SM and Ottewill RH: *Anal. Chim. Acta.*, submitted.
12. Winnik MA; Egan LS and Croucher MD: *Macromolecules*, in press 1986.
13. Winnik MA; Egan LS and Croucher MD: *J. Polym. Sci. Polym. Chem. Ed.*, 24, 1895 (1986).
14. Sperling LH: *Polym. Eng. Sci.*, 24, 1 (1984).
15. Pekcan O; Winnik MA and Croucher MD: *J. Polym. Sci. Polym. Lett.*, 21, 1011 (1983).
16. Winnik MA; Williamson B and Russell TR: Submitted.
17. (a) Barrett KEJ: *Dispersion Polymerization in Organic Media*, Wiley-Interscience, London,(1975).
    (b) Napper DH: *Polymeric Stabilization of Colloidal Dispersions*, Academic Press, New York, (1983).

DIELECTRIC SPECTROSCOPY OF MODEL POLYSTYRENE COLLOIDS

R.M. Fitch, L.S. Su and S.L. Tsaur

S.C. Johnson & Son, Inc.
Louis Laboratory
Racine, Wisconsin 53403

1. INTRODUCTION

When an oscillating electrical field of varying frequency is applied to an aqueous colloid, the (complex) impedance will be comprised of two parts, one in phase and one out of phase. These are the real and imaginary components. The processes involved will be (a) charge transfer at the electrodes, (b) diffusion or transference of ions, (c) distortions of the electrical double layers at the various interfaces in the system, and (d) orientations of the permanent dipoles present both in the continuous and discontinuous phases. As the frequency is scanned, various characteristic relaxations will be observed corresponding to these processes. Generally the data are analyzed in terms of equivalent electrical circuits comprised of resistors and capacitors connected in series and parallel, the characteristics of which indicate which processes produce which effects. In aqueous colloids comprised of dielectric particles it is the motions of the counterions in the Stern layer and diffuse layer surrounding the particle under the influence of the applied oscillating field which are of particular interest.

Recent advances in instrumentation have made the acquisition of data relatively rapid, as a result of computer control of the system along with extensive data analysis capabilities. Prior to this, each data point at each frequency was obtained by balancing a bridge; it was laborious and time-consuming work. Now a large amount of information can be obtained on a sample in only a few minutes time, scanning a broad frequency range from approximately 0.001 to 10,000 Hz.

2. REASONS FOR THIS WORK

Tsaur and Fitch[1] recently reported on the coagulation behavior of polystyrene colloids carrying strong electrolyte surface groups (sulfonate). In each series the particle size was the same, and the surface charge density was systematically varied. The particle diameters of each series were different. The coagulation kinetics in the presence of indifferent electrolytes showed that the calculated value of the surface potential, using the theory of Reerink and Overbeek, rose as surface charge increased. However, the critical coagulation concentration (CCC) rose and then fell with increased concentration of 2:2 electrolyte, but simply rose with increased concentration of 1:1 electrolyte. This kind of behavior is contrary to the predictions of the DLVO theory[2], but directionally consonant with the discrete ion effect theory of Levine, Mingins and Bell[3]. The results might also be explained by an electrostatic component to the attractive potential, as suggested by Ise and Sogami, although their theory deals with charged particles at some equilibrium distance of separation[4]. Patey has also suggested that an electrostatic term enters the attractive part of interparticle interactions ón the basis of his hypernetted chain theory[5]. Both of these latter theories predict an increasing attraction as surface charge density increases, contrary to DLVO.

There is a clear need for independent measurement of surface charge and potential. The former is now obtained by titration with alkali of ion-exchanged colloids in the acid form[6]. The latter is estimated, for example, by means of electrophoretic mobility[7], coagulation kinetics[8], time-average light scattering[9], and neutron scattering[10]. In all cases in the past there has been the problem of a lack of sufficient model colloids with independently variable physical characteristics to obtain statistically significant results. In the light- and neutron-scattering experiments cited above, for example, elegant measurements and calculations were performed on well characterized material, but only a single latex was examined.

Dielectric spectroscopy provides a means of obtaining rapidly an enormous amount of valuable information in a short time. The Fitch-Tsaur model colloids provide samples in which surface charge density and particle size are independent variables. To our knowledge, the work of Lyklema and coworkers is the only recent work employing the dielectric technique[11].

3. EXPERIMENTAL
3.1. Synthesis and Purification of Model Polystyrene Colloids

1. The surface-active monomer, sodium sulfophenyl vinyl benzyl ether (SSPVBE) was synthesized from vinyl benzyl chloride and sodium 4-hydroxybenzene sulfonate in aqueous alkali. It was crystallized from aqueous solution.

2. Model polystyrene colloids were synthesized in a two stage batch manner in which the particles of very uniform size were produced in the first stage, and the surface charge was determined by the amount of SSPVBE added in the second stage. Bisulfite/silver amine (BAg), which introduces only sulfonate end groups[12], was used as initiator. The resulting latexes were filtered, steam stripped to remove monomer, and dialyzed against D.I. water for 4 days. Subsequently they were treated with highly purified mixed bed ion exchange resin three times. The results are given in Table 1.

3.2. Dielectric Spectrometry
3.2.1. **Dielectric Spectrometer.** The dielectric spectrometer consists of three major components: the Solartron™ Frequency Response Analyzer (FRA), the Solartron™ 1186 Electrochemical Interface (ECI), and an HP9845B micro-computer for measuring the frequency-dependent complex a.c. impedance of a system (Figure 1). The Solartron™ 1172 FRA has three main functions: a programmable generator to provide a signal; a correlator to filter out harmonic oscillations and random noise as a function of frequency, and a display to present data points in Cartesian coordinates. The Solartron™ 1186 ECI, in conjunction with the FRA, controls the potential difference between electrodes and carries out a.c. characteristic measurements. The whole experiment is operated from the HP9845B keyboard, and data are stored on, and retrieved from tape for later analysis.

TABLE 1. Characteristics of Model Polystyrene Colloids

	C1	C2	C5	C4
Yield (%)	91.0	94.0	98.4	94.2
$\bar{D}_n$ (nm)	208	206	198	208
P	0.0335	0.0050	0.0070	0.0070
SCD ($\mu C/cm^2$)	6.6	12.5	16.5	21.1

Notes:

$\bar{D}_n$ number-average particle diameter, determined
        by quasi-elastic light scattering (QELS)

P   is the polydispersity parameter:

$$P = \frac{\bar{R}^6 \; \bar{R}^4}{\bar{R}^5{}^2} - 1$$

SCD   is surface charge density.

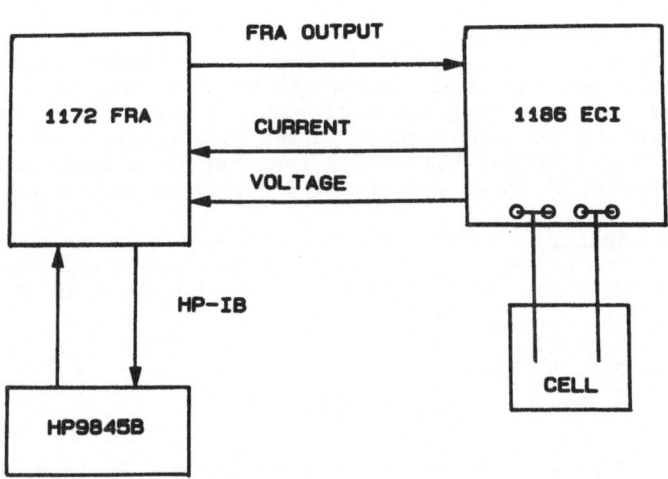

Figure 1: *Schematic diagram of colloid measuring apparatus.*

3.2.2. <u>Cell</u>. The cell is an assembly of two identical 316 stainless steel electrodes, nitrogen inlet and outlet, a sample container securely sealed with an O-ring to the Teflon™ stopper (Figure 2).

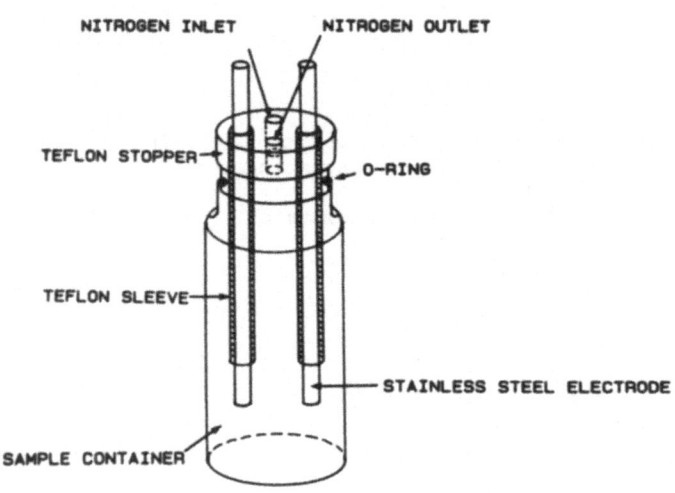

Figure 2: *Schematic diagram of two-electrode colloid cell.*

3.2.3. <u>Experimental Procedure</u>. Electrodes were polished with a fine silicon carbide emery cloth or fine steel wool, rinsed with milli-Q water and dried with acetone and nitrogen gas before each run. The entire cell assembly was continuously purged with nitrogen gas of high purity during measurements. The cell's response to a small amplitude sinusoidal potential of 10 mV was measured over a frequency range from 10^{-2} to 10^4 Hz.

3.2.4. <u>Data Analysis</u>. The measured frequency-dependent total impedance $Z_T(\omega)$ is represented by an equivalent electrical circuit of the modified Randel type, which is theoretically expressed in eq. [1]:

$$Z_T(\omega) = R_o + \frac{R_\infty}{1+(j\omega C R_\infty)^{1-\alpha}} \tag{1}$$

where α is the Cole-Cole distribution constant and R_o is a solution resistance connected in series with a subcircuit comprised of a resistance, R_∞, and an interfacial capacitance, C, coupled in parallel. Eq. [1] is a complex function, and its real and imaginary components are expressed in eqs. [2] and [3], respectively:

$$Z'(\omega) = R_o + \frac{R_\infty + \omega^\beta C^\beta R_\infty^{1+\beta} \cos(\beta\pi/2)}{1 + (\omega C R_\infty)^{2\beta} + 2(\omega C R_\infty)^\beta \cos(\beta\pi/2)} \tag{2}$$

$$Z''(\omega) = \frac{\omega^\beta C^\beta R_\infty^{1+\beta} \sin(\beta\pi/2)}{1 + (\omega C R_\infty)^{2\beta} + 2(\omega C R_\infty)^\beta \cos(\beta\pi/2)} \tag{3}$$

where $\beta = 1-\alpha$ and eq. [1] describes a complete semicircle on a complex plane, with its center lying on the real axis at a distance of $R_o + 1/2$ R_∞ from the origin when $\beta = 1$ or $\alpha = 0$.

For a successful least squares fitting of eq. [1], the experimental $Z''(\omega)$ should approach zero asymptotically at two limiting frequencies. The experimental $Z''(\omega)$ values of the system diverge at frequencies near 10^{-2} Hz. Therefore, eq. [1] is analytically inverted into a corresponding frequency-dependent total admittance function, $Y_T(\omega)$, to ensure perfect convergence of $Y''(\omega)$ in this frequency region. R_o, C, α, and f* are then obtained by means of non-linear complex least squares fitting of $Y_T(\omega)$. f* is defined as a dielectric relaxation frequency which is theoretically related to the frequency at which a maximum dielectric loss, $\varepsilon''(\omega)$, is attained.

3.2.5. <u>Dielectric Spectra</u>. The complex dielectric permittivity, $\varepsilon*(\omega)$, is calculated by substituting experimentally obtained a.c. impedance data as a function of frequency into eq. [4],

$$\varepsilon*(\omega) = [\omega C_o(\omega) \times |Z_T(\omega)|^2]^{-1}[Z''(\omega) + jZ'(\omega)] \tag{4}$$

or likewise substituting experimental a.c. admittance data as a function of frequency into eq. [5],

$$\varepsilon*_T(\omega) = -jY_T(\omega) [\omega Co(\omega)]^{-1} \tag{5}$$

where $C_o(\omega)$ is a frequency-dependent cell capacitance and was arbitrarily set equal to 10^5 for the calculation of relative dielectric permittivity data.

3.3. <u>Electrophoretic Mobility</u>

Measurements were kindly made by Dr. D. Fairhurst of the PenKem Company on the PenKem[TM] 3000, an instrument that automatically obtains the distribution of electrophoretic mobilities in each sample. The latexes were taken from an earlier series of PS model colloids prepared in almost the same fashion as the present series[1]. For the present study, even though the colloids were monodisperse in size, it was conceivable that surface charge was polydisperse.

4. RESULTS AND DISCUSSION
4.1. <u>Electrophoretic Mobilities</u>

Typical mobility distributions are shown in Figures 3 and 4. It is clear from these that there is only a single population according to surface (ζ) potential. The electrophoretic mobilities are unimodal and narrow.

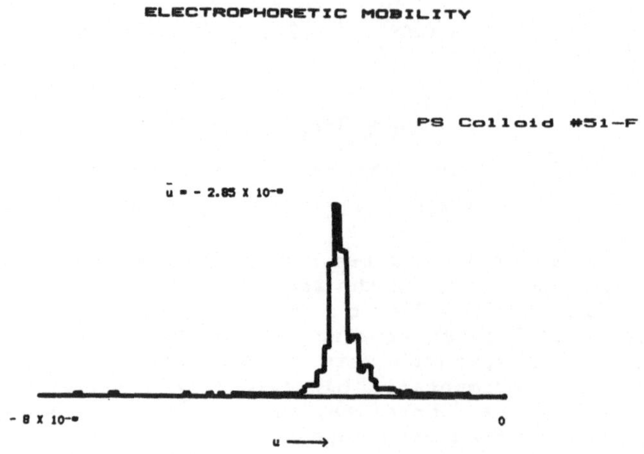

294

ELECTROPHORETIC MOBILITY

PS Colloid #51-F

$\bar{u} = -2.85 \times 10^{-8}$

-8×10^{-8} 0

u ⟶

Figure 3: *Electrophoretic Mobility Distribution of PS Colloid 51-F.*

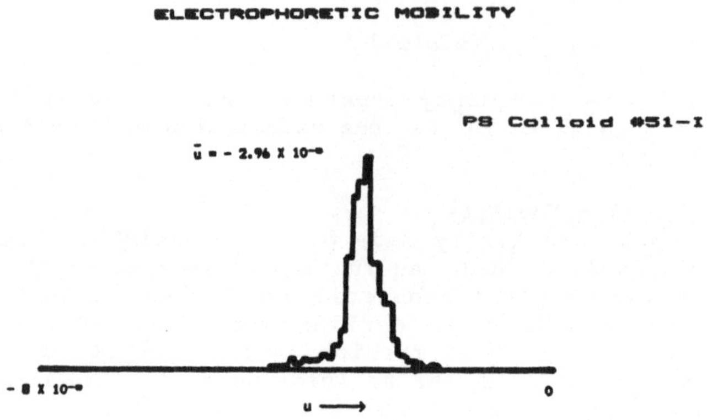

ELECTROPHORETIC MOBILITY

PS Colloid #51-I

$\bar{u} = -2.96 \times 10^{-8}$

-8×10^{-8} 0

u ⟶

Figure 4: *Electrophoretic Mobility Distribution of PS Colloid 51-I.*

4.2. Counterion Concentrations

When the imaginary component of the frequency-dependent total complex admittance, $Y''(\omega)$, is plotted as a function of the real part, $Y'(\omega)$, for one of our polystyrene colloids, a perfect semicircle is obtained. The position of the low frequency end of the curve near the origin indicates low ionic content, and the circularity, with its center lying near the real axis, indicates a narrow particle size distribution. The radius of the curve or the intercept of the curve on the real axis at high frequency is a direct function of the surface charge density at constant particle size, as shown in Figure 5.

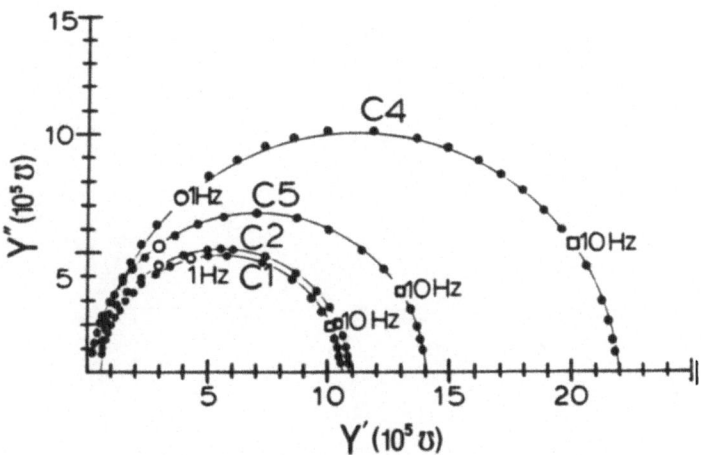

Figure 5: Frequency-dependent complex plane ac admittance curves of polystyrene latices with varied surface charge density (15% solids).

It has been shown experimentally that the ionic strength of an aqueous salt solution can be measured simply and rapidly by dielectric spectroscopy, regardless of the valence of the ions, over many decades of concentration[13]. The relationship is:

$$I_{ap} - I_o + b \log(c) \qquad (6)$$

where I_{ap} is the apparent ionic strength (AIS), and is the maximum value of $\theta \log \omega$ in a plot of this vs. $\log \omega$; θ is the phase angle determined from a.c. impedance data,

I_o is the apparent ionic strength of the pure solvent,
b is the slope, and
c is the true ionic strength (for 1:1 electrolyte,
    c equals the concentration of cations plus anions).
Typical results are shown in Figure 6.

296

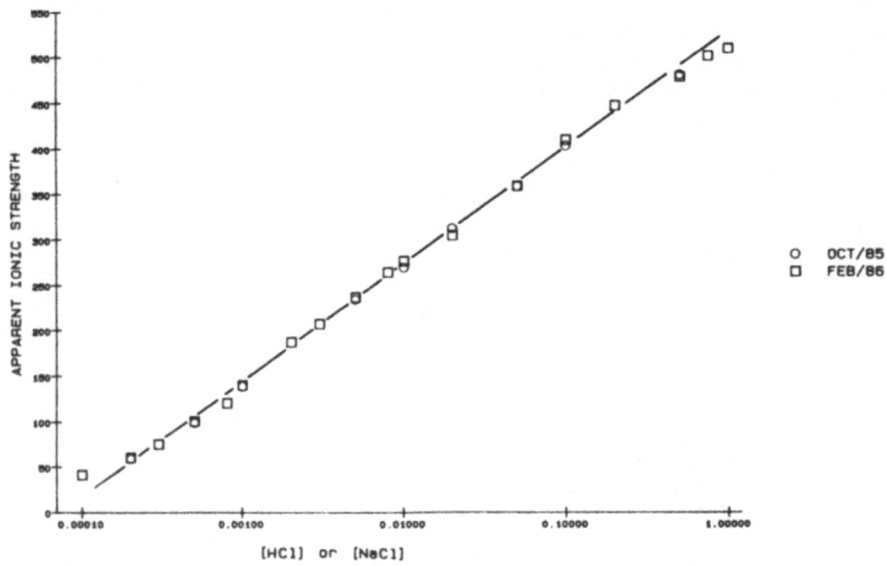

Figure 6: Relationship between "APPARENT IONIC STRENGTH" and logarithm concentration of NaCl in Milli-Q water.

In the case of polymer colloids which have been purified by ion exchange, the only ions contributing significantly to the ionic strength are those in the diffuse part of the electrical double layer. It is believed that the fixed ions (sulfonate in our case) on the surface of the latex particles contribute negligibly to the impedance, so that the slope, b, of such a colloid would be one half that of a simple 1:1 electrolyte. If the counterions are H^+, being small and highly mobile, most or all will be unadsorbed, so that they are the only ions which contribute to the apparent ionic strength. In Figure 7, curve C, the experimentally measured I_{ap} values are plotted against the concentration of the H^+ counterions determined by titration with NaOH. The slope is 0.50 times that of curve A, the theoretical value.

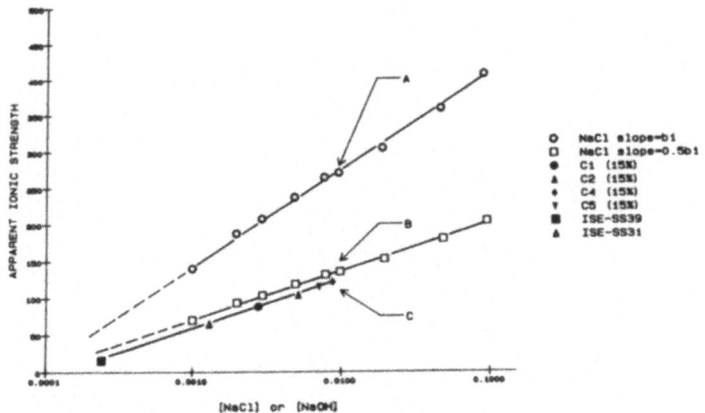

Figure 7: Correlation of "APPARENT IONIC STRENGTH" measured with FRA and end point NaOH concentration determined by conductivity titration.

Thus, if our premise is correct, one may estimate directly the concentration of H^+ counterions with a simple measurement of $\theta log\omega$ and thus the surface charge density, by use of curve B, Figure 7. Comparison of the counterion concentrations obtained by this means and by titration is given in Table II. The values obtained by dielectric measurements are between 0 and 0.4 x 10^{-3}M higher than those obtained by titration. Any extraneous electrolyte will cause the dielectric values to be high, whereas the titrations are specific to hydrogen ion. It is believed that with more careful attention to the possibilities of contamination during the experiment, these values could be brought closer together.

4.3. Degree of Ionization of the Electrical Double Layer

When the H^+ counterions are exchanged for Na^+ by titration with NaOH, the apparent ionic strength (AIS) decreases, as shown for each of the four PS colloids in Figure 8. It can be seen from this figure that there is a systematic increase in counterion concentration in both the H^+ and Na^+ forms according to the series Cl< C2< C5< C4. The difference in AIS from the beginning of titration to the end-point appears to be due to the adsorption of a substantial fraction of the Na^+ counterions into the Stern layer. The degree of ionization, f, should be a function of the apparent surface charge densities, before and after exchange of Na^+ for H^+. Furthermore, one would expect that the behavior would be such that f decreases with increasing σ_o. The results are presented in the lower half of Table 2. Values for f range from 0.12 to 0.085, corresponding to surface charge densities of 6.6 to 21.1 microcoulombs per square centimeter, in the same range that others have observed[14].

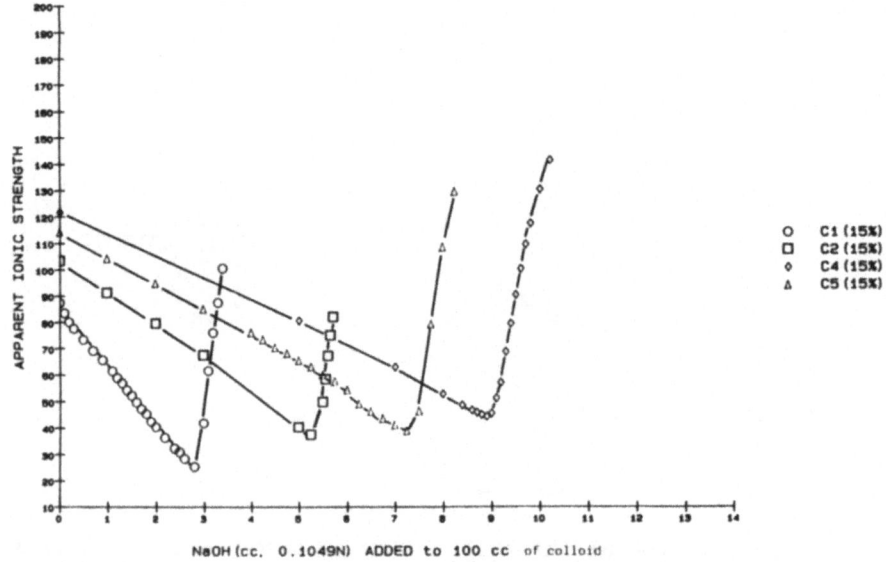

NaOH (cc. 0.1049N) ADDED to 100 cc of colloid

Figure 8: Relationship between "APPARENT IONIC STRENGTH" and NaOH titration concentration of polystyrene latices.

TABLE 2. Latex Characteristics: Comparison of Counterion Concentrations Determined by Two Methods

		C1	C2	C5	C4
Dia.(QELS)(nm)		208	206	198	208
Dia.(Dielect.spect.)(nm)		214	218	194	194
Surf. Charge ($\mu C/cm^2$)		6.6	12.5	16.5	21.1
AIS(H^+)		86	104	115	122
$[H^+]$	Curve B, Fig. 7	3.3	5.7	7.9	9.8
(M 10^3)	Titration	2.9	5.7	7.7	9.5
	AIS (Na^+)	23	37	36	42
$[Na^+]$ (M 10^4): Curve B, Fig. 7		4.5	7.1	6.8	8.3
f		0.14	0.12	0.086	0.085

Notes: f = $[Na^+]/[H^+]$

AIS = apparent ionic strength in $\theta \log \omega$ units

4.4. Particle Size

Schwarz[15] has studied the dielectric dispersion of colloidal polystyrene spheres in electrolyte solution. He proposes that the displacement of electrical charge within the double layer is attributable to the phenomenon of interfacial polarization observed in the low frequency region. The size of a sphere is theoretically derived as a function of the tangential mobility of ions along the surface of the sphere. There is a characteristic frequency of the dielectric dispersion process due to motion of the ions within the Stern layer responding to the externally applied electric field. Experimentally, a small amplitude d.c. electrical field is supplied by the 1186 ECI (a functional potentiostat) which is used in conjunction with the 1172 FRA. The output provides for semi-empirical determination of particle size, a, from the dimensionless apparent ionic mobility, $\Lambda/\theta\log\omega$ in which Λ is the solution conductance, and from the characteristic dielectric relaxation frequency, f*, obtained experimentally. This is expressed in the following equation:

$$a(A^\circ) = K[(\Lambda/(\theta\log\omega) \times (1/(2\pi f^*))]^{1/2} -3/\sqrt{c} \qquad (7)$$

where K is a cell-dependent constant and c is the concentration in moles/liter which can be accurately determined from the $\theta\log\omega$ versus log c curve in Figure 7. The second term on the right hand side of eq. (7) is the Debye thickness for 1:1 electrolyte.

The results for the determination of the radius, a, of polystyrene spheres with different surface charge density in this study calculated from eq. [7] (with K=31.342), along with those determined by QELS, are presented in Table 2. The agreement is remarkably good, with differences varying from 2.0% to 6.7% of the value determined by QELS. With further refinements of the technique it should be possible to get even better results. The advantage of the dielectric method lies in the fact that it can be applied to optically opaque and concentrated dispersions. The samples in this study were at 15% solids.

4.5. Characteristic Relaxation Frequency

It is well known from the literature that the characteristic relaxation frequencies of aqueous colloids of non-conducting particles will be low. In the samples examined here they are in the range of 1 to 2 Hz, as shown in Figure 9, in which the dielectric loss, ϵ'', is plotted against applied frequency. The maxima in these curves increase in frequency with increasing surface charge density, presumably because the counterions are held more strongly. During the course of a titration, not only are H$^+$ ions being replaced by Na$^+$, but also the latter are being adsorbed, effectively reducing the net surface charge. The result will be manifested in a progressive reduction in the characteristic relaxation frequency. This is shown in Figure 10. The end-point in this titration came at 8.9 cc of added NaOH.

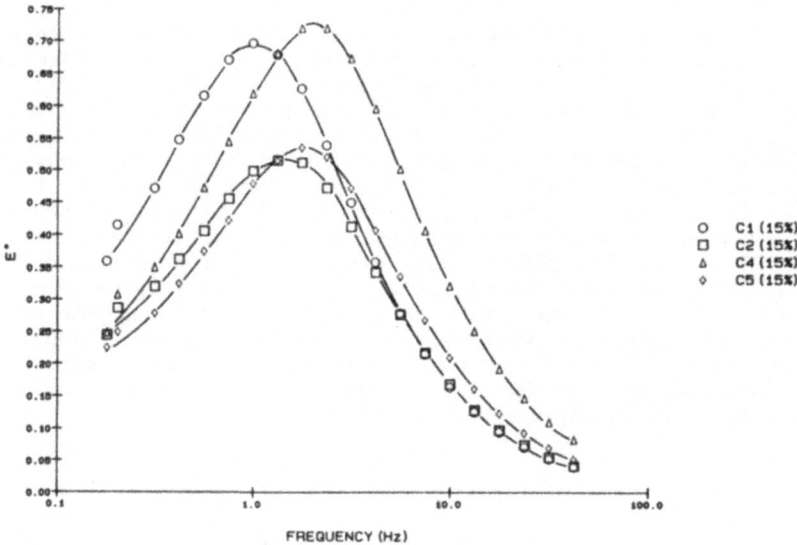

Figure 9: Dielectric loss permittivity vs frequency curves of polystyrene latices with different surface charge densities.

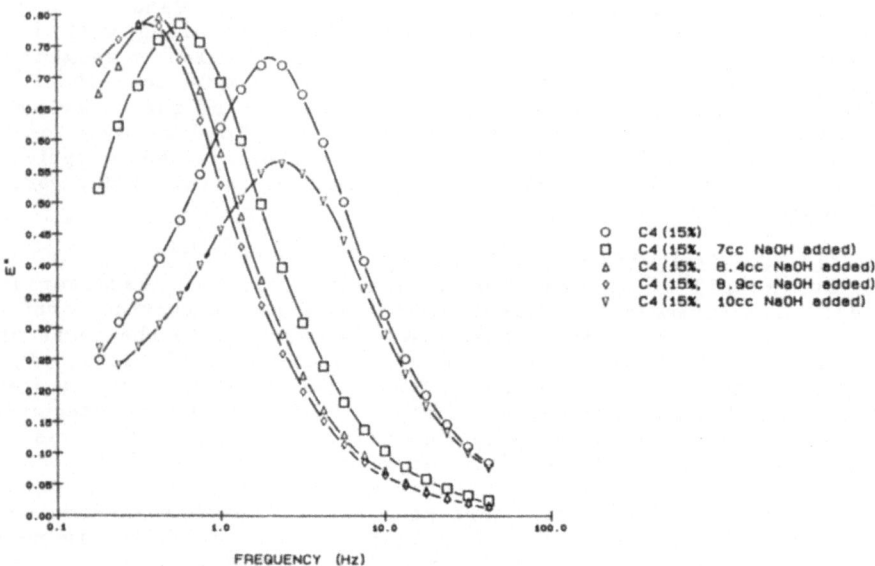

Figure 10: Dielectric loss permittivity vs frequency curves of C4 polystyrene latex titrated with NaOH solution.

301

4.6. Relationship to Coagulation Results

There is no evidence in this small series of PS colloids that there is any anomalous behavior in surface electrical properties at high levels of surface charge density as might have been deduced from the coagulation rate data of Tsaur and Fitch[1]. There is a regular progression in the apparent ionic strength in going up the series: C1, C2, C5, C4, the same as that for the titrated values of surface charge. The results in this study indicate that specific ion adsorption may play a more important role than earlier believed, so that coagulation studies with H^+ as well as other cations as "inert" electrolyte should be correlated with dielectric measurements.

4.7. Comparison of f-Values Independently Obtained

To determine the applicability of these techniques and ideas to colloids other than those of our own making, two highly purified polystyrene latexes were obtained from Prof. Norio Ise of Kyoto University. These were designated SS-31 and SS-39, and bore sulfonic acid groups on the particle surfaces. Their degree of dissociation had been determined by techniques which were quite laborious and entirely different from ours[14]. Upon titration with NaOH and measurement of AIS, the two samples lay directly on the extrapolated portion of curve C, Figure 7, as shown in Figure 11. Comparisons of the f-values determined by the two groups are shown in Table 3.

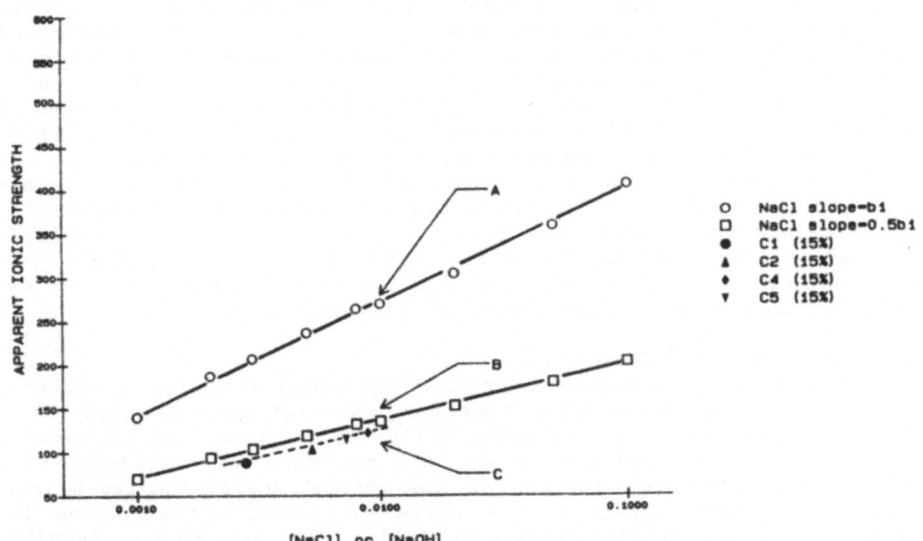

Figure 11: Correlation of "APPARENT IONIC STRENGTH" and end point NaOH concentration.

TABLE 3. Degree of Dissociation, f, determined by Two Methods

	SS-31		SS-39	
	Ise	This Work	Ise	This Work
% Solids				
5.3	0.23	0.228		0.098
7.9			0.09	

The alpha value (Equation [1]) of the C-Series colloids was almost zero, giving almost perfect semi-circles in Figure 5, and signifying very narrow particle size distribution. The alpha values for SS-31 and SS-39 were 0.119 and 0.117, respectively, indicating a somewhat broader dispersity. This may explain the small difference between our results and those of Ise and coworkers, although the agreement is remarkably good in any case.

4.8. Comments on Earlier Work

Lyklema and coworkers worked in a regime where $\kappa a \gg 1$, so that they were looking solely within the Stern layer at characteristic frequencies much higher than ours. We are working at $\kappa a \ll 1$, at low frequencies and are observing the diffuse part of the electrical double layer. Lyklema had to work at higher frequencies because of problems with polarization at the electrodes using their instrumentation at low frequencies. The Solarton^{TM} system employed by us can go down to 0.001 Hz without encountering such difficulties. Lyklema's results are highly model-dependent and require that the values of certain parameters be assumed. Even so, their results are somewhat unsatisfactory: as they say, "Quantitative discrepancies between theory and experiment have to do with surface- and colloid-chemical peculiarities of the latices..."[11]. Their theoretical analysis is nevertheless of great value and will serve as the basis for our own future studies on salt effects in model polymer colloids.

5. SUMMARY AND CONCLUSIONS

With modern, computer controlled equipment dielectric spectroscopy has become a powerful tool for the colloid scientist. In a matter of two minutes measuring time it is possible to have a determination of particle size, an estimate of particle size distribution, the ionic strength of the system, surface charge density and, by combining dielectric measurements with titration or ion exchange, to obtain the degree of ionization. The method requires samples that are fairly concentrated, so that properties of the system which are thermodynamically ideal cannot be studied. On the other hand, the technique is applicable to concentrated and optically opaque colloids.

Future studies in our laboratories will concentrate on determination of surface electrical potentials, salt effects and rheological properties.

REFERENCES

1. Tsaur SL and Fitch RM: *J. Colloid Interface Sci.*, in press.
2. Verwey EJW and Overbeek JThG: *The Theory of the Stability of Lyophobic Colloids"*, Elsevier, Amsterdam (1948).
3. Levine S; Mingins J and Bell GM: *J. Electroanal. Chem.*, 13, 280 (1967).
4. a) Ise N; Okubo T; Sugimura M; Ito K and Nolte H.J: *J. Chem. Phys.*, 78, 536 (1983).
   b) Sogami I: *Phys. Lett.*, A96, 199 (1983).
5. Patey GN: *J. Chem. Phys.*, 72, 5763 (1980).
6. van den Hul HJ and Vanderhoff JW: in *Polymer Colloids*; Fitch RM(ed) Plenum, NY (1971).
7. McCann GD; Vanderhoff JW; Strickler A and Sacks TI: *Separation and Purification Methods*, 2(1), 153 (1973).
8. Ottewill RH and Shaw JN: *Disc. Faraday Soc.*, 42, 154 (1966); *Kolloid Z.u.Z. Polym.*, 215, 161 (1966).
9. Ottewill RH and Richardson RA: *Colloid & Polymer Sci.*, 260, 708 (1982).
10. Cebula DJ; Goodwin JW; Jeffrey GC; Ottewill RH; Parentich A and Richardson RA: *Faraday Discuss. Chem. Soc.*, 76, 37 (1983).
11. Lyklema J; Dukhin SS and Shilov VM: *J. Electroanal. Chem*, 143, 1 (1983); Springer MM; Korteweg A and Lyklema J; ibid. 153, 55 (1983); Lyklema J; Springer MM; Shilov VM and Dukhin SS; ibid. 198, 19-26 (1986).
12. Traut GR and Fitch RM: *J. Colloid Interface Sci.*, 104, 216 (1985).
13. Su LS: private communication.
14. (a) Ito K; Ise N and Okubo T: *J. Chem. Phys.*, 82, 5732 (1985).
    (b) Wu WC; El-Aasser MS; Micale FJ and Vanderhoff JW: in *Emulsions, Latices & Dispersions*; Becher P and Yudenfreund MN(eds), Marcel Dekker, NY, 71-98 (1978).
15. Schwartz, G., *J. Phys. Chem.*, 66, 2636 (1962).

POLYMER COLLOIDS IN BIOMEDICAL FIELD

BIOMEDICAL APPLICATIONS OF POLYMER COLLOIDS: FUTURE DIRECTIONS

POSITION PAPER - #1

1. INTRODUCTION

It was the consensus of the participants that a number of important applications of polymer colloids in the biomedical field have emerged. These applications utilize recent advances in polymer science and new developments in biotechnology. The five most important research problems in the biomedical field, where the introduction of polymer particles has already or is anticipated to have a very significant impact are, in order of priority:

     (1) Bone marrow transplantation
     (2) Solid-phase immunoassays
     (3) Drug delivery systems
     (4) Extracorporeal and hemoperfusion systems
     (5) Identification and enumeration of lymphocyte
        populations.

These five applications as well as certain others of considerable significance are discussed in this position paper.

2. Magnetic particles in bone marrow transplantation

The major advantages of the immunomagnetic method of cell separation are time efficiency and the ability to handle large numbers of cells in a single separation. These characteristics make the immunomagnetic method highly suitable for:

2.1. Removal of T cells from human bone marrow cell suspensions to be used for allogeneic bone marrow transplantation across histo-compatibility barriers. Removal of these T cells results in prevention of graft-versus-host disease. Approximately 2,000 patients are treated with allogeneic bone marrow transplantation annually in the U.S. The number of such patients is severely limited by the availability of matched donors. In addition, graft-versus-host disease is observed in approximately 50% of the patients who received matched grafts, presumably because of partial matching and differences in minor histocompatibility complexes. T-cell depletion before bone marrow transplantation can provide marrow grafts for those patients who do not have matched donors and can reduce significantly the likelihood of graft-vs-host disease in those patients who have (partially) matched donors available.

2.2. Removal of tumor cells from human bone marrow cell suspensions to be used for autologous bone marrow transplantation. Autologous bone marrow transplantation has recently been widely used for the treatment of a number of cancers. Bone marrows from patients with cancer are quite often contaminated with metastasizing tumor cells. These tumor cells, although present usually at very low proportions are sufficient to cause early relapse of the disease. Removal of tumor cells from the bone marrow before reinfusion is required to avoid early relapse. Over two hundred patients with cancer have received autologous bone marrow grafts, after treatment with lethal chemotherapy and irradiation and purging of the tumor cells using magnetic monosized polymer particles developed by Ugelstad et al and appropriate monoclonal antibodies. Rapid and complete hemopoietic reconstitution has been observed in these patients demonstrating that bone marrow purged using microspheres can rescue their hemopoietic system. Although the results appear promising, final conclusions for the therapeutic effectiveness of the method to treat cancer require follow-up of the patients for longer time periods.

The immunomagnetic method for bone marrow purging appears to have a number of advantages over competing techniques, which are: (1) immunotoxins (monoclonal antibody-ricin conjugates); (2) antibody and complement treatment; (3) soybean lectin agglutination and sheep erythrocyte rosetting. These are discussed in previous papers of this volume.

Although available magnetic particles have been found satisfactory for bone marrow purging, further research is being carried out for optimizing the properties of the particles. These include: size, ability to bind antibodies covalently with high efficiency and addition of ligands for antibody coupling. Other improvements should involve instrumentation. Automated continuous flow or batch systems for purging of large numbers of cells should be explored.

3. Other Applications in Cell Separation

Cell separation experiments can be greatly benefited by the introduction of the immunomagnetic method. Although initial application of the method permitted only negative selection of cells (removal of a particular population), recently methods have been developed for positive selection of cell populations. This will facilitate cell separation experiments that are being carried out in the immunological laboratory using laborious and cumbersome techniques.

4. Polymer Particles for the Identification and Enumeration of Lymphocyte Populations

Identification and enumeration of lymphocyte populations are routinely carried out in the immunological laboratory for the diagnosis of various diseases. Extensive applications for the appropriate polymer particles are anticipated in this area. Monodispersity is a requirement that must be fullfilled by such particles. Identification and quantitative determination of lymphocyte populations using polymer particles requires only inexpensive instrumentation that is routinely available in all laboratories. In contrast, the methods that are used presently require complex and expensive instrumentation (fluorescence activated cell sorter) and highly trained personnel.

5. Drug Delivery Systems

Appropriate polymer particles are anticipated to have extensive applications in drug delivery. Initial studies using albumin-containing magnetic microspheres and an extracorporeal magnetic field have been promising. Polymer particles for drug delivery should be: biocompatible, bio- degradable, non-toxic, stable with a long shelf-life, of appropriate size to penetrate capillaries, allow the entrapment and release at controllable rates of drugs, and permit selective targeting to the tumor site using monoclonal antibodies or other acceptable means. Microspheres fullfilling these criteria when loaded with chemotherapeutic agents and targeted selectively to the tumor site are expected to be excellent drug delivery systems. Such particles are not available at the present but materials are available that may permit their development. Systems that will localize these particles to the tumor site appear to be essential.

6.   Extracorporeal and Hemoperfusion Systems
     Applications in the immediate future for polymer particles are
foreseen in the area of extracorporeal and hemoperfusion systems.
Appropriate polymer particles with chelating agents or antibodies can
provide selectivity to these procedures, which is completely lacking from
the systems currently in use. These systems are based on charcoal or ion
exchange columns.

7.   Solid-Phase Immunoassays
     7.1. Agglutination Assays. Most agglutination assays operate using
0.41-$1.2\mu m$ latex particles simply because the earliest latexes available,
with low variation in diameter size, were of these diameters. The
earliest latexes were polystyrenes. Most assays use physically-adsorbed
antibody, because hydrophobic surfaces could bind the Fc fragment of
antibodies (i.e. the portion of the molecule that is most remote from the
antigen-binding site). With what we now know, we would like to direct
polymer research in four principal areas to improve polymer colloids for
agglutination assays. These are as follows:

               Size and density of particles
               Surface characteristics
               Chemical binding spacers
               Shape

     These four areas of study are essential to achieve the relatively
simple objective of producing particles that, when coated with specific
antigen or antibody, have the properties of (a) strong and optimal
specific binding of particles via antigen-antibody bridges and (b) zero
non-specific binding or attraction between the particles. Ideally,
particles coated with specific protein should be indefinitely stable in
aqueous suspension until required for agglutination assays; and once
agglutinated, should form strongly bound oligomers.
     7.1.1. Size and Density of Particles. It is well known that a
particle density similar to the immunoassay buffer is important in order
to keep the particles in suspension. For both turbidimetric assays and
particle counting assays, monodispersity and size are important. The
number of Ag-Ab bridges between pairs of particles are about 2-8. With
larger particles, the shear forces across these bridges although they
have not been well characterized may result in disruption of agglutinates
when pumped at high speed in automatic machines. Thus particles of
smaller diameter may yield more robust assays. For turbidimetric assays,
optimum performance may be a function of the particle diameter/wavelength
ratio.
     7.1.2. Surface Characteristics. Protein-coated latexes are generally
unstable in aqueous suspension. They have a tendency to self-agglutinate
yielding a high background signal resulting in decreased sensitivity of
the assays and necessitating the use of calibrators with each assay run.
To overcome these problems with current generation hydrophobic particles,
the spaces between antibody or antigen molecules (coupled to particle
surfaces) have been filled with a ballast protein such as albumin. New
particles developed by Ugelstad and collaborators offer new approaches to
solving the problem of non-specific binding. This can be accomplished by
producing particles with a uniform distribution of hydrophobic or
functional group areas separated by hydrophilic areas. The former can be
used for attaching proteins; the latter for inhibiting non-specific
effects. Ideally, particles should be available with different

percentages of the two types of areas in order to be able to optimize assay concentration ranges. The use of chemical spacers on polymer particles ("hairy latexes") may be increasingly important. Spacer groups chemically coupled to the latex surface to which the protein is bonded have the ability to hold the protein away from the particle surface and allow greater flexibility with respect to orientation.

7.1.3. Shape Consideration. Polymer latexes with spherical geometry have hitherto been the norm. However, latexes with different shapes may offer interesting properties. For particle counting assays, if the antibody is located on one end of an ellipsoid particle, dimer agglutinates only could be formed. For turbidimetric assays, flat particles with functional groups (for protein binding) on flat areas could offer some interesting properties. Agglutinated compact stacks could be formed with very strong binding which could enhance light transmission or scatter depending upon the size to wavelength ratio.

In conclusion, the outlook for agglutination assays is optimistic. The problem of serum interference has been solved. With new particles with properties described above, plus other features (such as remanence-free magnetic properties), we can look forward to very fast assays (e.g. 4 minute assays) using simple stable reagents for very sensitive quantitative assays. One can envision inexpensive, high throughput automatic assays in the doctor's office on the one hand or veterinary "cow-side" tests on the other.

7.1.4. Marker-Based Immunoassays The advent of filtration and magnetic techniques in solid phase immunoassay has given latex particles a new lease on life as solid phases in solid immunoassays (radioimmunoassay, RIA, IRMA; enzyme immunoassay, EIA, ELISA; fluorescence immunoassay, FIA, DELFIA, PCFIA; chemiluminescence immunoassay, CIA; and others). The speed of reactions, (5-10 minutes) coupled with ease and speed of separation of the solid phase from the sample and labeled reagent, gives latex particles an enormous advantage over microtiter wells, test tubes and macrobeads. In addition, total automation becomes a realistic possibility. At the other end of the spectrum, single sample, rapid, over-the counter tests are on the horizon. The demands on latex particles in solid phase immunoassays are similar in many ways to those of latex agglutination assays. However, certain constraints are relaxed and new ones are introduced. High capacity binding with retention of antibody activity and antigenic determinants is still very important, if not critical, to the performance of the assay. However, the size of the particle, its shape and density, and to a certain extent, its tendency to flocculate are of secondary, non critical importance. The monodispersity of the particle preparation, although important for quality control purposes, is not an absolute requirement; however, the particle must not bind non-specifically to the labeled reagent. Therefore, the optimum particle must be developed which fulfills the first requirement of efficient binding of reagent but at the same time can be "switched off" to binding protein non-specifically. Alternatively, well defined, easily blocked binding sites have to be developed. These requirements are true whether covalent or physical adsorption is used for protein immobilization.

8. Chromatography

Polymer particles prepared by suspension polymerization have traditionally been applied in a number of chromatographic systems. Typically this has been styrene-divinylbenzene particles in the range of 10-1000 μm (porous, and nonporous) which have been used as such or modified to give ionic exchange resins. More recently strongly hydrophilic porous particles have been prepared by polymerizing highly hydrophilic monomeric and dimeric vinyl compounds in water and oil emulsions. These systems have been claimed to be superior to systems based upon natural polymers like dextran and agarose.

The methods for preparation of monodisperse particles developed by Ugelstad et al have up to now found their widest industrial application in chromatography. The system based upon these particles is called FPLC (fast protein liquid chromatography). The particles with a very uniform size have a very hydrophilic surface and are produced in the form of strong cationic and strong anionic exchangers and as polybuffer exchangers for chromatofocusing. The uniformity in particle size allows the preparation of very homogeneous beds resulting in very high efficiency and low resistance to flow. The FPLC systems have found wide applications in separation and analysis of biological materials and have proven to be superior to other systems in this area.

The industrial application of monosized polymer particles has up to now been restricted to ion exchange systems. The advantages of monodisperse over polydisperse particles with respect to speed and separation efficiency have been demonstrated also in other types of chromatography. It is anticipated that monosized particles will have also advantages in gel filtration, affinity chromatography, adsorption chromatography etc.

9. Quantitative Flow Cytometry Standards

Flow cytometry is a very important field in biomedical research, yet no good fluorescent latex particles are currently available for quantitative flow cytometry. Attempts have been made by several workers to supply the users of flow cytometers with fluorescent particles for this purpose but none of them fulfills all the requirements of good standard fluorescent particles for quantitative flow cytometry.

Following are some of the requirements for quantitative flow cytometry standard:

1. The fluorophore used in these particles should be the same as the one used to label the cells. Examples are fluorescein, rhodamine, phycoerythrin, allophycocyanin, Texas red, DAP, Hoechst 33258, 33342, and poropidium iodide.

2. Preferably the quantum yield of the fluorophore incorporated or attached onto the particles should be the same as the fluorophore attached to antibodies on the cells.

3. The size of the particles should range between 2.0μ to 10μ with a very narrow size distribution and a very low degree of aggregation.

4. The fluorescence intensity should be very stable for long term storage. The latex particles should have about the same refractive index as living cells to provide adequate light scatter signal.

5. The number of fluorophores per particle must be known accurately so that the number of antigenic sites on the cell can be calculated.

Within the development of tunable dye lasers across the UV-visible spectrum, another need exists for standard particles which may be employed for instrument alignment and quality control.

10. In vivo Imaging Applications

In vivo imaging applications is one potential use of microspheres. Radioactive compounds containing isotopes such as ^{99}Tc or ^{125}I can be covalently bound to the surface of microspheres. When injected i.v. they can be traced by means of gamma scintillation scan. One of the clinical

312

uses is to determine the site of internal bleeding. Currently, colloidal sulfur or denatured red blood cells are used as the carrier. One of the problems of this system is that the radioactive compound dissociates from the surface during the scan.

NMR imaging using liposomes has been developed recently. Nanospheres from emulsion polymerization and post-derivatization of such particles allow one to prepare particles containing fluorine molecules which can be detected by NMR with a fluorine-19 probe. When microspheres of these type are coupled with specific antibody, they can also be used for imaging of specific tumor cells or tissues. Microspheres for such applications must have the following properties:

1. Must be activated for chemical binding of fluorine or
   radioactive label containing molecules.
2. Must be biocompatible and non-toxic.
3. Must have proper size and size distribution
   that permits them to go through the capillaries.

11. DNA Probes

Although there are no reports in the literature on the use of polymer colloids in nucleic acid hybridization tests certain members of the Workshop group are actively involved in developing "latex particle"-based DNA/RNA hybridization assays and indeed have patents pending. It can be generally said that particle-based probe assays have many advantages over conventional nitrocellulose-membrane based techniques. Due to the nature of nucleotide chemistry, the particles involved are quite different from those used in immunoassays but must retain the same physical and functional properties. Some unique problems must be solved without optimized particles. At the present it is too early to say what the best coupling methodology is. Applications for such assays include diagnostics for infectious diseases, genetic disease screening, cancer detection, monitoring and classification, histocompatibility testing and many others.

SUMMARY

New and important applications of polymer particles in the biomedical field have emerged. These applications involve both the diagnosis and the treatment of a number of life-threatening or severe diseases and exploit recent advances in polymer chemistry and new developments in biotechnology (such as monoclonal antibodies). The most important applications of polymer particles in the biomedical field are the following, in order of priority: (1) Allogeneic or autologous bone marrow transplantation; (2) solid-phase immunoassays; (3) Drug delivery systems; (4) Extracorporeal and hemoperfusion systems; (5) Identification and enumeration of lymphocyte populations. Other applications involve cell separation methods, chromatographic separation of proteins, flow cytometry standards, DNA probes, in vivo imaging applications and immobilized enzymes. The recent development of improved polymer particles and emerging innovations in the development of particles with defined properties, will facilitate a broad range of basic and applied uses in the biomedical field.

12. **Panel**

Discussion Leader:
    C. D. Platsoucas (M. D. Anderson Hospital and Tumor Institute,
       USA)

Secretary:
    T. A. Wilkins (Acade Diagnostics Systems, Belgium)

Participants:
    F. N. Hansen (Dyno Industries A.S., Norway)
    M. E. Jolley (Pandex Laboratories, Inc., USA)
    K. Nustad (The Norwegian Radium Hospital, Norway)
    M. Papamichail (Hellenic Anticancer Institute, Greece)
    J. Ugelstad (Institutt for Industriell, Norway)
    C. J. Wang (Pandex Laboratories, Inc., USA)

FUTURE DIRECTIONS IN THE LATEX AGGLUTINATION ASSAYS

POSITION PAPER - #2

INTRODUCTION

The use of the latex particles in diagnostic immunoassays offers the following properties:

1. Uniform size particles in the ranges of 50-10,000nm.

2. A wide selection of surface functional groups which can be incorporated onto the latex particle.

3. Antigen, antibiodies, enzyme and lectins, can be physically adsorbed or covalently linked to the surface of the particles directly through a surface functional group or indirectly through a spacer arm.

4. Latex particles can be prepared with various polymer properties having various dispersion medium properties e.g. ionic strength, pH and electrical charges. High density latex particles can be used.

5. It is possible to introduce radioisotopes, enzymes, fluorophores, magnetic material, dyes, photographic synthesizer, chemi and biolumin-escent lectins, either directly or into the protein adsorbed.

6. Latex particles used in the diagnostic immunoassay may be classified in two overlapping groups based on their size and stability. The first class is uniform sized particles of 0.1μ to 1.5μ in diameter; these particles are used as carriers for antigen and antibody reaction in latex agglutination tests; the second class of microsphere polystyrene or other polymeric structure, ranges in size from $1.5-2\mu$ to 100μ or larger in the form of particles and beads, of limited stability and used in all immunoassay, ELISA, immunospheres, cell separation and chromatography.

Latex particles used as a tool in agglutination reactions is the subject of this position paper.

Latex are a colloidal dispersion of polymeric spheres which derive their stability from one of two mechanisms: 1) double layer repulsion, 2) steric stabilization. Monodisperse polystyrene latex particles, particularly larger sizes tend to settle upon standing because of the density of the polymer (1.05g/cc).

The approximate time for complete settling of the particles yielding a clear supernatant layer is 1-3, 3-6 and 6-12 months per particle diameter of 10,000, 8,000, and 5000 Å respectively. Particles of 3500 Å are smaller in diameter showing no settling upon standing for several years. The rather strong attachment of hydrophile protein e.g. human serum albumin, immunoglobulin to low energy surfaces such as polystyrene are due to van der Waals interaction and interfacial forces, hydrogen bonds and the long range forces of the London-Keesom-Debye varieties.

Uncoated and protein-coated latex particles under 1.5μ are metastable in regard to their sensitivity to electrolyte, pH, buffers, and protein concentration. Since 1956 there have been 170 papers which describe the different uses of latex particles in immunoassays. From all these research papers only 80 commercial kits have been developed. The main reason that half of the tests were unsuccessful was due to the difficulties in keeping the latex system stable. Instability was observed prior, during and after the protein coating procedure.

The following variables contribute to instability of latex particles:

1. Particles larger than 0.6μ tend to settle on prolonged standing. For these particles gentle shaking may return them to a uniform

dispersion. However, other particles settle permanently because they form aggregates. These particles cannot be used for protein coating procedures. In a dilution of 10ppm (parts per million) the particles should be visible in a microscope as mostly single particles, with only an occasional doublet.

2. Colloidal stabilization is provided by anionic and nonionic detergent. While stability of some latex particles may be improved by the addition of detergents, some others may be destabilized.

3. Most of the latex particles are destabilized by electrolyte in higher ionic strengths. Latex particles with carboxyl and amino groups differ markedly in their sensitization to electrolytes from polystyrene latex containing SO_4 groups. Latex particles have a critical coagulation concentration to electrolytes. The study of this concentration is a practical way to determine colloidal stability.

4. Buffers are critical for the performance of each test, because there is a pH dependency for each protein coated latex. Organic buffers have a tendency to destabilize most of the particles coated with protein.

5. Concentration of latex particles is critical in developing a new test. There are great difficulties in extending practical approaches of dilute solution to concentrated suspension and vice versa.

6. Too many functional groups onto the surface can interfere with physical as well as chemical bindings of proteins. The presence of two different types of functional groups or the hydrophobic spacers would complicate the colloidal stability.

7. A molecular flat orientation of antibiodies on the surface of polystyrene particles is critical and could be accomplished by keeping a low Ab or Ag surface concentration.

8. Mixture of different proteins decreases the binding of each individual protein onto the surface of the latex. However, some commercial kits utilize polyvalent antisera adsorbed physically onto the surface of the latexes.

9. There is no apriori way to predict the stability of covalent attachment technique. It must be investigated for each particle and information obtained from one particle cannot be extrapolated to another particle. New information must be obtained when particles are of same size but different manufacturers.

10. Denaturation of antigen or antibody epitopes during chemical coupling is critical. Polysaccharides, glycoproteins are less denatured by chemical coupling than protein antigen in their reactivity with their respective antibody.

11. Coupling procedure should not produce inter and intra molecular bridging. Carbodiimide, glutardehyde used in chemical coupling may inhibit the sensitivity of the latex particle tests.

12. Filtration, high speed centrifugation, ultrasonication, drying, would destabilize particles. Very few particles can be lyophilized.

13. There is a great need for the developing of techniques of cleansing coated latex particles. Dialysis, serum replacement used in cleansing procedure have not as yet been successful.

For the performance of a practical latex particle, testing a number of variables should be considered: particle size, particle concentration, surface charge density and distribution, the number of functional group of each type, density of the particle, concentration of adsorbed emulsifier, composition of copolymer latexes, the composition of aqueous phase (residual monomer, electrolyte, ionic strength, emulsifier, pH, products of side reactions), critical coagulation. All this information is needed to be included in the manufacturer products. There are far too many latex particles on the market today each particle having its own characteristic, for example, size and functional groups. There is a great need in this diversity to produce fewer latex particles with well defined colloidal properties. Adsorption of proteins onto the surface of latex particles forms a flat monolayer. There is still 8% of the surface area uncovered by ligand molecule. This uncovered surface area between molecular is too small to accomodate antibody molecules in a flat orientation. Any adsorption which takes place in this area does so by head on orientation of the same molecule. In the latex fixation test the surface of 0.8μ latex particle is covered by approximately 75,000 gamma globulin molecules. Only 4-5 molecules of IgM rheumatoid factor are necessary to destabilize 2 latex particles. It appears that there is no need for so many antibody molecules to be adsorbed onto the surface. A smaller surface area of the polystyrene for specific protein adsorption will limit the nonspecific adsorption of proteins.

To eliminate the nonspecific adsorption of the proteins onto the surface of latex particles, a large number of methodologies have been proposed. None of these methods have been standardized and which method is best is a matter of trial and error.

It has been established that adsorption of protein onto the surface of latex particles differs from one particle to another and in each case the condition of adsorption should produce maximum sensitivity and high specificity. For this purpose it is assumed that antigen or antibody can be attached to the solid phase support yet maintain immunological activity with the lowest nonspecific adsorption.

The rational design of specific chemical coupling of immunoglobulins require careful examination of structural features associated with antibody specificity and of the antigen structure. The specificity of the antigenic binding is determined by 3 hypervariable segments of both light and heavy chains of immunoglobulin. The amino acids residue in this region tends to be aromatic but usually interspaced with polar residue. Some of the binding sites of the antibody are no longer than 35 Å x 11 x 65 Å, consisting of no more than 17 amino acids. The antigenic sites of some antigens are about 20 x 25 Å and do contain 10-20 amino acids. The binding forces between antigen and antibody are: electrostatic, hydrogen bonding or van der Waals forces. These bonds are nonspecific lying in their spatial complimentary configuration of the reactive atoms or atomic group. Chemical coupling reaction are specific for a particular amino acid residue and can lead to the antigenic binding sites being blocked by the binding acids. Many of the coupling procedures produce multiple inter and intra molecular cross-linking, especially where more than one amino group is present in the reactant. Extensive multipoint binding takes place during chemical coupling of the protein and latex particle, because there is a very large number of functional groups on the particle surface. This multiple binding may render either the antigen or the antibody inactive.

An idealized particle should have a small surface area for protein physical binding and a larger surface area of hydrophilic polymer with few functional groups (COO⁻, NH⁺ and SH groups).

Two moieties in immunoglobulins may provide linkage sites which are not likely to interfere with the antigenic binding sites. One is the carbohydrate prosthetic group which is usually present in the C region of heavy chain and occasionally in the V region. It can be simple or highly branching and containing fructose, galactose, mannose and salicylic acid. The other linkage can be provided by free sulfhydryl groups formed on reduction cleavage of interchain disulphide bridges in immunoglobulin. The disulphide bridge in the Fab_2 fragment can be split to yield FabSH monomer.

The development of techniques for the isolation of the antibody population of defined characteristic is of importance in their binding to polymer surface. Serum antibodies are heterogeneous with regard to their immunochemical characteristic, isotopic class and subclass allotype, idiotype and with regards to their affinity for the particular antigen.

Antigen should be as pure as possible. Monoclonal antibodies secreted by hybridoma are useful because they recognize single antigenic determinant and are homogeneous in respect to their specificity and affinity. The commercial production of monoclonal antibodies, precisely defined concerning their affinity, of lower no cross reactivity, of more than one antigenic specificity will lead to improvements in the immunoassay technique. This antibody may be used for maping antigenic structures to complex antigens such as tumor markers, or specific cell antigens. Bacteria possess many structures that may be antigenic, such as capsular polysaccharides, glycoproteins, flagellar proteins, pili, genes, wall components, and intra and extracellular products. Monoclonal antibodies to all these bacterial antigens are expected to be produced. Traditionally clinical microbiology has been dependent upon the isolation of individual microbes by their visualization on the microscopic slide test and their identification by growth dependent test. Rapid automated microbiology systems promises faster turn around time but usually at considerably increased expense.

Tests that take minutes to perform like the latex agglutination test lie predominantly in antigen detection in normal sterile fluid as well as in the tissue specimen. The speed and simplicity of these tests allow prompt and appropriate therapy. Serological identification directly from specimen (throat, urine, stool, pus, blood) or directly from culture plates (at genus or species level) is around the corner. Spot tests combined with colony morphology, gram stain, and serology can have a significant clinical impact. The ELISA techniques are utilizing polystyrene plates for immunoassay. The replacement or a combination of these plates with polymer microsphere and polystyrene beads is not far away. In the diagnosis of parasitic infection, and more importantly in epidemiologic studies, serological techniques have considerable practical value. Antibodies to parasites are composed of heterogeneous mixture not well defined. The necessity of purified specific antigen is particularly evident for the new generation of highly sensitive immunoassays.

There is a continued interest in the direction of viral antibodies which could serve as a reliable marker of an active viral infection. Serological markers in virology have provided insights regarding the phases of infection, degree of infectivity, prognosis and immune status of the patient. Great attention has been focused on the assay of viral IgM antibodies since antibodies of this class characteristically have an early appearance and limited persistance in viral infection. With the events of hybridoma technology major problems associated with the use of polyvalent antisera in diagnostic tests has been eliminated. Antibody to viral glycoprotein, virus genomes, and polypeptides have also demonstrated their potential use in specific serological tests. Latex particle agglutination as well as the use of polymeric spheres may successfully replace the ELISA technique in virology.

Many fully automated spectrophotometric techniques have been developed for the quantitation of the latex particle agglutination. The reagent used in various kits can be easily standardized by spectrophotometry.

A theory of antigen-antibody reaction and of the latex particle coated with antigen and flocculated by antibody needs to be formulated. The kinetics of flocculation of latex protein dispersions by electrolyte should also be examined. While there are many theories of endpoint versus kinetics of antigen-antibody reactions using immunonephelometer assay there are only few theories of flocculation of antigen antibody reaction using latex particles.

Panel:

Discussion Leader:
J. M. Singer (Montefiore Hospital & Medical Center, USA)

Participants:
M. Chang (Jet Propulsion Laboratory, USA)
J. C. Daniel (Centre de Recherches D'Aubervilliers, France)

BIOMEDICAL APPLICATIONS OF POLYMER PARTICLES WITH EMPHASIS ON CELL SEPARATION

Chris D. Platsoucas

Department of Immunology
M.D. Anderson Hospital and Tumor Institute
Houston, Texas 77030

1. INTRODUCTION

Applications of polymer particles in the biomedical field have been concentrated in the past primarily in the areas of blood flow determination and in vitro immunoassays. Microspheres have been employed for the determination of myocardial, renal, cerebral and other blood flow and perfusion rates (Phibbs, et al, 1967; Hollenberg, 1975; Hoffbrand and Forsyth, 1969). Polymer particles, and latexes in particular have been extensively used in immunoassays, starting in 1956 with the development of the Latex Agglutination Test (Singer and Plotz, 1956). Recently a significant number of additional applications of polymer particles in the biomedical field have emerged. These applications exploit advances in polymer chemistry in combination with new developments in the field of biotechnology. Several of these applications are well underway, whereas others are at an early developmental stage. Several of these applications are described in the following paragraphs of this review, with emphasis on the use of microspheres in cell separation.

2. SOLID-PHASE IMMUNOASSAYS.

Polymer particles have extensive applications in the development of immunoassays, which are based on the antigen-antibody reaction. The following are the most frequently used types of immunoassays (reviewed by Hunter, 1978; Rose and Friedman, 1980; Gribnau et al, 1986) employing polymer particles or polymer materials: agglutination tests (Singer and Plotz, 1956); radioimmunoassays (Yalow and Benson, 1970; Walsh, et al, 1970); enzyme-linked immunoadsorbant assays (ELISA)(Engvall and Perlmann, 1972); fluorescence immunoassays (Jolly, et al, 1984). These in vitro diagnostic tests are widely used for the determination of the concentrations of various proteins, hormones, drugs, and other substances in body fluids (Singer, 1961; Heymer, et al, 1973; Malin and Edwards, 1972; Milgram and Goldstein, 1962; Guesdon, et al, 1978; Hunter, 1978; Rose and Friedman, 1980; Gribnau, et al, 1986). The development of highly specific monoclonal antibodies and recombinant antigens (such as viral antigens) will facilitate further development of these immunoassays. The use of polymer particles in immunoassays is discussed in this volume by Dr. Singer (Singer, 1987).

3. LABELING AND IDENTIFICATION OF LYMPHOCYTE POPULATIONS.

Microspheres prepared from various polymerization products such as polygluteraldehyde (Rembaum, et al, 1978), mixtures of methyl methacrylate, hydroxyethyl methacrylate and methacrylic acid (Molday, et al, 1974; 1977), and polyacrylamide (Ljungstedt, et al, 1978), have been employed for the identification and enumeration of distinct populations of lymphoid cells. These microspheres have been coated with antibodies specific for cell surface antigens selectively expressed on functionally distinct lymphocyte populations. Both direct methods of labeling lymphocytes with antibody coated microspheres and indirect methods have been used. Indirect methods of labeling employ two antibodies (such as goat anti-rabbit or anti-mouse on the particles and rabbit or mouse anti-human on the cells) or appropriate ligands, such as biotin (Kaplan, et al, 1983; Wojchowski and Sytkowski, 1986). Also, fluorescent microspheres as well as radioactive microspheres have been used (Molday, et al, 1975; Gordon, et al, 1977; Merivuori, et al, 1980; Mirro, et al, 1981; Higgins, et al, 1981; Ross and Lambris, 1982; Lambris and Ross, 1982; Oonishi and Uyesaka, 1985; Tomaszewski, et al, 1986; Bonnefoy, et al, 1986). Both light and electron microscopy (scanning and transmission) have been employed for the identification of populations of lymphoid cells.

4. EXTRACORPOREAL AND HEMOPERFUSION SYSTEMS.

Extracorporeal and hemoperfusion systems are potentially important areas of application of microspheres. Hemoperfusion is a procedure by which molecules usually of small molecular weight (such as poisons) are removed from the circulation. The cleared plasma and the blood cells are returned back to the patient. Today this is accomplished by non-selective means, by passing the plasma through activated charcoal columns or ion exchange columns that remove the compound desired together with a number of other useful elements. Blood cells and plasma are separated before the procedure using a blood separator. Appropriately designed polymer particles with chelating agents or antibodies can provide selectivity to the procedure. This approach can be very useful for the treatment of patients that have taken overdoses of drugs or those subject to poisoning with heavy metals. These systems may be useful for the removal of iron from the circulation of patients with thalassemia major without affecting other soluble blood components (Margel and Marcus, 1986). These patients have highly increased iron in the circulation and abnormally high iron deposits as a result of multiple transfusions. Another application utilizes microspheres with covalently attached monoclonal antibodies or appropriate antigen(s) for the removal from the circulation of antibodies in patients with autoimmune disease (Margel, et al, 1984; Margel and Marcus, 1986).

Extracorporeal perfusion systems over formalin-treated S. aureus or staphyloccocal protein A columns have been used with some success for the treatment of dogs with spontaneous breast adenocarcinoma and patients with breast cancer (Terman, et al, 1980; 1981). However, the substantial toxicity associated with the treatment has somewhat decreased its popularity. Polyacrolein or polyglutaraldehyde microspheres encapsulated in agarose matrix (Marcus, et al, 1982; Margel and Offarim, 1983; Margel, et al, 1983) and covalently coupled to bovine serum albumin (BSA), have been successfully used for the removal of anti-BSA antibodies (Marcus, et al, 1984). Crosslinked agarose microspheres of diameter in the range of 0.5 to 0.8 mm were found to provide an optimal matrix for hemoperfusion. Ligands can be covalently attached to these particles through the reactive aldehyde groups of the polyacrolein (Margel and Marcus, 1986). The complement component C1q bound to these particles permitted quantitative removal of BSA-anti-BSA immunocomplexes. Also, these particles with covalently attached anti-digoxin antibodies have been successfully used for the removal of digoxin from the circulation of digoxin intoxicated dogs. The procedure was well tolerated and all treated animals survived the intoxication. In contrast, none of the untreated animals, or those whose blood was hemoperfused over particles without antidigoxin antibodies, survived the intoxication (Margel, et al, 1984; Marcus, et al, 1985).

5. DRUG DELIVERY SYSTEMS.

Polymer particles have been studied extensively over the last decade for use as drug delivery systems for controlled release and targeting (reviewed by Widder and Green, 1985; Lim, 1984; Davis, et al, 1984). The advantages of controlled release systems (Langer and Folkman, 1976; Murray, et al, 1984) have long been recognized and include: (1) continuous release of constant levels of drugs at specific sites or in systemic circulation; (2) reduction of toxicity; and (3) reduction of the need for frequent administration of drugs.

Targeting drugs to specific sites has also important advantages: (1) specific targeting of tumor cells can be accomplished; (2) High local concentrations of the chemotherapeutic agent can be achieved. These concentrations cannot be achieved by systemic administration of equivalent amounts of the chemotherapeutic agent because of high toxicity; (3) The efficacy (therapeutic index) of the treatment is significantly enhanced, whereas the toxicity is contained.

The need to administer high concentrations of drugs is most urgent in chemotherapy. Most tumors exhibit a dose-related response to chemotherapy and irradiation (Skipper, et al, 1964; Frei and Canellos, 1980) and high levels of drugs or irradiation can completely eliminate the tumor cells from the body in certain cancers. However, administration of such high levels of chemotherapy are prohibited because they will cause death of the patients due to toxicity. The use of targeting systems permit the achievement of high local concentrations of drugs that can eradicate the tumor completely under acceptable toxicity levels.

A large number of materials have been investigated as polymer matrixes for drug delivery. These include: albumin (Scheffel, et al, 1972; Evans, 1972a; 1972b; Widder, et al, 1978; Widder and Senyei, 1983; Yapel, 1985; Longo and Goldberg, 1985; Tomlinson and Burger, 1985; Senyei, et al, 1985); starch (Mosbach and Schroder, 1979; Russell, 1983); ethyl cellulose (Kato, et al, 1981; Ohnishi, et al, 1984; Kato, et al, 1985); ethyl oleate based emulsions (Morimoto, et al, 1983); polylactates and polyglycolates (Miller, et al, 1977; Kitchell and Wise, 1985; Cowsar, et al, 1985); dextran (Edman, et al, 1980; Sjoholm and Edman, 1984; Schroder, 1985) and others (Tokes, et al, 1982; Sjoholm and Edman, 1979; Edman and Sjoholm, 1979).

Polymer particles to be used in drug delivery systems must be biodegradable, biocompatible, non-toxic, stable, have sufficient shelf-life, allow the entrapment and release, at controllable rates, of drugs (including those of relatively high molecular weight) and have appropriate size to penetrate capillaries. Also, polymer particles in drug delivery must permit selective targeting to the tumor cells or organ of interest, by using monoclonal antibodies (Illum and Jones, 1985; Freeman and Mayhew, 1986) or other suitable means such as an extracorporeal magnet (Widder, et al, 1981a). Biocompatible microspheres loaded with chemotherapeutic agents and selectively targeted to tumor cells using appropriate means such as monoclonal antibodies, are expected to be effective delivery systems for the treatment of certain tumors. Drugs released from the microspheres will permit the achievement of high local drug concentrations able to destroy the tumor cells to which the microspheres bind through the antibody, while the systemic concentrations will be kept within acceptable limits.

Although a number of polymer particles containing drugs have been administered in vivo for evaluation of their properties for drug delivery, only a few have been actually tested for therapeutic efficacy. Earlier attempts involved magnetic starch polymers (Mosbach and Schroder, 1979), albumin microspheres (Senyei, et al, 1978; Widder, et al, 1978, Sugibayashi, et al, 1979), acrylic polymers (Sjoholm and Edman, 1979; Edman and Sjoholm, 1979) and dextran particles (Sjoholm and Edman, 1984). L-asparaginase immobilized to polyacrylamide microspheres was found to be significantly more effective in treating L-asparagine-dependent tumors than free enzyme, in both mice and rats (Edman and Sjoholm, 1979; 1983). Poly (alkyl cyanoacrylate) particles (size distribution 10 to 1000 nm)(Kreuter, 1983; 1985) have been used as carriers of chemotherapeutic agents. Brasseur, et al, (1980) demonstrated considerable decrease in tumor growth of the S250 soft tissue carcinoma in rats with actinomycin D containing particles in comparison to free drug. Similar findings were made by Kreuter and Hartmann (1983) against the Crocker sarcoma S180 using 5-fluorouracil containing microspheres. Mizushima, et al, (1986) used lipid microspheres as carriers for delta $7-PGA_1$ to treat L1210 leukemia in mice. Significantly prolonged survival was observed in animals treated with drug-containing microspheres in comparison to those receiving free drug.

Hsieh, et al, (1981) developed a system for modulated sustained release of drugs using a magnetic field. Bovine serum albumin and magnetic steel beads were incorporated into an ethylacrylate vinyl acetate copolymer matrix. Albumin was released slowly and continuously by the particles in aqueous medium. Application of an oscillating magnetic field increased the rate of release of albumin by 100%. Application of the magnetic field for 6 hr period over a period of five days resulted in increases and decreases of the rate of release, demonstrating a pattern of modulated sustained release.

Albumin-coated magnetic microspheres prepared by emulsion polymerization, have been studied for drug delivery by Widder and Senyei using on extracorporeal magnet as a targeting system (Senyei, et al, 1978; 1985; Widder, et al, 1978; 1979a; 1979b; 1981a, 1981b; Widder and Senyei, 1983). Albumin-coated magnetic microspheres containing doxorubicin and magnetite (Fe_3O_4) were targeted selectively using an extracorporeal magnet to Yoshida Sarcoma tumors transplanted in the tails of rats (Widder, et al, 1981a). All animals that received the doxorubicin containing microspheres and targeting exhibited either total remission (75% of the animals) or marked tumor regression (remaining 25%) with no deaths or metastases. In contrast, all groups of animals that received either free doxorubicin or drug-bearing microspheres without targeting by the external magnet or placebo microspheres, exhibited significant increase in tumor size with widespread metastases and most rats died (Widder, et al, 1981a). Similar observations were made by Morris et al (1984), in treating Yoshida sarcoma-bearing rats with albumin magnetic microspheres containing vindesine sulfate. Sugibayashi, et al (1982) compared free doxorubicin and albumin microspheres containing doxorubicin for the treatment of AH7974 lung carcinoma in the rat. They observed that: (1) magnetically targeted doxorubicin-microspheres exhibited greater anti-tumor effect than free doxorubicin, as determined by histopathological examination; (2) tumor growth (assessed by determining lung weight) was significantly delayed in the animals treated with doxorubicin-microspheres than in those treated with free drug; (3) the concentration of doxorubicin in the lungs of animals treated with doxorubicin-microspheres was eight times higher than that in the animals treated with free drug; (4) the mean survial time of animals treated with doxorubicin-microspheres was significantly longer, than those treated with free drug.

Ovadia, et al, (1983) investigated the selective localization of albumin-coated magnetic microspheres in different anatomic sites. They reported that even with high magnetic field gradients, effective localization of microspheres appeared to be restricted to tissues relatively close to body surfaces. Complex vasculature networks, tissue-vascular barriers and subsurface depth are factors restricting the specific localization of microspheres by the external magnetic field (Ovadia, et al, 1983). In addition, albumin-coated magnetic microspheres were an effective carrier for myelin basic protein and were used in the induction of experimental allergic encephalomyelitis in the rat (Ovadia, et al, 1982; Carbone, et al, 1983). Goldberg, et al, (1984) reported that significantly higher concentrations of adriamycin could be administered in intratumor or intravenous injections in mice, if the drug was bound to human serum albumin microspheres. Equal concentrations of free adriamycin killed 80% of the animals. Also, when microsphere-bound adriamycin was administered, the severe necrotic lesion that develops at the site of injection when free drug is injected, was not observed.

Microspheres labeled with radioactive isotopes have been used primarily for diagnostic purposes in medicine. Albumin microspheres labeled with ^{99m}Tc (reviewed by Burger, et al, 1985) have been extensively used for diagnostic in vivo imaging in brain scans (Burdine, et al, 1970), determination of the structure and function of the reticuloendothelial system (Scheffel, et al, 1972), studies on phagocytic properties of Kupffer cells (Reske, et al, 1981), determination of the rate and pattern of gastric emptying (Theodorakis, et al, 1980), blood flow (Hoffbrand and Forsyth, 1969; Hof, 1982), lung ventilation (Millar, et al, 1982) and cerebral perfusion imaging (Etani, et al, 1983). Bartlett, et al, (1984) studied the localization of magnetic microspheres in 36 canine osteogenic sarcomas using human albumin magnetic microspheres labeled with ^{99m}Tc and an extracorporeal magnet as the targeting system. They observed a definite improvement in tumor localization and retention of the labeled microspheres in the presence of extracorporeal magnet, when the microspheres were administered intra arterially. In contrast, the magnet had no effect when the particles were given intravenously. An additional potential of microspheres in cancer treatment will be delivery of radioactive isotopes of high specific activity to the tumor cells, in efforts to eradicate the tumor. For this application effective targeting systems will be needed.

Another application of polymer particles in drug delivery is that of chemoembolization (Kato and Nemoto, 1978; Kato, et al, 1980; 1981; 1985). Microspheres containing mitomycin C are injected intra arterially by selective catheterization of the appropriate artery that brings blood supply to the tumor. Ethyl cellulose is an inert and water-soluble material. The mean particle diameter of these microspheres is in the range of 100 to 250 μm, depending on the method of the preparation. When given intra arterially they embolize in the small vessels at the target site. Mitomycin C is released slowly in the tumor and the surrounding tissue resulting in high local concentration of the drug, whereas the systemic blood levels are kept low. The therapeutic effect of these mitomycin C containing microspheres have been shown to be a function of the sustained drug release and of the infarction (Kato, et al, 1980). Over the last few years, approximately 300 patients with advanced carcinoma of the kidney, liver, prostate, bladder and lung were treated with chemoembolization using mitomycin C ethyl cellulose microspheres. Approximately, thirty five percent of the patients exhibited a response consisting of tumor reduction greater than 50% in area, 37% of the patients showed a tumor reduction less than 50% and 28% did not respond (Kato, et al, 1985). Side effects consisting of fever, anorexia, bone marrow depression and local pain appeared to be manageable. Similar results were obtained (responses in approximately 58% of the patients) when 20 patients with inoperable hepatocellular carcinoma were treated by chemoembolization using ethyl cellulose microspheres containing mitomycin C (Ohnishi, et al, 1984).

6. MAGNETIC POLYMER PARTICLES IN CELL SEPARATION.
6.1. General Considerations

Cell Separations on the basis of magnetic fields were first introduced by Rous and Beard (1934) for the removal of Kupffer cells that phagocytized iron particles. Subsequently, cells were allowed to phagocytize magnetite particles coated with sensitizing agents, and then were removed using a magnet (Lichtenstein, 1973). Recently magnetic microspheres coated with appropriate antibodies have been introduced by several groups for the separation of different populations of cells (immunomagnetic separations). These particles are described in Table 1. Almost all these particles are ferromagnetic and contain substantial amounts of iron (15 - 40%) either as magnetite (Fe_3O_4) or maghemite(γFe_2O_3). Most of them (discussed below) have been generated by coating magnetite (Fe_3O_4) particles (that are available commercially under the name Ferrofluid) with various polymers, using a number of polymerization reactions (Table 1). Others formed magnetite particles from ferrous and ferric chloride solutions (Khalafalla and Reimers, 1980; Molday and MacKenzie, 1982). Non-covalent coating with albumin permitted the attachment of various ligands (Molday and MacKenzie, 1982; Molday, 1984; Owen and Sykes, 1984; Owen and Liberti, 1986; Poynton, et al, 1983; 1985). A different approach has been employed by Ugelstad (Ugelstad, et al, 1983; Treleaven, et al, 1984) who first formed porous polystyrene monosized polymer particles and then precipitated the magnetic material inside the particles from iron salt solutions.

All magnetic microspheres to be used in cell separation must fulfill the following criteria: (1) they should be stable in media or isotonic buffers used in cell separation; (2) they should not aggregate in physiological media; (3) they should bind relatively strongly to immunoglobulin molecules or appropriate ligands by adsorption or form covalent bonds under appropriate reaction conditions; (4) they should not bind to cells non-specifically. Binding to cells should be directed exclusively by the antigen-antibody reaction; (5) they should allow satisfactory physical separation (in magnetic fields) of the cells labeled with the particles, from the unlabeled cells.

Both direct and indirect labeling methods can be used in immunomagnetic separations. However, it is becoming clear that best results can be obtained using indirect methods. According to the direct methods antibody or antibodies (or other ligands) are placed on the surface of the microspheres, and are reacted directly with the cells. This antibody(ies) recognizes a cell surface antigen or structure selectively expressed on the population of cells that it is desired to be separated. Alternatively, when indirect labeling methods are used, the cells are coated with antibodies recognizing a cell surface antigen(s) and then are conjugated with the magnetic microspheres, which have been previously coated with a second antibody recognizing the antibody

TABLE I

MAGNETIC MICROSPHERES EMPLOYED IN CELL SEPARATION

Microspheres	Chemical Composition	Polymerization Method	Diameter	Representative Cell Types Separated	Magnetic Separation Method	References
Methacrylate	Methyl methacrylate; 2-hydroxyethyl methacrylate; meth-acrylic acid; ethylene glycol dimethacrylate; magnetite.	^{60}Co gamma irradiation	30 to 50 nm	Mouse T and B cells	Laboratory magnet	Molday, et al, 1977.
Polyglutaral-dehyde	Glutaraldehyde; magnetite.	Chemical polymerization of glutaral-dehyde at high pH.	0.1 to 0.5 μm	Mixtures of red blood cells from different species.	Laboratory magnet	Margel, et al, 1979; Rembaum, et al, 1982.
Polyacrolein grafted polystyrene	Polystyrene; acro-lein; magnetite.	^{60}Co gamma irradiation.	10 μm	Mixtures of red blood cells from different species.	Laboratory magnet	Rembaum, et al, 1982.
Polyacrolein	Acrolein; magnetite.	Anionic polymeri-zation of acrolein at pH 10.5.	4 μm or other sizes.	Mixtures of red blood cells from different species.	Laboratory magnet	Margel, et al, 1982.
Agarose-poly-acrolein	Acrolein; magnetite; agarose.	Anionic polymeri-zation of acrolein. Encapsulation with agarose.	150-250 μm	Mixtures of red blood cells from different species. Mouse T and B lymphocytes.	Laboratory magnet	Margel, et al, 1983.
Hydrogel	2-hydroxy methacry-late; N-N'-methylene bisacrylamide; methacrylic acid magnetite.	Redox polymeri-zation system	>50 nm	Neuroblastoma; cells expressing G_{M1}	Electro-magnet	Kronick, et al, (1978); Kronick, 1980.

TABLE I CONTINUED

Name	Composition; Method	Size	Application	Separation	Reference
Magnogel	Acrylamide; agarose; magnetite. Chemical polymerization.	50 to 160 µm	Mouse and rat T and B lymphocytes.	Laboratory magnet	Antoine, et al, 1978.
Albumin	Albumin; protein A; magnetite. Emulsion polymerization.	0.2 to 1.5 µm	Red blood cells from various species; rat T and B lymphocytes; removal of human T cells; removal of HLA-BW 6+ lymphocytes.	Laboratory magnet	Senyel et al, 1978; Widder, et al 1979a; 1979b; Kandzia, et al, 1981; 1984, Ohman, et al, 1985.
Iron-dextran	Dextran; magnetite. Precipitation	30 to 40 nm	Red blood cells from myeloma cells.	Laboratory magnet; High gradient magnetic separation.	Molday and MacKenzie, 1982; Molday, 1984.
Cobalt/magnetite albumin immunocolloid.	Magnetite; albumin; cobalt. Precipitation/reduction.	150 nm or 20 to 60 nm	Removal of leukemic cells from human bone marrow cell suspension. Autologous bone marrow transplantation.	Samarium/cobalt magnets and high gradient magnetic separation.	Poynton, et al, 1983; 1985.
Protein-magnetite	Magnetite; albumin. Precipitation	30 to 40 nm	Red blood cells from different species; Mouse T and B lymphocytes.	High gradient magnetic separation.	Owen and Sykes 1984; Owen, 1986.

TABLE I CONTINUED

Highly porous monosized polystyrene particles.	Styrene divinyl-benzene; magnetite.	Chemical	3 μm	Removal of neuroblastoma cells from human bone cell suspensions. Autologous bone marrow transplantation.	Samarium/cobalt magnets; electromagnet.	Ugelstad, et al, 1979; 1980. Treleaven, et al, 1984.
M-450 and M-450 tosylated monosized polystyrene particles.	Styrene divinyl-benzene; magnetite.	Chemical	4.5 μm	Removal of T cells from human bone marrow cell suspensions. Human T lymphocyte populations.	Samarium/cobalt magnets.	Platsoucas, et al, 1987a. Gaudernack, et al, 1986; Lea, et al, 1985.

molecules on the cells. A frequently employed combination of two different antibodies in indirect labeling methods is polyclonal rabbit or goat anti-mouse immunoglobulin for coating of the microspheres and mouse monoclonal antibodies recognizing a cell surface antigen selectively expressed on the cell population to be separated, for coating of the cells. Additional ligands in combination with immunoglobulin have been also used. Widder, et al, (1978; 1979a; 1979b) used staphylococcal protein A (SpA) as a ligand on magnetic microspheres. SpA binds to the Fc portion of most immunoglobulins of mammalian origin, which bind to the cells through the Fab portions of the molecule. Owen (1986) and Owen and Liberti (1986) employed avidin-conjugated monoclonal antibodies for labeling of the cells and biotin-conjugated particles for the separation of mouse spleen T lymphocytes. The specificity of the immunomagnetic separations is based on the affinity of the antibody for the cell surface antigen and not on the physical properties (magnetism) of the microspheres. On the basis of this criterion all immunomagnetic methods are biological and not physical methods for cell separation (Platsoucas and Catsimpoolas, 1980).

Cells labeled with magnetic microspheres are retained using a magnetic field. Various kinds of instruments have been used, including (Table I): conventional laboratory magnets; laboratory electromagnets (Kronick, et al, 1978; Kronick, 1980); samarium-cobalt magnets in a closed flow system able to handle large numbers of cells (up to 10^{11} at a time) (Treleaven, et al, 1984); and columns for high gradient magnetic separations (Melville, et al, 1975; Owen, 1978; 1983; Molday, 1984)(The high gradient magnetic method has been reviewed by Obertenffer (1973); Oder, (1976); and Watson, (1973)). All these separation instruments are simple and easy to use. However, there is substantial room for improvement of the instrumentation, a need that is becoming apparent in large scale cell separations (see below bone marrow purging). Furthermore, although certain immunomagnetic separations are effective for depletion of individual cell types from mixed populations of cells (as it will be discussed in other paragraphs of this review), in general the retained cell population by the magnetic field cannot be recovered, with few exceptions (Gaudernack, et al, 1986).

6.2. Magnetic microspheres

Magnetic affinity beads, or magnetic microspheres, conjugated with an antimouse immunoglobulin antibody were first introduced by Molday, et al (1977) for the separation of mouse spleen T and B lymphocytes. The magnetic microspheres were prepared by ^{60}Co gamma irradiation-induced polymerization of methyl methacrylate, hydroxyethyl methacrylate, methacrylic acid and ethylene glycol dimethacrylate in the presence of colloid iron oxide particles (Fe_3O_4). The microspheres were purified by density gradient centrifugation derivatized using 1, 7-diaminoheptane and conjugated to the appropriate immunoglobulin using the glutaraldehyde reaction. Mouse spleen lymphocytes labeled with this reagent were allowed to sediment on a column containing phosphate buffered saline and 5% bovine serum albumin. A horseshoe magnet was placed for two hours on the side of the column to separate the immunoglobulin-positive cells. However, only 1x10^6 lymphocytes were processed at a time. The time period required for the magnetic separation step was rather long (2 hours). The method for preparation of these microspheres was rather complex and their tendency to form aggregates have limited their use in cell separation (Molday, 1984).

The same research group (Margel, et al, 1979; Rembaum, et al, 1982), reported the synthesis of polyglutaraldehyde magnetic microspheres, by polymerization of glutaraldehyde at high pH in the presence of magnetite. These microspheres possess aldehyde functional groups and can be easily conjugated with antibodies. These reagents which are easier to prepare were employed for the separation of mixtures of human and turkey red blood cells. Ten million red blood cells could be handled at a time, in a procedure requiring two hours for the magnetic separation step. However, aggregation of these particles in physiological buffers or media have limited their applications in cell separation. Rembaum, et al, (1982) has grafted polyacrolein onto polystyrene magnetic microspheres using ^{60}Co gamma irradiation. Margel, et al, (1982) developed polyacrolein magnetic microspheres by anionic polymerization. Both these microspheres were used for the separation of red blood cells from different species. Margel, et al (1983) encapsulated polyacrolein or polyglutaraldehyde magnetic microspheres with agarose. These microspheres have diameter in the range of 150-250 micrometers. These microspheres coupled with the appropriate antibodies or lectins were used for the separation of human from turkey erythrocytes and T and B mouse spleen lymphocytes. However, the limited application of these particles in immunomagnetic separations does not permit complete evaluation of their properties.

Kronick, et al (1978) reported a method for the separation of a subpopulation of neuroblastoma cells, based on hydroxyethyl methacrylate microspheres (hydrogel). These cells expressed in their membrane the ganglioside G_{M1}, which binds to the cholera toxin (choleragen). Preparation of the magnetic hydrogel microspheres with 99% yield and predictable size, was achieved by polymerizing 2-hydroxyethyl methacrylate, N,N'-methylene bisacrylamide and methacrylic acid in the presence of magnetite (50nm particles), using a ferrous ion-persulfate redox polymerization system, instead of a conventional initiator system. Choleragen was then coupled to the magnetic beads by the carbodiimide and glutaraldehyde reactions. A subpopulation of C-1300 neuroblastoma cells (approximately 10%) that expressed ganglioside G_{M1} in their membranes which binds to cholera toxin, were successfully removed by this procedure. Cell suspensions containing G_{M1}-negative cells of purity higher than 99% that were able to grow in culture were obtained using this approach. The separation which was carried out using a small electromagnet was rapid and it was completed in six minutes. Two to five million cells were processed at a time (Kronick, 1980). However, the use of these magnetic microspheres in cell separation is limited to allow evaluation of their performance.

Antoine, et al (1978) used large (50 to 160μm) magnetic polyacrylamide agarose beads (magnogel) coated with anti-Ig antibodies for the separation of mouse or rat T and B lymphocytes. Approximately 99% of the cells eluted through these columns were Ig-negative, however, the retained cells by these anti-Ig beads, were 62-79% Ig-bearing cells and 21-38% immunoglobulin negative cells.

Widder, et al, (1978, 1979a, 1979b; 1981b) and Senyei, et al, (1985) reported the synthesis of magnetic microspheres containing human serum albumin and magnetite by an emulsion polymerization procedure originally developed by Scheffel, et al, (1972). Albumin and staphylococcal protein A (SpA) solution containing magnetite were mixed with cottonseed oil and the suspension was sonicated and subsequently added to cottonseed oil preheated to 110-165°C for hardening. The microspheres were washed extensively with phosphate-buffered saline supplemented with a nonionic detergent to remove the cottonseed oil and were coupled to immunoglobulin molecules via protein A. Protein A-magnetic microspheres were coated by incubation with either rabbit anti-chicken erythrocyte antibody or rabbit anti-rat immunoglobulin antibody and were used, respectively, for the separation of artificial mixtures of chicken and sheep erythrocytes, or removal of Ig-bearing lymphocytes from rat spleen cell suspensions. Cells bearing antibody-coated SpA microspheres were removed by applying a 4000 gauss bar magnet (gradient:1500 gauss/cm) for 1 min on the side of the tube. The removal of these cells was complete. However, very small numbers of lymphoid cells were handled at a time (4×10^5). Kandzia, et al, (1981) employed albumin-coated SpA-containing magnetic microspheres for the removal of HLA-Bw 6 positive human peripheral blood lymphoctes from mixtures of HLA-Bw 6 and -Bw 4 cells. However, only 0.5×10^6 cells have been separated at a time. In another report these investigators (Kandzia, et al, 1984) reported certain modifications of the polymerization procedure resulting in the formation of microspheres with more uniform size. Ohman, et al, (1985) employed albumin-coated magnetic microspheres and the T101 and anti-Leu 5 anti-T cell monoclonal antibodies for the removal of T cells from suspensions of human peripheral blood lymphocytes. At a ratio of 160:1 of microspheres to cells they achieved nearly complete removal of T cells, on the basis of immunofluorescence analysis using the fluorescence activated cell sorter.

We investigated the effectiveness of albumin-coated SpA-containing magnetic microspheres, prepared as described by Widder, et al (1979a; 1979b) for the removal of T lymphocytes, coated with the anti-Leu 5 and anti-Leu 2a monoclonal antibodies, from human bone marrow cell suspensions. First, erythrocytes, certain other mature erythroid cells and granulocytes were removed from human bone marrow cell suspensions by centrifugation on a ficoll/hypaque density cushion, as previously described (Platsoucas, et al, 1981). Bone marrow cells were removed from the interface, washed three times with phosphate-buffered saline and incubated on ice for one hour with anti-Leu 5 and anti-Leu 2a monoclonal antibodies (0.2 micrograms each per 20×10^6 cells) with frequent agitation and subsequently were washed to remove unbound antibody. Appropriate numbers of albumin-coated microspheres (to achieve a microsphere: cell ratio of 100:1) were added to affinity-purified goat anti-mouse IgG (400 micrograms per 10×10^6 cells) and incubated at room temperature for 1.5 hours on a rotator. The microspheres were extensively washed with phosphate-buffered saline by centrifugation. Bone marrow cells coated with the anti T-cell monoclonal antibodies were mixed with goat anti-mouse IgG conjugated magnetic microspheres and incubated at 4°C for 2 hours. T lymphocytes labeled with microspheres were removed by passing the cells through a chamber (Figure 1) containing samarium cobalt magnets using a peristaltic pump (Platsoucas, et al, 1987a), similar to the one orginally described by Treleaven, et al, (1984). A representative experiment of depletion of human bone marrow cells from T lymphocytes using albumin coated SpA containing magnetic microspheres is shown in Table 2. E-rosette forming cells were reduced in the eluted cells to 5% from 42% in the initial fraction. Proliferative response to phytohemaglutinin (PHA) was practically eliminated in the eluted fraction. Proliferative response to a pool of allogeneic mononuclear leukocytes in mixed lymphocyte culture was significantly reduced in comparison to the controls (Table 2). These results suggest that quantitative depletion, although apparently non-complete, of T cells from human bone marrow cell suspensions can be achieved using albumin-coated SpA containing magnetic microspheres (Platsoucas, et al, unpublished results). However, there were certain problems associated with the albumin-coated SpA-containing magnetic microspheres and their use in cell separation: (1) the microspheres were not uniform in size and their diameter distribution ranged from 0.2 to 1.5 µm; (3) they formed microaggregates (three or more microspheres) in media or physiological buffers.

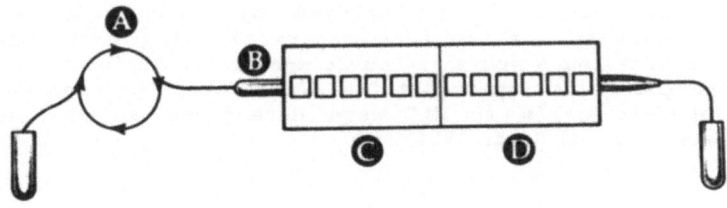

Figure 1: Apparatus used for immunomagnetic removal of T cells from human bone marrow cell suspensions using anti-T cell monoclonal antibodies and either albumin-coated SpA-containing magnetic microspheres or M-450 magnetic monosized polymer particles. This apparatus consists of two polycarbonate chambers with six samarium cobalt permanent magnets attached to the bottom (1 cm x 1 cm x 0.4 cm; Magnet Sales and Manufacturing Company, Culver City, CA) and a peristaltic pump (A). To minimize nonspecific trapping of the cells that may occur because of the formation of aggregates at the beginning of the chamber due to strong magnetic fields, the following design was adopted. In the first chamber (C), the magnets were separated from the interior space by 4 mm of polycarbonate polymer material and the distance between two successive magnets was 1 cm. In the second chamber (D), the magnets were separated from the interior by 2 mm of polycarbonate and the distance between two successive magnets were 6 mm. A sterile disposable 2 ml pipet (B) was placed in the chambers and was connected to the peristaltic pump (A) and to the collection tube with plastic tubing. The actual separation took place in this sterile pipet (B). The separation system was sterilized by washing with 70% alcohol and UV irradiation and washed with phosphase buffered saline supplemented with 1% fetal calf serum.

TABLE 2.   Removal Of T Lymphocytes From Human Bone Marrow Cell
Suspensions Using Albumin-Coated SpA-Containing Magnetic
Microspheres Coated With Anti-Leu 5 and Anti-Leu 2a Monoclonal
Antibodies.

| | E-rosettes | Proliferative Responses (Counts per minute) | | |
	%	Medium	PHA	MLC
Initial Sample[*]	42%	1,540 ± 359	142,654 ± 16,721	9,300 ± 2,227
Eluted Cells	5%	144 ± 67	267 ± 92	1,057 ± 218

[*]10^6 human bone marrow cells purified by centrifugation on
ficoll/hypaque density cushion were separated using microspheres
coated with the anti-Leu 5 and anti-Leu 2a monoclonal antibodies.   The
separation procedure was repeated twice. Proliferative responses to
PHA and allogeneic cells in MLC were determined as previously
described (Platsoucas and Good, 1981).

Molday and MacKenzie (1982) prepared ferromagnetic iron dextran
particles by adding to a mixture of ferrous and ferric chloride (in
25% (w/w) dextran, M.W. 40,000) ammonium hydroxide until pH 10-11 was
achieved. The resulting small colloidal magnetite particles were
stabilized by the dextran. Aggregated iron dextran and free dextran
were removed by centrifugation and gel filtration. Aldehyde groups
were generated by oxidation of the dextran with periodic acid and
proteins were covalently attached to the microspheres via the aldehyde
groups through a Schiff base which was subsequently stabilized by
reduction with sodium borohydride. Human erythrocytes, labeled with
anti-erythrocyte antibody conjugated iron-dextran microspheres, were
separated by this procedure from a mixture of erythrocytes and
unlabeled myeloma cells. A 500 G permanent laboratory magnet was used
in these experiments. Eluted cells were exclusively myeloma cells,
whereas retained cells contained labeled erythrocytes and 5-10%
myeloma cells (Molday and MacKenzie, 1982). The same research group
recently used the high gradient magnetic separation method to retain
human erythrocytes labeled with anti-human erythrocyte antibody
conjugated-protein A-iron dextran particles. Retention of 96 ± 2% of
the labeled cells was achieved.
    Poynton, et al (1983; 1985) prepared cobalt/magnetite albumin
immunomagnetic colloid particles by forming magnetite from ferrous and
ferric chloride solutions at high pH (Molday and MacKenzie, 1982) and
then coating it with cobalt metal (precipitated from cobalt chloride
solution by reduction with sodium borohydrate) and human serum
albumin. The albumin is then crosslinked around the particles with
glutaraldehyde and goat anti-mouse immunoglobulin is attached using
parabenzoquinone (Poynton, et al, 1985). Using this procedure, these

investigators removed CALLA-positive cells from the bone marrow of four patients with acute lymphoblastic leukemia. The patients then received high doses of chemotherapy followed by reinfusion of their "cleansed" marrow. These experiments are discussed in another part of this paper.

Owen and Sykes (1984) prepared magnetite-protein colloid particles by precipitating magnetite from solutions of ferrous and ferric chloride (Khalafalla and Reimers 1980; Molday and MacKenzie, 1982) and coating the particles with albumin (Poynton, et al, 1983). Goat anti-rabbit immunoglobulin or other ligands were then coupled to the albumin using the glutaraldehyde reaction or N-succinimidyl-3(2-pyridyldithio) proprionate (SPDP). These particles were used for the separation of red blood cells from different species and of T from B mouse spleen lymphocytes (Owen and Sykes, 1984; Owen, 1986; Owen and Liberti, 1986). The limitation of these particles and of the iron-dextran particles (Molday and MacKenzie, 1982) is that they bind non-specifically to populations of leukocytes (Owen, 1986; Owen and Liberti, 1986).

Truly monosized and monodisperse magnetic microspheres were developed by Ugelstad (Ugelstad, et al, 1979; 1980; Treleaven, et al, 1984; Nustad, et al, 1986). Porous monosized polystyrene particles were prepared by polymerization of styrene divinylbenzene and the magnetic material was precipitated inside from iron salt solutions. The following particles were made:

(1) highly porous polystyrene magnetic monosized particles of 3 microns in diameter with a surface area of approximately 100 m^2 per gram (approximately 10^8 microspheres/mg). These polystyrene particles were modified to make them hydrophylic (Ugelstad, et al, 1980). Under appropriate conditions these particles bind noncovalently protein to their surface. Proteins are known to bind nonspecifically to polystyrene (reviewed by Voller, et al, 1980). These magnetic microspheres were used for purging of neuroblastoma cells from human bone marrow cell suspensions for autologous bone marrow transplanatation (Treleaven, et al, 1984). These studies will be discussed in another part of this review.

(2) M-450 polystyrene magnetic monosized particles. These monodispersed particles have been recently developed by Ugelstad and collaborators (Ugelstad, et al, unpublished results) from preformed highly porous polystyrene magnetic particles by filling the pores with a polymeric material. This resulted in reduction of the surface of the microspheres to 3 to 5 m^2 per gram from about 100 m^2 per gram. The diameter of these particles is 4.5 microns. These monodispersed particles have a rather hydrophobic character which permits noncovalent binding of protein to the microspheres by physical adsorption. The characteristics of these microspheres are discussed in another chapter of this volume (Ugelstad, et al, 1987).

(3) Tosylated M-450 polystyrene magnetic monosized particles. The presence of sufficient numbers of hydroxyl groups on the surface of the M-450 particles allow covalent binding of proteins by the tosyl chloride method of Nilsson and Mosbach (1980), in addition to non-covalent binding. Therefore, these particles bind immunoglobulin both by covalent and non-covalent means (Ugelstad, et al, 1987).

Both M-450 and the tosylated M-450 particles have been used for the depletion of T-cells from human bone marrow cell suspensions (Platsoucas, et al, 1987a; 1987b). In addition, Lea et al (1985) and Gaudernack, et al (1985; 1986) have succeessfully used the M-450 particles for the separation of human T lymphocyte subpopulations. Also, Danielsen, et al (1986) used these particles to study interactions between cell surface antigens and their corresponding antibodies bound to the particles.

6.3. Immunomagnetic methods in bone marrow transplantation.

The major advantages of the immunomagnetic methods over already existing techniques is time efficiency and capacity to handle large numbers of cells (up to 1×10^{11}) at a time. These advantages make the immunomagnetic methods very suitable for: (1) the removal of tumor cells from human bone marrow cell suspensions to be used in autologous bone marrow transplantation (bone marrow purging); (2) the removal of T cells from human bone marrow cell suspensions to be used for allogeneic bone marrow transplantation across histocompatibility barriers (T-cell purging).

6.4. Removal of tumor cells from human bone marrow cell suspensions. Autologous bone marrow transplantation.

Autologous bone marrow transplantation is becoming an increasingly popular method of treatment of patients with certain cancers including leukemia, lymphoma, neuroblastoma, testicular carcinoma, ovarian cancers, and others (Epstein, et al, 1979; Buckner, et al, 1970; Tobias, et al, 1977; Graze and Gale, 1978; Diesseroth and Abrams, 1979; Dicke, et al, 1978; 1979, 1984; Graze, et al, 1979; Appelbaum, et al, 1978a; 1978b). It is well known that most tumors exhibit a dose-related response to chemotherapy and irradiation (Skipper, et al, 1964; Frei and Canellos, 1980). Therefore, administration of high levels of chemotherapy and irradiation in certain cancers can indeed eradicate all tumor cells in the body. However, these high levels of chemotherapy and irradiation are prohibited because they will result in the death of the patient due to failure of the hemopoietic system. This is prevented by removing bone marrow from the patient before treatment and then reinfusing the marrow back to the patient to achieve hemopoietic reconstitution. This is usually achieved within two to four weeks and death because of the permanent distruction of the hemopoietic system is prevented. Removal of contaminating tumor cells in vitro is highly recommended before infusion of the marrow. It is likely that these cells are present in the bone marrow, because metastatic spread of tumor cells at this site of the body is observed at early stages of most cancers. Reinfusion with the bone marrow even of small numbers of tumor cells that often may be below the detectable limits of the available methods, dramatically increases the likelihood of relapse. Removal of tumor cells from the bone marrow in vitro is known as bone marrow purging. A number of methods have been developed for the removal of tumor cells from human bone marrow cell suspensions, including immunotoxins (monoclonal antibodies conjugated with Ricin) (Thorpe, et al, 1982), monoclonal antibody plus complement treatment (Ritz, et al, 1982) and in vitro treatment of the bone marrow with chemicals (such as 4-hydroperoxycyclophosphamide; Korbling, et al, 1982) that lyse preferentially the tumor cells without affecting the normal cells.

The bone marrow purging procedure has the following steps: (1) Removal from the patient usually in first or second remission of bone marrow from the bilateral posterior iliac crest regions. Approximately 10% of the patient's bone marrow is usually removed. Mononuclear cells are prepared and they are cryopreserved. Bone marrow cells can be cryopreserved for prolonged time periods, without loss of their hemopoietic activity (Graze and Gale, 1978; Diesseroth and Abrams, 1979); (2) Treatment of the patient with lethal or near lethal irradiation and high doses of chemotherapy; (3) Thawing of the bone marrow cells and preferably removal of the tumor cells using in vitro cell separation techniques (purging); (4) Reinfusion of the bone marrow (preferably after purging) to the patient. The reinfusion of the bone marrow cells (untreated or "cleansed" of tumor cells) is needed in order to rescue the patient from death due to complete destruction of his hemopoietic system by the chemotherapy and irradiation treatments.

Poynton, et al, (1983; 1985) used cobalt magnetite albumin immunomagnetic colloid particles coated with goat antimouse immunoglobulin to remove CALLA-positive cells (coated with the J5 monoclonal antibody) from the bone marrow of four patients with acute lymphocytic leukemia. The purged bone marrow was reinfused into the patients after treatment with high doses of chemotherapeutic agents. Rapid and complete hemopoietic reconstitution was observed in three patients, whereas the fourth patient died on day 17 after transplantation due to infection. Remission of three months or longer was observed in the remaining patients (Dicke, et al, 1984). Additional research and long term follow up is required to evaluate the value of this method for the treatment of patients with leukemia.

Treleaven, et al (1984) used highly porous 3 μm polystyrene magnetic particles and six monoclonal antibodies recognizing neuroblastoma-associated differentiation antigens for the removal of tumor cells from bone marrow of patients with neuroblastoma. The "cleansed" bone marrow was reinfused to the patients. A flow system using permanent samarium-cobalt magnets, an electromagnet and a peristaltic pump permitted removal of tumor cells labeled with the magnetic microspheres from 10^9 to 10^{10} bone marrow cells within a few hours. Approximately 65% of the initial nucleated cells were recovered. These bone marrows were initially contaminated with 3-8% tumor cells. At least three log depletion was achieved, as determined by indirect immunofluorescence using twelve monoclonal antibodies recognizing neuroblastoma-associated differentiation antigens as well as by histological and cytological procedures. Approximately 95% of the hemopoietic progenitor cells (CFU-e, BFU-e, BFU-e, CFU-gem and CFU-c) were recovered. The viability of the cells passing through the separation chamber was higher than 98%. Hemopoietic reconstitution was observed in the patients that received purged bone marrow within 2-4 weeks after treatment. Over twenty five patients have been transplanted using this procedure (Treleaven, et al, 1984; Kemshead, et al, 1985). Long-term follow-up of these patients is needed to determine the value of the method for treatment of patients with neuroblastoma.

Seeger, et al, (1985) and Reynolds, et al, (1985) investigated in vitro removal of cells from neuroblastoma tumor lines from mixtures with human bone marrow cells from normal donors (50 to 100x10^6 bone marrow cells contaminated with 10-20% cells from neuroblastoma tumor cell lines). Three to four logs of tumor cells were consistently removed from the bone marrow using two cycles of separation with monoclonal antibodies and magnetic microspheres.

6.5. Depletion of T cells from human bone marrow cell suspensions. Allogeneic bone marrow transplanatation across histocompatibility-barriers.

Bone marrow transplantation is a method of choice for the treatment a number of fatal diseases including severe combined immunodeficiency, aplastic anemia, leukemia, and others. This procedure involves elimination of the ailing bone marrow (and other hemopoietic cells) of the patient by high doses of chemotherapy and irradiation. The patient's bone marrow is replaced by infusion of marrow cells from a healthy donor, which within a few weeks reconstitute the destroyed hemopoietic system of the patient. This healthy donor must be matched with the recipient for the major histocompatibility antigens. Otherwise, the donor's T cells react and proliferate to the "alien" histocompatibility antigens of the host, resulting in graft-versus-host disease (GvHD) (Simonsen, 1957). GvHD limits the applicability of allogeneic bone marrow transplantation to those patients for which matched donors (siblings or others) are available. However, even when matched donors are used GvHD develops in approximately 50-60% of the recipients, because of differences in matching in the major or minor histocompatibility loci. It is fatal in up to 25% of the patients affected (Thomas, et al, 1975; Weiden, et al, 1980; Storb, et al, 1983; Gale, et al, 1983). Immunosuppressive therapy (anti-thymocyte globulin, cyclosporin A) has been used to decrease or eliminate the danger of the GvHD in certain patients. However, patients subjected to immunosuppressive therapy are at high risk because of fatal complications due to infection (Thomas, et al, 1975; Winston, et al, 1979; Ramsay, et al, 1982; Powles, et al, 1978; Deeg, et al, 1983).

It has been clearly demonstrated in man (Reisner, et al, 1980; 1981; 1983; Rodt, et al, 1981; Blacklock, et al, 1983; Sharp, et al, 1983; Prentice, et al, 1984; Ash, et al, 1984; Filipovich, et al, 1984) and experimental animals (Dicke, et al, 1968; Rodt, et al, 1974; Tyan, 1973; Onoe, et al, 1980; Korngold and Sprent, 1983; Vallera, et al, 1983; Reisner, et al, 1978) that mature T cells and immediate committed T-cell precursors present in the graft are responsible for the generation of both acute and chronic GvHD. Removal of T cells from the graft resulted in elimination or partial reduction of both the incidence and the severity of GvHD in man. Various methods have been used for this purpose including soybean lectin agglutination followed by rosetting with sheep erythrocytes (Reisner, et al, 1980; 1981; 1983), anti-T cell immunotoxins (Filipovich, et al, 1984) and treatment of the bone marrow before reinfusion with anti-T cell monoclonal antibodies and complement (Rodt, et al, 1981; Blacklock, et al, 1983; Sharp, et al, 1983; Prentice, et al, 1984; Ash, et al, 1984).

We employed M-450 magnetic monosized particles, as well as M-450 tosylated particles developed by Ugelstad, et al for the depletion of T lymphocytes from human bone marrow cell suspensions. Human bone marrow was obtained from healthy young adult volunteers by multiple aspirations from the posterior iliac crest. Bone marrow cells were separated on a Ficoll/Hypaque density cushion as previously described (Platsoucas, et al, 1981). The cells were incubated for 2 hrs at 4°C with mixtures of two (anti-Leu 5 and anti-Leu 2a) or four (anti-Leu 1, anti-Leu 2a, anti-Leu 3a, anti-Leu 5) anti-T cell monoclonal antibodies (approximately 2.5 mcgs of each monoclonal antibody was added to 10×10^6 cells) and were washed three times. Appropriate numbers of the M-450 microspheres, to achieve microsphere:cell ratio of 100:1, were coated with affinity purified goat or rabbit anti-mouse immunoglobulin for 14 hrs at 4°C. These microspheres bind immunoglobulin noncovalently by adsorption. In the experiments reported here 1.75 mg of antibody per gram of particles were used and were sufficient for optimal cell separation. The particles were washed and incubated with monoclonal antibody-coated bone marrow cells for 2 hrs at 4°C. The cells were separated in an apparatus (Figure 1) similar to the one described by Treleaven, et al (1984), at a rate of 6 ml/min (Platsoucas, et al, 1987a). All microspheres were retained by the magnet. To access the level of T cell depletion proliferative responses to PHA and to allogeneic mononuclear leukocytes in mixed lymphocyte culture were determined in the eluted fraction and the unseparated cells by methods described previously (Platsoucas and Good, 1981). Representative results are shown in Table 3.

TABLE 3. Removal Of T Lymphocytes From Human Bone Marrow Cell Suspensions Using Anti-T Cell Monoclonal Antibodies[+] And 4.5 Micron Magnetic Monosized Polymer Particles Coated By Physical Adsorption With Goat Anti-Mouse Immunoglobulin

	E-rosettes %	Proliferative Responses (Counts per minute)		
		Medium	PHA[++]	MLC[++]
Initial Sample[*]	15%	2,451 ± 542	26,109 ± 1,673	17,288 ± 459
Eluted Cells[++]	1%	120 ± 55	280 ± 98	952 ± 190

[+]Monoclonal antibodies used: anti-Leu 5 and anti-Leu 2a.

[++]Proliferative response to PHA and to allogeneic cells MLC were determined as previously described (Platsoucas and Good, 1981). Cultures of unfractionated human bone marrow cells and cells depleted of T lymphocytes were incubated for 6 days at 37°C, pulsed with tritiated thymidine during the last 18 hours of the culture and harvested using an automated cell harvester.

[*]$6X10^6$ Human bone marrow cells purified by centrifugation on fiicoll/hypaque density cushion.

[**]$2.4X10^6$ cells weere recovered (40%).

All these responses were absent in the T cell depleted fractions. In contrast, unseparated cells responded vigorously to both PHA and allogeneic cells (Table 3). E-rosette forming cells were determined by rerosetting with sheep erythrocytes as previously described (Platsoucas, et al, 1980) and were 0.25% ± 0.42% in the eluted fractions versus 22.5 ± 10.4% in the unseparated cells. These results represent the mean ± one standard deviation of six experiments. Recovery of eluted cells was in the range of 30 to 60%. In addition, tosylated M-450 particles coated with sheep anti-mouse immunoglobulin (Nustad, et al, 1984; 1986) by physical adsorption and covalent binding were evaluated for T-cell depletion. These particles were kindly provided for these studies by Dr. Nustad of the Norwegian Radium Hospital, Oslo, Norway. They were found to be equally effective to the M-450 particles, which were coated with anti-mouse immunoglobulin by adsorption only, in removing T lymphocytes from human bone marrow cell suspensions. Approximately 3.5 mcgs of anti-mouse immunoglobulin/mg of particles were attached to these microspheres. Removal of T cells from the bone marrow resulted in complete elimination of proliferative responses of the eluted cell fractions to PHA and allogeneic cells in mixed lymphocyte culture. In contrast, the separated bone marrow cells exhibited a vigorous proliferative response (Platsoucas, et al, 1987a; 1987b). Quantitation of T cell removal from the bone marrow was achieved using a limiting dilution microculture assay (Kernan, et al, 1985). Approximately, 2.5 log reduction of T cells was achieved in a single step using these magnetic particles (Platsoucas, et al, 1987a; 1987b), demonstrating that this method allows significant quantitative depletion of T cells from human bone marrow. Studies of the recovery of committed hemopoietic progenitor cells of the granulocyte-macrophage lineage showed that selective loss of hemopoietic progenitors was not observed in the immunomagnetic separation (Platsoucas, et al, 1987a; 1987b). Although in these experiments small numbers (up to $20x10^6$) of human bone marrow cells were depleted of T cells at the time, it is anticipated that the method can be scaled up to separate large numbers of cells sufficient for bone marrow transplantation across histocompatibility barriers without major difficulties. Depletion of T cells greater than 2.5 logs can be achieved in one cycle of magnetic separation and greater degrees of depletion can be achieved by repeating the procedure for the second time or by combining the procedure with other methods of depletion of T cells. It is anticipated that the introduction of magnetic monosized particles and of appropriate flow systems will facilitate bone marrow transplantation across major histocompatibility barriers.

7. CONCLUSIONS

Removal of tumor cells or T cells from human bone marrow to be used, respectively, for autologous or allogeneic bone marrow transplantation, appears to be the most promising application of magnetic polymer particles. The advantages of the immunomagnetic methods for bone marrow purging over existing techniques are significant: (1) Time efficiency and capacity to handle large numbers of cells in limited time periods. Up to $1x10^{11}$ bone marrow cells can be purged by the immunomagnetic method within a few hours. Complete hemopoietic reconstitution has been achieved in most of the patients transplanted (Kemshead, et al, 1984; 1985; Poynton, et al, 1983; 1985). Laborious cell separation methods are not employed. (2) Only depleted cell populatiosns are reinfused to the patients. Lysed or inactivated T cells are not reinfused as it is the case with cell elimination methods such as antibody plus complement treatment or immunotoxin treatment in vitro. (3) The method takes place in a sterile closed continuous flow system, which diminishes the danger of infection. (4) Agents of which the viral or other contamination cannot be effectively controlled (such as sheep erythrocytes) are not employed. (5) The method can be better standardized and its end point is well defined. From all the magnetic particles developed for cell separation, the magnetic monosized polystyrene particles of Ugelstad, et al have been sufficiently tested in actual bone marrow transplantation. These particles have various attractive futures in comparison to those prepared by other methods, including uniformity in size, lack of non-specific binding to other cells, satisfactory coating with immunoglobulins and high magnetic susceptibility. However, the cell separation methods based on these magnetic microspheres are at a rather early stage of development. Significant improvements are needed on the design of the equipment and the continuous flow systems, to allow complete capture of all labeled cells and completion of the separation in a few hours. Furthermore, methods for the enumeration of the remaining tumor cells in the graft are needed. These methods should be able to detect very low numbers of tumor cells.

In conclusion, immunomagnetic methods of cell separation appear to be very promising. However, additional studies are needed to determine the therapeutic value of the immunomagnetic method in bone marrow purging, including long-term follow up of the transplanted patients.

8. ACKNOWLEDGMENTS

This work was supported in part by grant PCM-8119178 from the National Science Foundation, grants CA 23766 and CA 41699 from the National Institute of Health, grant IM 409D from the American Cancer Society and a grant from the Whitaker Foundation. I thank Ms. Helen Farr for typing this manuscript.

REFERENCES

1. Antoine JC; Ternyac T; Rodisgot M and Auremeas S: Lymphoid cell fractionation on magnetic polyacrylamide-aganese beads. *Immunochemistry* 15:443, (1978).

2. Applebaum FR; Herzig GP and Ziegler JL: Successful engraftment of cryopreserved autologous bone marrow in patients with malignant lymphoma. *Blood* 52:98, (1978a).

3. Applebaum FR; Diesseroth AB and Graw GR Jr.: Prolonged complete remission following high dose chemotherapy of Burkitt's lymphoma in relapse. *Cancer* 41:1059, (1978b).

4. Ash RC; Serwint M; Doukas M; Romond E; Bradley P; Metcalf M; Marshall E; Geil J; Greenwood M; MacDonald JS and Thompson JS: Marrow T cell depletion with anti-human T cell antibodies is effective for graft versus host disease (GvHD) prophylaxis in human allogeneic marrow transplantation. *Blood* 64(suppl-1):744 (1984).

5. Bartlett JM; Richardson RC; Elliot GS; Blevins WE; Janas W; Hale JR and Silver RL: Localization of magnetic microspheres in 36 canine osteogenic sarcomas. In: Microspheres and Drug Therapy. Pharmaceutical, Immunological and Medical Aspects, Davis SS; Illum L; McVie JG and Tomlinson E(eds) Elsevier, Amsterdam, p.413 (1984).

6. Blacklock HA; Prentice HG and Gilmore M: Attempts at T cell depletion using OKT3 and rabbit complement to prevent GvHD in allogeneic BMT, *Exp. Hematol.*, 11 (suppl 13):37, (1983).

7. Bonnefoy JY; Banchereau J; Aubry JP and Wijdenes J: A flow cytometric micromethod for the detection of Fc epsilon receptors and IgE binding factors using fluorescent microspheres. *J. Immunol. Meth.* 88:25, (1986).

8. Brasseur F; Couvreur P; Kante B; Deckers-Passau L; Roland M; Deckers C and Speiser P: Actinomycin D absorbed on polymethyl-cyanoacrylate nanoparticles: increased efficiency against an experimental tumor. Eur. *J. Cancer* 16:1441, (1980).

9. Buckner CD; Storb R and Dillingham LA: Low temperature preservation of monkey marrow in dimethylsulfoxide. *Cryobiology* 7:136, (1970).

10. Burdine JA; Sonnemaker RE; Ryder LA and Spjut HJ: Perfusion studies with technetium-99m human albumin microspheres. *Radiology* 95:101, 1970.

11. Burger JJ; Tomlinson E; DeRoo JE and Palmer J: Technetium-99m labeling of albumin microspheres intended for drug targeting. In: Drug and Enzyme Targeting, Widder KJ and Green R(eds) *Meth. Enzymol.* 112:43, (1985).

12. Carbone AM; Ovadia H and Paterson PY: Role of macrophage myelin basic protein interaction in the induction of experimental allergic-encephalomyelitis in Lewis Rats. *J. Immunol.* 131:1263, (1983).

13. Cowsar DR; Tice TR; Gilley RM and English JP: Poly(lactide-coglycolide) microcapsules for controlled release of steroids. In: Drug and Enzyme Targeting, Widder KJ and Green R(eds), *Meth. Enzymol.* 112:101, (1985).

14. Danielson H; Funderud S; Nustad K; Reith A and Ugelstad J: The interaction between cell surface antigens and antibodies bound to monodisperse polymer particles in normal and malignant cells. Scand. *J. Immunol.* 24:179, 1986.

15. Davis SS; Illum L; McVie JG and Tomlinson E(eds): Microspheres and Drug Therapy. Pharmaceutical, Immunological and Medical Aspects, Elsevier, Amsterdam, New York, pp. 448, (1984).

16. Deeg HJ; Storb R; Thomas ED et al: Marrow transplantation for acute, nonlymphoblastic leukaemia in first remission: Preliminary results of a randomized trial comparing cyclosporin and methotrexate for the prophylaxis of graft-versus-host disease. Transplant. Proc. 15:1385, (1983).

17. Dicke KA; Van Hooft J and Van Bekkum DW: The selective elimination of immunologically competent cells from bone marrow and lymphatic cell mixtures. Transplantation 6:562, (1968).

18. Dicke KA; McCredie KB; Spitzer G et al: Autologous marrow transplantation in patients with adult acute leukemia. Transplantation 26:169, (1978).

19. Dicke KA; Zander A; Spitzer G et al: Autologous bone marrow transplantation in adult leukemia in relapse. Lancet 1:514, (1979).

20. Dicke KA; Jagnnath S; Spitzer G; Poynton C; Zander A; Vellekoop L; Reading C; Jehn UW and Tindle S: The role of autologous bone marrow transplantation in various malignancies. Sem. Hematol. 21:109, (1984).

21. Diesseroth A and Abrams RA: The role of autologous stem cell reconstitution in intensive therapy for resistant neoplasms. Cancer Treat. Rep. 63:461, (1979).

22. Edman P and Sjoholm I: Acrylic microspheres in vivo. II. The effect in rat of L-asparaginase given in microparticles of polyacrylamide. J. Pharmacol. Exp. Ther. 211:663, (1979).

23. Edman P; Ekman B and Sjoholm I: Immobilization of proteins in microspheres of biodegradable polyacryldextran. J. Pharm. Sci. 69:838, (1980).

24. Edman P and Sjoholm I: Acrylic microspheres in vivo VI: antitumor effect of microparticles with immobilized L-asparaginase against 6C3HED lymphoma. J. Pharm. Sci. 72:654, (1983).

25. Engvall E and Perlmann P: ELISA. III. Quantitation of specific antibodies by enzyme linked antiimmunoglobulin in antigen coated tubes. J. Immunol. 109, 129, (1972).

26. Epstein RB; Storb R; Clift RA: Autologous bone marrow grafts in dogs treated with lethal doses of cyclophosphamide. Cancer Res. 29:1072, (1969).

27. Etani H; Kimura K; Yoneda S; Tsuda Y; Nakamura M; Kataoka K; Iwata Y and Abe H: Demonstrating patency of STA-MCA anastomosis with Tc-99m albumin microspheres. J. Nucl. Med. 24:136, (1983).

28. Evans RL: U.S. Patent, 3,663,685, (1972a).

29. Evans RL: U.S. Patent, 3,663,687, (1972b).

30. Filipovich AH; Youle RJ; Neville Jr. DM; Vallera DA; Quinones RR and Kersey JH: Ex-vivo treatment of donor bone marrow with anti-T-cell immunotoxins for prevention of graft-versus-host disease. Lancet 1:469, (1984).

31. Freeman AI and Mayhew E: Targeted drug delivery. Cancer 58:573, (1986).

32. Frei E III; Canellos GP: Dose: a critical factor in cancer chemotherapy. Am. J. Med. 69:586, (1980).

33. Gale RP; Kersey JH; Bortin MM; Dicke KA; Good RA and Zwaan FE: Bone marrow transplantation for acute lymphoblastic leukaemia. Lancet II:639, (1983).

34. Gaudernack G; Leivestad T; Qvigstad E; Ugelstad J and Thorsby E: Isolation of functionally active T cell subsets by monoclonal antibodies conjugated to monodisperse magnetic microspheres M-450. Scand. *J. Immunol.* 22:444, (1985).

35. Gaudernack G; Leivestad T; Ugelstad J; Thorsby E: Isolation of pure functionally active CD8$^+$ T cells. Positive selection with monoclonal antibodies directly conjugated to monosized magnetic microspheres. *J. Immunol. Meth.* 90:179, (1986).

36. Goldberg EP; Iwata H and Longo W: Hydrophilic albumin and dextran ion-exchanged microspheres for localized chemotherapy. In: Microspheres and Drug Therapy. Pharmaceutical, Immunological and Medical Aspects, Davis, SS; Illum JG; McVie JG and Tomlinson E(eds): Elsevier, Amsterdam, p.309, (1984).

37. Gordon IL; Dreyer WJ; Yen RCK and Rembaum A: Light microscope identification of murine B and T cells by means of functional polymeric microspheres. *Cellular Immunology* 28:307, (1977).

38. Graze PR and Gale RP: Autotransplantation for leukemia and solid tumors. *Transplant. Proc.* 10:177, (1978).

39. Graze RP; Wells JR and Ho W: Successful engraftment of cryopreserved autologous bone marrow stem cells in man. *Transplantation* 27:142, (1979).

40. Gribnau TC; Leuvering TH; van Hell H: Particle-labelled immunoassays: a review. *J. Chromatogr.* 376:175, (1986).

41. Guesdon JL; Thierry R and Avrameas S: Magnetic enzyme immunoassay for measuring human IgE. *J. Allergy Clin. Immunol.* 61:23, (1978).

42. Heymer B; Schachenmayer W; Buttmann B; Spank R; Haterkamp O and Schmidt W: A latex agglutination test for measuring antibodies to streptococcal micropeptides. *J. Immunol.* 11:478, (1973).

43. Higgins TJ; O'Neill HC and Parrish CRA: A sensitive and quantitative fluorescence assay for cell surface antigens. *J. Immunol. Meth.* 47:275, (1981).

44. Hof RP: Measuring regional blood flow with tracer microspheres: a method, its problems and its application. *Blood Circulation* 21:29, (1982).

45. Hoffbrand BI and Forsyth RP: Validity studies of the radioactive microsphere method for the study of the distribution of cardiac output, organ blood flow and resistance in the conscious rhesus monkey. *Cardiovascular Research* 3:426, (1969).

46. Hollenberg NK: The Renal Circulation. In: The Peripheral Circulation, Zelis R; (ed)Grune Stratton, New York, (1975).

47. Hsieh DS; Langer R and Folkman J: Magnetic modulation of release of macromolecules from polymers. *PNAS* 78:1863, (1981).

48. Hunter WM: Radioimmunoassay. Handbook of Experimental Immunology, Weir DM(ed), Blackwell Scientific Publications, p.14.1, (1978).

49. Illum L and Jones PDE: Attachment of Monoclonal Antibodies to Microspheres. In: Drug and Enzyme Targeting, Widder KJ and Green R(eds): *Meth. Enzymol.* 112:67, (1985).

50. Jolley ME; Wang CH; Ekenberg SJ; Zuelke MS and Kelso DM: Particle concentration fluorescence immunoassay (PCFIA): A new rapid immunoassay technique with high sensitivity. *J. Immunol. Meth.* 67:21, (1984).

51. Kandzia J; Anderson MJD and Mueller-Ruchholtz W: Cell separation by antibody coupled magnetic microsphere and their application in conjunction with monoclonal HLA antibodies. *J. Cancer Res. Clin. Oncol.* 101:165, (1981).

52. Kandzia J; Scholz W; Anderson MJD and Mueller-Ruchholtz W: Magnetic albumin protein A immunomicrospheres. I. Preparation, antibody binding capacity and chemical stability. *J. Immunol. Meth.* 75:413, (1985).

53. Kaplan MR; Calef E; Bercovici T and Gitler C: The selective detection of cell surface determinants by means of antibodies and acetylated avidin attached to highly fluorescent polymer microspheres. *Biochim. Biophys. Acta.* 728:112, (1983).

54. Kato T and Nemoto R: Microencapsulation of mitomycin C intra-arterial infusion chemotherapy. *Proc. Jpn. Acad. Ser. B* 54:413, (1978).

55. Kato T; Nemoto R; Mori H and Kumagai I: Sustained-release properties of microencapsulated mitomycin C with ethylcellulose infused into the renal artery of the dog kidney. *Cancer* 46:14, (1980).

56. Kato T; Nemoto R; Mori H; Takahashi M; Harada M: Arterial chemoembolization with mitomycin C microcapsules in the treatment of primary or secondary carcinoma of the kidney, liver, bone and intrapelvic organs. *Cancer* 48:674, (1981).

57. Kato T; Unno K and Goto A: Ethylcellulose microcapsules for selective drug delivery. In: Drug and Enzyme Targeting, Widder KJ and Green R(eds). *Meth. Enzymol* 112:139, (1985).

58. Kemshead JT; Treleaven J; Gibson F; Ugelstad J; Rembaum A and Philip T: Monoclonal antibodies and magnetic microspheres used for the depletion of malignant cells from bone marrow. In "Autologous bone marrow transplantation: Proceedings of the First International Symposium" Dicke KA; Sptzer G and Zander A (eds.) U. of Texas Press, Houston, TX, p.409, (1985).

59. Kernan NA; Burns MJ; Collins NH; O'Reilly RJ and Dupont B: Limiting dilution microculture assay for quantitation of T lymphocytes in bone marrow. *Transpl. Proc.* 17:437, (1985).

60. Khalafalla SE and Reimers GW: Preparation of dilution-stable aqueous magnetic fluids. *IEEE Trans. Magnetics* 16:178, (1980).

61. Kitchell JP and Wise DL: Poly(lactic/glycolic acid) biodegradable drug-polymer matrix systems. In: Drug and Enzyme Targeting, Widder KJ and Green R(eds.) *Meth. Enzymol* 112:436, (1985).

62. Korbling M; Hess AD; Tutschuka PJ; Kaiser H; Colvin MO and Santos GW: 4-Hydroperoxycyclophosphamide: a model for eliminating residual tumour cells and T-lymphocytes from the bone marrow. *Br. J. Haematol.* 52:89, (1982).

63. Korngold R and Sprent J: Surface markers of T cells causing lethal graft-versus-host disease in mice, Recent Advances in Bone Marrow Transplantation: Gale RP and Liss AR(eds), New York, p.199, (1983).

64. Kreuter J and Hartmann HR: Comparative study on the cytostatic effects and the tissue distribution of 5-fluorouracil in a free form and bound to polybutylcyanoacrylate nanoparticles in sarcoma 180 bearing mice. *Oncology*, 40:363, (1983).

65. Kreuter J: Evaluation of nanoparticles as drug-delivery systems. II: Comparison of the body distribution of nanoparticles with the body distribution of microspheres (diameter greater than 1 micron), liposomes and emulsions. *Pharm. Acta. Helv.* 58:217, (1983).

66. Kreuter J: Poly(alkyl acrylate) Nanoparticles. In: Drug and Enzyme Targeting, Widder KJ and Green R(eds). *Meth. Enzymol.* 112:129, (1985).

67. Kronick PL; Campbell G and Joseph K: Magnetic microspheres prepared by redox polymerization used in a cell separation based on gangliosides. *Science* 200:1074, (1978).
68. Kronick PL: Magnetic microspheres in cell separation. *Methods Cell Separation* 3:115, (1980).
69. Laakso T; Anderson J; Artursson P; Edman P; Sjoholm I: Acrylic microspheres in vivo. X. Elimination of circulating cells by active targeting using specific monoclonal antibodies bound to microparticles. *Life Sci.* 28:183, (1986).
70. Laubins JD and Ross GD: Assay of membrane complement receptors (CR1 and CR2) with C3b - and (3d-coated florescence microspheres. *J. Immunol.* 128:186 (1982).
71. Langer R and Folkman: J. Polymers for the sustained release of proteins and other macromolecules. *Nature* 263:797, (1976).
72. Langer R; Brown L and Edelman E: Controlled Release and Magnetically Modulated Release Systems for Macromolecules. In: Drug and Enzyme Targeting, Widder KJ and Green R(eds) *Meth. Enzymol.* 112:399, (1985).
73. Lea T; Vartdal F; Davies C and Ugelstad J: Magnetic monosized polymer particles for fast and specific fractionation of human mononuclear cells. Scan. *J. Immunol.* 22:207, (1985).
74. Lichtenstein B: U.S. Patent No. 3,709,791 (1973).
75. Lim F(ed): Biomedical Applications in microencapsulation, CRC Press, Boca Raton, (1984).
76. Lindberg B; Lote K and Teder H: Biodegradable starch and microspheres: A new medical tool. In: Microspheres and Drug Therapy, Pharmaceutical, Immunological and Medical Aspects: Davis SS; Illum JG; McVie and Tomlinson E(eds) Elsevier, Amsterdam. p.153, (1984).
77. Ljungstedt I; Ekman B and Sjoholm I: Detection and separation of lymphocytes with specific surface receptors by using microparticles. *Biochem. J.* 170:161, (1978).
78. Longo WE and Goldberg EP: Hydrophilic albumin microspheres. In: Drug and Enzyme Targeting, Widder KJ and Green R(eds): *Meth. Enzymol.* 112:18, (1985).
79. Malin S and Edwards: J. Detection of hepatitis associated antigen by latex agglutination. *Nature* (Lond.) 235:182, (1972).
80. Marcus L; Offarim M and Margel S: A new immunoadsorbent for hemoperfusion: agarose-polyanolein microspheres beads. I. In vitro studies. *Biomater. Med. Devices Artif. Organs.* 10:157 (1982).
81. Marcus L; Mashiah A; Offarim M and Margel S: Extracorporeal removal of specific antibodies by hemoperfusion through immunosorbent agarose-polyacrolein microsphere beads: removal of antibovine serum albumin in animals. *J. Biomed. Mater. Res.* 18:1153, (1984).
82. Marcus L; Margel S; Savin H; Offarim H and Ravid M: Therapy of digoxin intoxication in dogs by specific hemoperfusion through agarose polyacrolein microsphere beads-antidigoxin antibodies. *Am. Heart J.* 110:30, (1985).
83. Margel S and Offarim M: Novel effective immunoadsorbents based on agarose-polyaldehyde microsphere beads: synthesis and affinity chromatography. *Anal. Biochem.* 128:342, (1983).
84. Margel S and Marcus L: Specific hemoperfusion through agarose acrobeads. *Appl. Biochem. Biotechnol.* 12:37, (1986).

85. Margel S; Zisblatt S and Rembaum A: Polyglutaraldehyde: A new reagent for coupling proteins to microspheres and for labeling cell surface receptors. II. Simplied labeling methods by means of non-magnetic and magnetic polyglutaraldehyde microspheres. *J. Immunol. Meth.* 28:341, (1979).

86. Margel S; Beitler U and Ofarim M: Polyacrolein microspheres as a new tool in cell biology. *J. Cell Sci.* 56:157, (1982).

87. Margel S; Ofarim M; Eshar Z: Cell fractionation with affinity ligands conjugated to agarose-polyacrolein microsphere beads. *J. Cell Sci.* 62:149, (1983).

88. Margel S; Marcus L; Savin H; Offarim H and Mashiah A: Specific removal of digoxin by hemoperfusion through agarose polyacrolein microsphere polyacrolein microsphere beads and antidigoxin antibodies. *Biomater. Med. Devices Artif. Organs* 12:25, (1984).

89. Melville D; Paul F and Roath S: Direct magnetic separation of red cells from whole blood. *Nature* 255:706, (1975).

90. Merivuori H; DeLa-Chapella A and Schroder J: Cell labeling and separation with polyglutaraldehyde microspheres. *Exp. Cell. Res.* 130:464, (1980).

91. Milgram F and Goldstein R: Agglutination of sensitized red blood cells by latex particles. *Vox. Sang.* 7:86, (1962).

92. Miller AM; McMillan L; Hannan WJ; Emmett PC and Aitken RJ: The preparation of dry, monodisperse microspheres of 99m Tc-albumin for lung ventilation imaging. *Int. J. Appl. Radiat. Isot.* 33:1423, (1982).

93. Miller RA; Brady JM and Cutright DE: Degradation rates of oral resorbable implants (polylactates and polyglycolates): rate modification with changes in PLA/PGA copolymer ratios. *J. Biomed. Mat. Res.* 11:711, (1977).

94. Mirro J; Schwartz JF and Civin CI: Simultaneous analysis of cell surface antigen and cell morphology using monoclonal antibodies conjugated to fluorescent microspheres. *J. Immunol. Meth.* 47:39, (1981).

95. Mizushima Y; Shoji Y; Kato T; Fukushima M and Kurozumi S: Use of lipid microspheres as a drug carrier for antitumor drugs. *J. Pharm. Pharmacol.* 38:132, (1986).

96. Molday RS; Dryer WJ; Rembaum A and Yen RCK: Latex spheres as marker for studies of cell surface receptors by scanning electron microscopy. *Nature* 249:81, (1974).

97. Molday RS; Yen RCK and Rembaum A: Application of magnetic microspheres in labeling and separation of cells. *Nature* 2268:437, (1977).

98. Molday RS and MacKenzie KD: Immunospecific ferromagnetic iron-dextran reagents for the labeling and magnetic separation of cells. *J. Immunol. Meth.* 52:353, (1982).

99. Molday RS: Cell labeling and separation using immunospecific microspheres. In: Cell Separation, Methods and Selected Applications, 3, Pretlow II, TG and Pretlow TP(eds); Academic Press, NY, p.237, (1984).

100. Molday RS; Dreyer WJ; Rembaum A and Yen RCK: New immunolatex spheres: Visual markers of antigens on lymphocytes for scanning electron microscopy. *J. Cell Biol.* 164:75, (1975).

101. Morimoto Y; Sugibayashi K and Akimoto M: Magnetic guidance of Ferro-colloid-entrapped emulsion for site-specific drug delivery. *Chem. Pharm. Bull.* 31:279, (1983).

102. Morris RM; Poore GA; Howard DP and Sefranka JA: Selective targeting of magnetic albumin microspheres containing vindesine sulfate: Total remission in Yoshida sarcoma-bearing rats. In: Microspheres and Drug Therapy. Pharmaceutical, Immunological and Medical Aspects, Davis SS; Illum L; McVie JG and Tomlinson E (eds): Elsevier, Amsterdam, p.439, (1984).

103. Mosbach K and Schroder UI: Preparation and application of magnetic polymers for targeting of drugs. FEBS Lett. 102:112, (1979).

104. Murray J; Brown L and Langer R: Controlled release of microquantitites of macromolecules. Cancer Drug Delivery, 1:119, (1984).

105. Nilsson K and Mosbach K: p-Toluensulfonyl chloride as an activating agent of agarose for the preparation of immobilized affinity ligands and proteins. Eur. J. Biochem. 112:397, (1980).

106. Nustad K; Johansen L; Ugelstad J; Ellingsen T and Berge A: Hydrophilic monodisperse particles as solid-phase material in immunoassays: Comparison of shell-and-core particles with compact particles. Eur. Surg. Res. 16:Suppl 2, 80, (1984).

107. Nustad K; Danielsen H; Reith A; Funderud S; Lea T; Vartdal F and Ugelstad J. Monodisperse polymer particles in immunoassays and cell separation, in Microspheres: Medical and Biological Applications, Rembaum A and Tokes ZA(eds), CRC Publications, in press (1987).

108. Obertenffer JA: High gradient magnetic separation. IEEE Trans. Magnetics 9:303, (1973).

109. Oder RR: High gradient magnetic separation theory and applications. IEEE Trans. Magnetics 12:428, (1976).

110. Ohman M; Garovoy MR; Widder KJ and Streigel JA: Depletion of T lymphocytes from peripheral blood using monoclonal antibodies and protein magnetic microspheres. Hum. Immunol. 14:128, (1985).

111. Ohnishi K; Tsuchiya S and Nakayma T: Arterial chemoembolization of hepatocellular carcinoma with mitomycin C microcapsules. Radiology 152:51, (1984).

112. Oonishi T and Uyesaka N: A new standard fluorescence microsphere for quantitative flow cytometry. J. Immunol. Meth. 83:1143, (1985).

113. Onoe K; Fernandes G and Good RA: Humoral and cell-mediated immune responses in fully allogeneic bone marrow chimeras in mice. J. Exp. Med. 151:115, (1980).

114. Ovadia H; Carbone AM and Paterson PY: Albumin magnetic microspheres a novel carrier for myelin basic protein. J. Immunol. Meth. 53:109, (1982).

115. Ovadia H; Paterson PY and Hale JR: Magnetic microspheres as drug carriers: Factors influencing localization at different anatomical sites in rats. Isr. J. Med. Sci. 19:631, (1983).

116. Owen CS: High gradient magnetic separation of erythrocytes. Biophys. J. 22:171, (1978).

117. Owen CS: Magnetic cell sorting. In: Cell Separation Methods and Selected Applications. Pretlow II, TG and Pretlow TP(eds) Academic Press, NY, 2, p.127, (1983).

118. Owen CS: Magnetic sorting of leukocytes. Cell Bioph. 8:287 (1986).

119. Owen CS and Liberti PA: Magnetic-Protein conjugates for the separation of cells by high gradient magnetic filtration. In: Cell Separation: Methods and Selected Applications. Pretlow II, TG and Pretlow TP(eds) Academic Press, NY, 4 in press (1986).

120. Owen CS and Sykes NL: Magnetic labeling and cell sorting. *J. Immunol. Meth.* 73:41, (1984).

121. Phibbs RH et al. Rheology of microspheres injected into circulation of rabbits. *Nature* 216:1339, (1967).

122. Platsoucas CD and Catsimpoolas N: Biological methods for the separation of lymphoid cells. *Meth. Cell Separation* 3:157, (1980).

123. Platsoucas CD and Good RA: Inhibition of specific cell mediated cytotoxicity by monoclonal antibodies to human T-cell antigens. *Proc. Natl. Acad. Sci.* USA 78:4500, (1981).

124. Platsoucas CD; Good RA and Gupta S: Separation of human lymphocyte subpopulations by density gradient electrophoresis. I. Different mobilities of T (Tmu, Talpha) and B lymphocytes from human tonsils. *Cell. Immunol.* 51:238, (1980).

125. Platsoucas CD; Beck JD; Kapoor N; Good RA and Gupta S: Separation of human bone marrow cell populations by density gradient electrophoresis: Differential mobilities of myeloid (CFU-C), monocytoid, and lymphoid cells. *Cellular Immunol.* 59:345, (1981).

126. Platsoucas CD; Chae F; Collins N; Kerman N; Laver J; Ellinger T; Steristad P; Bjorgum J; Rembaum A; Good RA; O'Reilly R and Ugelstad J: The use of magnetic monosized polymer particles for the removal of T cells from human bone marrow cell suspensions. Microspheres: Medical and Biological Applications: Rembaum A and Tokes Z(eds) CRC Press, Inc., Boca Raton, Florida, in press, (1987a).

127. Platsoucas CD; Kerman N; Collins N; Chae F; Ugelstad J and O'Reilly R: Depletion of T cells from human bone marrow cell suspension using the M-450 magnetic monosized polymer particles and anti-T cell monoclonal antibodies, Submitted (1987b).

128. Powles RL; Clink H; Sloane J; Barrett AJ; Kay HE and McElwain TJ: Cyclosporin A for the treatment of graft-versus-host disease in man. *Lancet* 2:1327, (1978).

129. Poynton CH; Dicke KA; Culbert S; Frankel L; Jagannath S and Reading CL: Immunomagnetic removal of CALLA positive cells from human bone marrow. *Lancet* 1:524, (1983).

130. Poynton CH; Reading CL and Dicke KA: Colloidal immunomagnetic fluids for cells separation. In "Autologous bone marrow transplantation; Proceedings of the Fifth International Symposium". Dicke KA; Spitzer G and Zander A(ed), U. of Texas Press, Houston, TX p.433, (1985).

131. Prentice HG; Janossy G; Price-Jones L; Trejdosiewicz LK; Panjwani D; Graphakos S; Ivory K; Blacklock, HA; Gilmore ML; Tidman N; Skeggs DBL; Ball S; Patterson J and Hoffbrand AV: Depletion of T lymphocytes in donor marrow prevents significant graft-versus-host disease in matched allogeneic leukaemic marrow transplant recipients. *Lancet* 1:472, (1984).

132. Ramsey NKC; Kersey JH; Robison LL et al: Prevention of acute graft-versus-host disease: a randomized study demonstrating the influence of treatment regimen and age. *N. Engl. J. Med.* 306:392 (1982).

350

133. Reisner Y; Itziscovitch L; Meshorer A and Sharon N: Hemopoietic stem cell transplantation using mouse bone marrow and spleen cells fractionated by lectins. *Proc. Natl. Acad. Sci. USA.* 75:2933, (1978).

134. Reisner Y; Kapoor N; O'Reilly RJ and Good RA: Allogeneic bone marrow transplantation using stem cells fractionated by lectins: VI. In vitro analysis of human and monkey bone marrow cells fractionated by sheep red blood cells and soybean agglutinin. *Lancet* 2:1320, (1980).

135. Reisner Y; Kapoor N; Kirkpatrick D; Pollack M; Dupont B; Good RA and O'Reilly RJ: Transplantation for acute leukaemia with HLAA and B non-identical parental marrow cells fractionated with soybean agglutinin and sheep red blood cells. *Lancet* 2:327, (1981).

136. Reisner Y; Kapoor N; Kirkpatrick D; Cunningham-Rundles S; Dupont B; Hodes MZ; Good RA and O'Reilly RJ: Transplantation for severe combined immunodeficiency with HLA-A, B, D, DR incompatible parental marrow cells fractionated by soybean agglutinin and sheep red blood cells. *Blood* 61:341, (1983).

137. Rembaum A, Margel S and Levy J: Polyglutaraldehyde: A new reagent for coupling proteins to microspheres and for labeling cell surface receptors. *J. Immunol. Meth.* 24:239 (1978).

138. Rembaum A; Yen RCK; Kempner DH and Ugelstad J: Cell labeling and magnetic separation by means of immunoreagents based on polyacrolein microspheres. *J. Immunol. Meth.* 52:341, (1982).

139. Reske SN; Vyska K and Feinendegen LE: In vivo assessment of phagocytic properties of Kupffer cells. *J. Nucl. Med.* 22:405, (1981).

140. Reynolds CP; Seeger RC; Vo DD; Ugelstad J and Wells J (eds)Vince UA; Spitzer G and Zander A: Purging of bone marrow with immunomagnetic beads: Studies with neuroblastoma as a model system. In: *Autologous Bone Marrow Transplantations"*, Proceedings of the Fifth International Symposium, U. of Texas Press, Houston, TX, p. 439 (1985).

141. Ritz J; Bast RC and Clavell LA: Autologous bone-marrow translantation in CALLA-positive acute lymphoblastic leukaemia after in vitro treatment with J5 monoclonal antibody and complement. *Lancet* ii:60, (1982).

142. Rodt H; Kolb HJ; Netzel B; et al. Effect of anti-T cell globulin on GvHD in leukaemic patients treated with BMT. *Transplant. Proc.* 13:257, (1981).

143. Rodt H; Thierfelder S and Eulitz M: Anti-lymphocytic antibodies and marrow transplantation. III. Effect of heterologous anti-brain antibodies on acute secondary disease in mice. *Eur. J. Immunol.* 4:25, (1974).

144. Rose NR and Friedman H: Manual of Clinical Immunology, Second Edition, American Society for Microbiology, Washington, D.C., pp.1105, (1980).

145. Ross GD and Lambris JD: Identification of a C3bi-specific membrane complement receptor that is expressed on lymphocytes, monocytes, neutrophils and erythrocytes. *J. Exp. Med.* 1155:96, (1982).

146. Rous P and Beard JW: Selection with the magnet and cultivation of reticuloendothelial cells (Kupffer cells). *J. Exp. Med.* 59:577, (1934).

147. Russell GFJ: Starch microspheres as drug delivery systems. *Pharm. Int.* 4:260, (1983).

148. Scheffel V; Rhodes BA; Natarjan TK and Wagner HN: Albumin microspheres for study of the reticuloendothelial system. *J. Nuclear Med.* 13:498 (1972).

149. Schroder U: Crystallized Carbohydrate Spheres for Slow Release and Targeting. *Meth. Enzymol.* 112:1116, (1985).

150. Seeger RC; Reynolds PD; Vo J; Ugelstad J and Wells J: Depletion of neuroblastoma cells from bone marrow with monoclonal antibodies and magnetic immunobeads. In "Advances in Neuroblastoma Research", Evans AE; D'Angio G and Seeger RC(eds): Alan R. Liss, Inc. NY p.443, (1985).

151. Senyei A; Widder K and Czerlinksi G: Magnetic guidance of drug-carrying microspheres. *J. App. Phys.* 49:3578, (1978).

152. Senyei AE; Driscoll CF and Widder KJ: Biophysical Drug Targeting: Magnetically Responsive Albumin Microspheres. In: Drug and Enzyme Targeting: Widder KJ and Green R(eds) *Meth. Enzymol.* 112:56, (1985).

153. Sharp TG; Sachs DH; Fauci AS; Messerschmidt GL and Rosenberg SA T cell depletion of human bone marrow using monoclonal antibody and complement-mediated lysis. *Transplantation* 35:112, (1983).

154. Simonsen M: The impact on the developing embryo and newborn animal of adult homologous cells. *Acta Path. Miocrobiol. Scand.* 40:480, (1957).

155. Singer JM and Plotz CM: The latex fixation test. I. Application to the serological diagnosis of rheumatoid arthritis. *Am. J. Med.* 21:888, (1956).

156. Singer JM: The latex fixation test in rheumatic disease. A review. *Am. J. Med.* 31:766, (1961).

157. Singer JM: The use of polystyrene latexes in Medicine, in: *Future Directions in Polymer Colloids*, El-Aasser MS and Fitch R(eds) Martinus-Nijhoff Publishing Company, Bordrecht, The Netherlands. This volume (1987).

158. Singh M; Silver RL; Lin C; Chowdhury L and Post M: Localization with magnetic microspheres reduces the systemic toxicity of doxorubicin. *Orthop Trans* 8:324, (1984).

159. Sjoholm I and Edman P: Acrylic microspheres in vivo. I. Distribution and elimination of polyacrylamide microparticles after intravenous and intraperitoneal injection in mouse and rat. *J. Pharmacol. Exp. Ther.* 211:656, (1979).

160. Sjoholm I and Edman P: The use of biocompatibile microparticles as carriers of enzymes and drugs in vivo. In: Microspheres and Drug Therapy. Pharmaceutical, Immunological and Medical Aspects, S. S. Davis, L. Illum, McVie JG and Tomlinson E(eds) Elsevier, Amsterdam, p.245, (1984).

161. Skipper HE; Schnabel FM, Jr; Wilcox WS: Experimental evaluation of potential anticancer agents. XIII. On the criteria and kinetics associated with "curability" of experimental leukemia. *Cancer Chemother. Rep.* 34:1, (1964).

162. Storb R; Prentice RL; Buckner CD; Clift RA; Appelbaum F; Deeg J; Doney K; Hansen RP; Mason M; Sanders JE; Singer J; Sullivan KM; Witherspoon RP and Thomas ED: Graft-versus-host disease and survival in patients with aplastic anemia treated by marrow grafts from HLA-identical siblings. Beneficial effect of a protective environment. N. Engl. *J. Med.* 308:302, (1983).

163. Sugibayashi K; Morimoto Y; Nadai T; Kato Y; Hasegawa A; Arita T: Drug-carrier property of albumin microspheres in chemotherapy. II. Preparation and tissue distribution in mice of microsphere-entrapped 5-fluorouracil. *Chem. Pharm. Bull* 27:204, (1979).

164. Sugibayashi K; Okumura M and Morimoto Y: Biomedical applications of magnetic fluids III. Antitumor effect of magnetic albumin microspheres-entrapped adriamycin on lung metastasis of AH-7974 in rats. *Biomaterials* 3:181, (1982).

165. Terman DS; Young JB; Sheaver WT; Ayus C; Lehane D; Mattioli C; Espada R; Howell JF; Yamamoto T; Zaleski HI; Miller L; Frommer P; Feldman L; Henry JF; Tilquist R; Cook G and Daskal Y: Preliminary observations of the effects on breast adenocarcinoma of plasma perfused over immobilized protein A. *N. Engl. J. Med.* 305:1195, (1981).

166. Terman DS; Yamamoto T; Mattioli M; Cook G; Tilquist R; Henry J; Poser R and Daskal Y: Extensive necrosis of spontaneous canine mammary adenocarcinoma after extracorporeal perfusion over Staphylococcus aureus Cowans I. Description of acute tumoricidal response: Morphologic, Histologic, Immunohistochemical, Immunologic, and Serologic Findings. *J. Immunol.* 124:795,1 (1980).

167. Theodorakis MC; Digenis GA; Beihn RM; Shambhu MB and DeLand FH: Rate and pattern of gastric emptying in humans using 99mTc-labeled triethylenetetraamine polystyrene resin. *J. Pharmc. Sci.* 69:568, (1980).

168. Thomas E; Storb R; Clift R; et al. Bone marrow transplantation. *N. Engl. J. Med.* 292:895, (1975).

169. Thorpe PE; Mason DW; Brown ANF; Simmonds SJ; Ross WCJ; Cumber AJ and Forrester JA: Selective killing of malignant cells in leukemic rat bone marrow using an antibody-ricin conjugate. *Nature* 297:594, (1982).

170. Tobias JS; Weiner RS, Griffiths CT; Richman CM; Panier LM and Yourvee RA: Cryopresened autologous marrow infusion following high dose cancer chemotherapy. *Eur. J. Cancer*, 13:269 (1977).

171. Tokes ZA; Rogers KE and Rembaum A: Synthesis of adriamycin-coupled polyglutaraldehyde microspheres and evaluation of their cytostatic activity. PNAS 79:2026, (1982).

172. Tomaszekwski JE; Good DB and Zmijewski CM: Cell surface antigen identification by a modified fluorescein immunosphere method. *Am. J. Clin. Pathol.* 85:219, (1986).

173. Tomlinson E and Burger JJ: Incorporation of water-soluble drugs in albumin microspheres. In: Drug and Enzyme Targeting, Widder KJ and Green R(eds). *Meth. Enzymol.* 112:27, (1985).

174. Treleaven JG; Ugelstad J; Philip T; Gibson FM; Rembaum A; Caine GD and Kemshead JT: Removal of neuroblastoma cells from bone marrow with monoclonal antibodies conjugated to magnetic microspheres. *Lancet* 1:70, (1984).

175. Tuma RF: The use of degradable starch microspheres for transient occlusion of blood flow and for drug targeting to selected tissues. In: Microspheres and Drug Therapy. Pharmaceutical, Immunological and Medical Aspects, Davis SS; Illum L; McVie JG and Tomlinson E(eds) Elsevier, Amsterdam, p.1 189, (1984).

176. Tyan M: Modification of severe graft-versus-host disease with antisera to the theta antigen or to whole serum. *Transplantation* 15:601, (1973).

177. Ugelstad J; Kaggerud KH; Hansen FK et al: Absorption of low molecular weight compounds in aqueous dispersions of polymer-oligomer particles: a two step swelling process of polymer particles giving an enormous increase in absorption capacity. *Makromol Chem.* 180:737, (1979).

178. Ugelstad J; Mork PC; Kaggerud KH et al: Swelling of oligomer particles. New methods of preparation of emulsions and polymer dispersion. *Advances in Colloid Interface Science* 13:101, (1980).

179. Ugelstad J; Soderberg L; Berge A and Bergstrom J: Monodisperse polymer particles-a step forward for chromatography. *Nature* 303:96, (1983).

180. Ugelstad J; Berge A; Ellingsen T; Bjorun J; Schmid R; Stenstad P; Aume O; Nilsen TN; Funderud S and Nustad K: Biomedical applications of monodisperse magnetic polymer particles, In: *Future Directions in Polymer Colloids*, ElAasser MS and Fitch R(eds): Martinuns-Nijhoff Publishing Company, Dordrecht, The. Netherlands. This volume (1987).

181. Vallera DA; Youle RJ; Neville DM and Kersey JH: Bone marrow transplantation across major histocompatibility barriers. V. Protection of mice from lethal graft-vs-host disease. Pretreatment of donor cells with monoclonal anti-Thy-1-2 coupled to the toxin ricin. *J. Exp. Med.* 155:949, (1982).

182. Voller A; Bidwell D; Bartlett A: Double antibody sandwich method for detection and measurement of antigen. In: Rose NR; Friedman H(eds) Manual of Clinical Immunology, Chapter 29, Washington, D.C., American Society for Microbiology, (1980).

183. Walsh JH; Yalow R and Benson SA: Detection of Australia antigen and antibody by means of radioimmunoassay techniques. *J. Infect. Dis.* 121:550, (1970).

184. Watson JHP: Magnetic filtration. *J. Appl. Phys.* 44:4209, (1973).

185. Weiden PL and the Seattle Marrow Transplant Team. Graft-vs-host disease in allogeneic marrow transplantation. In: Gale RP; Fox CF(eds). Biology of bone marrow transplantation. New York, Academic Press, (1980).

186. Wells JR; Sullivan A and Cline MJ: A technique for the separation and cryopreservation of myeloid stem cells from human bone marrow. *Cryobiology* 16:201, (1979).

187. Widder KJ; Senyei AE and Scarpelli DG: Magnetic microspheres: A model system for site specific drug delivery in vitro. *Proc. Soc. Exp. Biol. Med.* 58:141, (1978).

188. Widder KJ; Flauret G and Senyei A: Magnetic microspheres: Synthesis of a novel parenteral drug carrier. *J. Pharm. Sci.* 68:79, (1979a).

189. Widder KJ; Senyei AE; Ovadia H and Peterson PY: Magnetic protein A microspheres: A rapid method for cell separation. *Clin. Immunol. Immunopathol.* 14:395, (1979b).

190. Widder KJ; Morris RM; Poore G; Howard Jr. DP and Senyei AE: Tumor remission in Yoshida Sarcoma bearing rats by selective targeting of magnetic albumin microspheres containing doxorubicin. *Proc. Natl. Acad. Sci. USA* 78:579, (1981a).

191. Widder KJ; Sendyei AE; Ovadia H and Paterson P: Specific cell binding using staphylococcal protein A magnetic microspheres. *J. Pharm. Sci.* 70:387, (1981b).

192. Widder KJ and Senyei AE: Magnetic microspheres: A vehicle for selective targeting of drugs. *Pharmac. Ther.* 20:377, (1983).

354

193. Widder KJ and Green R: Drug and Enzyme Targeting, Methods in Enzymology, 112, Academic Press, Inc. Press, Inc., Orlando, New York, (1985).
194. Winston D; Gale R; Meyer D and Young L: Infectious complications of human bone marrow transplantation. *Medicine* 58:1, (1979).
195. Wojchowski DM and Sytkowski AJ: Detection of low-density cell-surface molecules using biotinylated fluorescent microspheres. *Biochim. Biophysi. Acta.* 857:61, (1986).
196. Yalow RS and Benson SA: Immunoassay of endogenous plasma insulin in man. *J. Clin. Invest.* 39:1157, (1960).
197. Yapel Jr, AF: Albumin Microspheres: Heat and Chemical Stabilization. In: Drug and Enzyme Targeting, Widder, KJ and Green R(eds): *Meth. Enzymol.* 112:3, (1985).

355

BIOMEDICAL APPLICATIONS OF MONODISPERSE MAGNETIC POLYMER PARTICLES

J. UGELSTAD, A. BERGE, T. ELLINGSEN, J. BJORGUM, R. SCHMID
P. STENSTAD, O. AUNE and T.N. NILSEN

SINTEF Applied Chemistry Division, Trondheim, Norway

S. FUNDERUD

Apothekernes Laboratorium
Oslo, Norway

K. NUSTAD
The Norwegian Radium Hospital
Oslo, Norway

1. INTRODUCTION

The method developed by Ugelstad and coworkers[1-3] for preparation of monosized polymer particles allows preparation of particles of any size from 1 to 100 μm with standard deviation in diameter ~ 1%. Also the method allows preparation of particles from a large number of monomers. Particles with highly crosslinked polymer of high mechanical strength, porous macroreticular particles and core and shell particles are prepared. The application of the porous macroreticular particles in fast protein liquid chromatography (FPLC)[4] as marketed by Pharmacia is already well established and has in many cases led to very significant improvements in analysis and separation of complex protein mixtures.

Crosslinked core and shell particles containing a low density core material and a shell of polymer with functional groups are prepared in the size range of 3 μm. Their relatively large size means that they are easy to handle and yet they settle very slowly in water. The fact that they are crosslinked allows their use in nonaqueous solvents. These particles are used in a number of immunoassays[5-8].

This paper deals with monosized magnetic polymer particles and their applications, especially in various cell separation procedures.

In most biochemical and biomedical applications of magnetic polymer particles a major problem lies in the preparation of polymer particles which are well characterized with respect to size as well as content of magnetizable material. The ease of handling during binding of antibody, washing etc. requires that the particles should show very little magnetic remanence after having been subjected to the magnetic field.

Most procedures for preparation of magnetic particles for biochemical and medical application have involved the use of finely divided magnetite. Particles where the antibody is directly adsorbed to the magnetic bead have found use in immunoassays[9].

Polymerization of monomers by emulsion polymerization[10] or by controlled anionic polymerization of acrolein[11] in the presence of finally divided magnetite have resulted in magnetizable polymer particles.

Other methods for preparation of polymer particles containing magnetite involve suspension processes where the Fe_3O_4 is made hydrophobic and added to the monomer mixture containing oil soluble initiators. This mixture is then suspended in water and polymerized[12,13]. Preparation of natural polymers containing magnetizable materials has involved mixing of an aqueous solution of natural polymers such as starch[14], dextran, agarose[15,16] and albumin[17,18] with Fe_3O_4. Formation of particles is then achieved by preparation of a suspension of these mixtures in a water insoluble organic liquid. Insolubilization of the polymer may take place by crystallization (dextran, agarose), heat denaturation (albumin) or crosslinking.

It has been reported that very small Fe_3O_4 particles (<0.01 μm) covered with dextran may be obtained from precipitation of Fe^{2+} and Fe^{3+} salts from an aqueous solution containing dextran[20]. Apparently the dextran becomes deposited on the precipitated iron compound. The dextran covered Fe_3O_4 particles were tested for isolation of legionella from other water bacterias.

The particles described above have been used in different immunoassays, for biophysical drug targeting combined with slow release and to a limited degree in cell and bacteria separation, but they have not yet been applied clinically. In cancer therapy removal of cancer cells by magnetic means has been tried, using very fine colloidal dispersions of cobalt[19]. In this case the secondary antibody was bound directly to the metal surface by physical adsorption.

The type of magnetizable particles which have found the widest application in various cell separation procedures are the monosized magnetic polymer particles developed by Ugelstad et al[21] which are described in the present paper.

2. MONOSIZED MAGNETIC POLYMER PARTICLES

Several types of magnetic particles have been prepared. The methods have in common that one starts with monosized polymer particles which in a second step are made magnetic by formation of Fe_3O_4 or other magnetic iron compounds in the particles[21]. One type which involved compact hydrophilic particles was prepared by introducing polymer particles with iron complexing groups in a solution of Fe^{2+} and Fe^{3+} and precipitating Fe_3O_4 in the particles. Other types involved the use of monosized porous macroreticular particles. Iron complexing agents may be introduced on the surface of the pores and the formation of precipitated Fe_3O_4 inside the pores may be carried out as described above. Another method involved that one has oxidizing groups bound to the surface of the pores. Under appropriate conditions Fe^{2+} is continuously transported from an outer phase to the inside of the pores where it precipitates as very fine grains of Fe_3O_4 which to a large extent is transformed to the magnetic γ-Fe_2O_3 form.

Magnetic particles prepared from macroreticular polymer particles have a large surface area ($50 - 200 m^2/g$). One may reduce this surface by different procedures. If the particles are suspended in a solution of monomers with an oil soluble peroxide initiator one will get a catalyzed polymerization in the pores of the particles at temperatures where the polymerization in the bulk is negligible. Crosslinking monomers may be used as well as monomers which will give polymer with functional groups.

With the porous particles we may end up with covalently bound functional groups (-XH) at the surface of the pores, such as OH, -COOH, -NH_2. In this case it is possible to fill the pores with polymeric substances bound covalently through reaction with the -XH groups. The method applied in this case may involve reaction with monomeric, oligomeric and polymeric substances, containing groups that react with

-XH, such as anhydride groups, ester groups, epoxy groups, isocyanate groups, aldehyde groups, lactones etc. These treatments which aim at filling up the pores may at the same time provide functional groups for chemical coupling of the protein.

Unreacted groups in the polymer coating like anhydride, epoxy and isocyanate groups may moreover be used to bind hydrophilic compounds to the surface of the beads with functional groups suitable for chemical coupling of proteins. Likewise the unreacted groups may be used to bind molecules to the surface which may act as a link or spacer arm between the particles and the protein.

3. MAGNETIC PROPERTIES OF THE PARTICLES

The magnetic particles which up to now have been most widely used in cell separation are prepared from macroporous particles followed by filling the pores with polymeric substances.

Scanning electron micrograph of such particles denoted M-450 is shown in Figure 1. The particles are 4.5 μm in diameter. The specific surface area is 3-5 m^2/g and the iron content about 20% by weight.

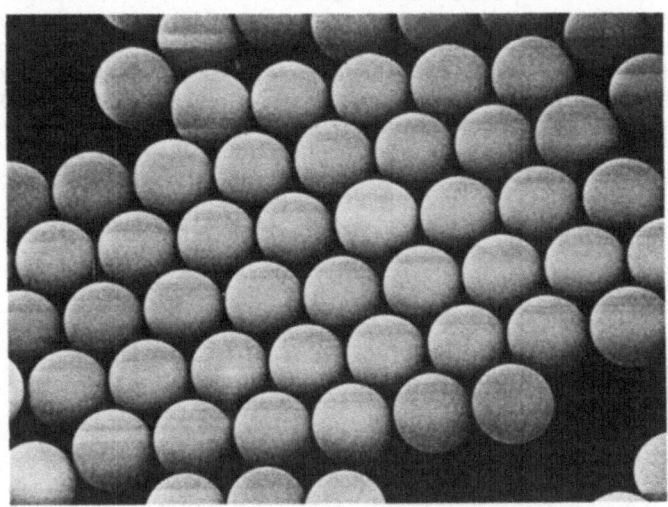

Figure 1: SEM of 4.5 μm magnetic particles (M-450).

In Figure 2 is shown an electron micrograph of an ultra-thin section of such particles. The iron compound is evenly distributed throughout the particles as very fine deposits. The composition of the iron deposits in the final particles has been analyzed by Mossbauer spectroscopy[22] and powder diffraction measurements[23]. From these analysis it is evident that the iron in the actual particles is present as γ-Fe_2O_3, maghemite, which is of the same spinel lattice as magnetite. The magnetic Fe ions in γ-Fe_2O_3 are identical, and ferrimagnetism arises from an unequal distribution of these ions in the so called A and B sites[24].

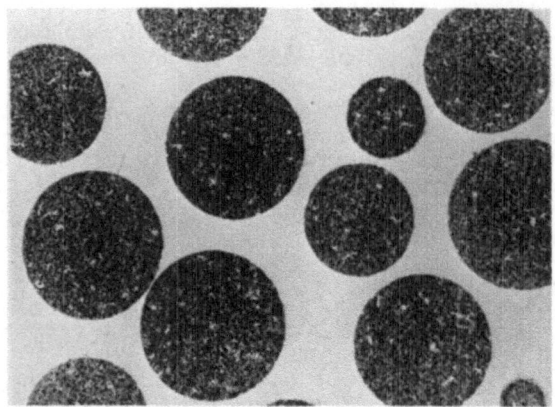

Figure 2: Electron micrograph of an ultrathin section of magnetic particles.

Figure 3 shows a typical magnetization curve for M-450 particles at room temperature. The measurements were carried out with a vibrating sample magnetometer (VSM, Princeton Applied Research) using a Ni probe as standard[23]. As shown in Figure 3 there is no hysteresis and consequently both the remanence and coercivity are zero. The probable reason for this is that the single maghemite crystals within the particles are sufficiently small to behave as single domains and moreover they are of a size where the thermal effects are strong enough to spontaneously demagnetize a previously magnetized or saturated sample. Figure 2 indicates that the maghemite grains may be of a size about 50 - 100Å.

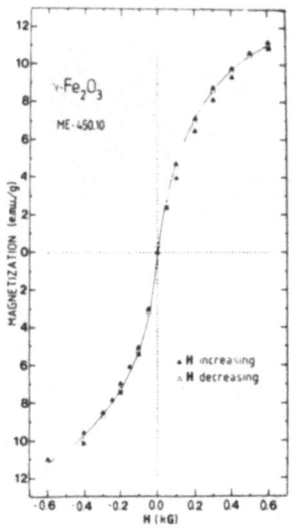

Figure 3: Magnetization curve for M-450 particles.

Maghemite may be almost as "magnetic" as magnetite. The saturation moment of finely precipitated magnetite is about 92 emu/g depending to some extent upon the preparation method, whereas pure maghemite has a saturation moment of 76 emu/g at room temperature[24].

The lack of magnetic remanence is of the utmost importance for the application of the particles in cell separation and in immunoassays. The use of the particles involves that they are repeatedly collected in a magnetic field and thereafter have to be redispersed. Any remanent magnetism would reduce the ease of redispersion.

4. BINDING OF ANTIBODY TO THE PARTICLES

The first magnetic particles from SINTEF to be used in cell separation were porous particles where no after-treatment of the particles after introduction of iron was applied. These particles which had a surface area of 200 m^2/g, did as expected bind large quantities of antibody in what appeared to be an irreversible reaction. Although the total amount of protein bound was high expressed as mg protein per gram particles, the amount expressed as mg protein per m^2 surface was low as was the activity per unit weight of adsorbed antibody. Apparently not all of the pores were accessible to the protein molecules and likewise a major part of the adsorbed antibody was buried in the pores and therefore functionally inactive due to steric hindrance.

Figure 4 shows a result from an adsorption experiment carried out with the M-450 particles which as mentioned are 4.5 μm magnetic particles with filled pores. The adsorption was carried out in 1 ml phosphate buffer (0.1 M, pH 7.7). 100μg sheep anti-rabbit IgG was added to different amounts of particles (2 - 20mg). The reaction mixture was rotated for 16 hours. The particles were then washed three times with 1ml of Tris buffer (0.05 M Tris, 0.1 M NaCl, 0.01 % BSA, 0.01 % merthiolate, pH 7.7). In each washing cycle the particles were rotated for about 24 hours. The amount of antibody adsorbed after 3 times washing is given as a function of amount of antibody added per gram of particles.

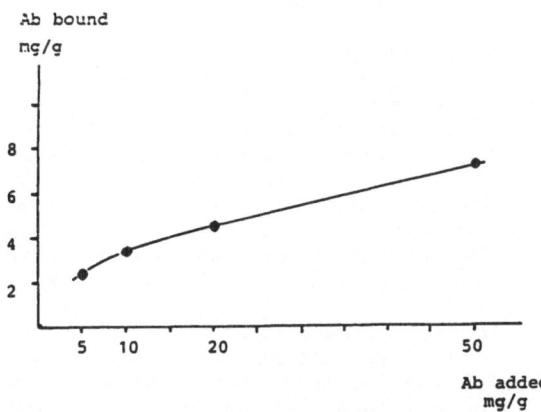

Figure 4: Amount of antibody (Ab) adsorbed on M-450 particles as a function of the amount of Ab added. (The values are given as mg Ab per gram of particles).

An increase in the number of washing steps did not remove more antibody from the particles, indicating an irreversible binding. Even so, adsorption experiments as illustrated in Figure 4 gave a gradual attainment of saturation, typical for low-energy reversible adsorption of small molecules. Similar behaviour has been found for adsorption of proteins on polystyrene beads[25].

The amount of antibody bound to the M-450 beads by physical adsorption is about 3 mg per gram, corresponding to 1 mg per m^2 when applying 10 mg of antibody per gram particles during incubation. With completely smooth surface one would expect 2-5 mg per m^2.

The M-450 particles are sufficiently hydrophobic to allow a relatively high and strong physical binding of the antibody. The coating applied for these particles does, however, give a substantial amount of hydroxyl groups on the surface. It may therefore appear that also with these particles one should preferably bind the antibody by chemical coupling. The method used for coupling of antibodies chemically has most often been performed by activation of the hydroxyl groups with tosyl or tresyl chloride[26]. With this method the amount of antibody bound to M-450 particles increases about 50% as compared to that obtained with physical adsorption of the antibody. The number of OH-groups on the surface of the M-450 beads is sufficiently high to allow a manifold covalent binding of the antibody molecules covering the beads.

Quite recently we have prepared particles which are so highly hydrophilic that they bind relatively small amounts of antibody by physical adsorption compared to that which may be bound by chemical coupling. These particles may be prepared from the same type of porous particles as the M-450 particles, however, the outer layer consists of very hydrophilic oligomeric or polymeric substances of synthetic or natural origin, preferably with a high content of functional groups. In this case all the antibody molecules on the particles are obviously attached by covalent bonds.

The orientation of the antibodies on the particle surface is very important for their ability to bind mouse monoclonal antibodies attached to a cellular surface. We find that antibodies bound by physical adsorption and by sulfonyl chloride activation are better orientated when the antibody is partially denaturated before use.

5. BINDING OF PARTICLES TO TARGET CELLS

To achieve specific binding of the magnetic particles to the target cells three procedures may be used:

1. Direct coupling of the monoclonal antibody (MoAb) to the magnetic particles. With IgG MoAb this method has always led to unsatisfactory results in the way that the cell binding efficiency has been low. This seems to indicate that steric hindrance is present which may be due to distance or to poor orientation of the antibody. Accordingly with the much larger IgM MoAb (a pentamer) one may get very effective and selective coupling of beads to cells even if the MoAb is directly coupled to the beads.

2. A secondary antibody is attached to the beads and the MoAbs are added to the cell mixture in a separate step. The secondary antibody is a polyclonal one (PAb) which may for instance be sheep or goat anti-mouse antibody. In this case the monoclonal antibodies are the mouse MoAbs. The secondary antibody is attached in a separate step to the beads by physical adsorption or chemical coupling and may be looked upon as a spacer arm. Obviously excess of the MoAbs have to be effectively washed out of the cell mixture before addition of the beads with the secondary

antibody. This makes the method somewhat time consuming. Also the extensive washing to remove excess of MoAbs normally leads to loss of cells.

3. Secondary and primary antibodies are attached to the beads in subsequent steps. In this case a PAb is attached to the beads in the first step, where it is bound by physical adsorption or by chemical coupling. Excess secondary Ab is removed by washing and then the MoAbs are added. After a given incubation time the excess of MoAbs are removed by washing. The secondary antibody may in this case also function as a spacer arm and in addition lead to a favorable orientation of the MoAbs.

The advantage of method 3 compared to method 2 is that the washing procedures in method 3 always involve cleaning of magnetic beads with attached antibodies, which is much simpler than removal of excess MoAbs from a cell mixture.

In many cases one uses a panel of different MoAbs to ensure an effective removal of target cells. The target cells may have different antigen expressions on the surface recognized by different MoAbs. It may therefore be argued that comparing methods 2 and 3, method 2 will, due to steric considerations, be the most effective one from the point of getting as strong as possible binding between particles and cells.

The binding of a magnetic particle with sheep antimouse IgG antibodies to a cell covered with monoclonal mouse IgG has been studied by scanning and transmission electron microscopy for normal mouse hepatocytes and Raji cells[27]. The initial binding of particle bound antibody to the cellular antigen follows a zipper mechanism as multiple antigen-antibody complexes are formed. This seems to trigger a cellular mechanism whereby movement of the cellular surface takes place as revealed by membrane folds, pseudopodes and lamellae extending from the surface and enveloping large parts of the particles.

6. CELL SEPARATION BY MEANS OF MONODISPERSE MAGNETIC PARTICLES
6.1. Purging of bone marrow for malignant cells.

Many malignant diseases initially respond well to chemotherapy and/or radiotherapy, but may subsequently recur. When treated by further drug therapy in conventional doses they may become resistant, thus making a cure for the disease very unlikely. It may therefore be desirable in some cases to use very high dose chemotherapy combined with radiotherapy to bring about tumor regression. The agents used will, however, effect not only the tumor cells present in the body but also normal dividing cells, in particular bone marrow stem cells. Over the last few years autologous bone marrow transplantation combined with high dose chemotherapy and/or radiotherapy has become an accepted form of treatment for patients suffering from some malignant diseases. This mode of therapy removes the very chemo- and radiosensitive bone marrow stem cells from the body while therapy is being given, thus protecting them, and they may be reinfused into the patient once therapy is completed. The stem cells then re-engraft in the bone marrow compartment. Since they are the patients own cells (autologous transplantation) the problems associated with receiving bone marrow from a donor (allogeneic transplantation), as graft versus host diseases, are avoided.

This form of therapy is, however, limited by the fact that bone marrow cells removed for autologous transplantation may be contaminated with tumor cells. If this is the case, retransfusion of bone marrow following therapy may lead to re-seeding and engraftment of tumor cells along with the normal bone marrow cells. Several methods of removing tumor cells from bone marrow prior to reinfusion have been described.

These include the use of monoclonal antibodies and complement to kill cells, or the use of toxic substances such as ricin linked to monoclonal antibodies.

The method for binding of magnetic beads to tumor cells for cleaning of bone marrow described above leads to selective binding of magnetic beads to the tumor cells. After incubation the tumor cells, coated with magnetic beads, and the excess of beads are removed from the mixture by a magnetic device and the normal cells are reinjected into the patient. This method seems to be the most effective one for removal of cancer cells from bone marrow. It is, however, sensitive to the type and quality of the antibodies applied. Moreover, the success of the method obviously depends upon that no cancer cells survive the chemo- and/or radiotherapy given to the patient before reinjecting the cleaned bone marrow.

6.2. Removal of Neuroblastoma Cells From Bone Marrow

Clinical application of our magnetic beads for removal of tumor cells was first carried out by John Kemshead and coworkers[28-30] for removal of neuroblastoma cells from bone-marrow. Originally they used the porous magnetic beads. Experiments with separation of other types of cells indicate that compact particles or porous particles where the surface area had been drastically reduced were superior to the porous ones[31,32] so that such particles, for the moment mostly the M-450 beads, are the ones used for selective removal of cells.

Kemshead et al demonstrated that direct coupling of the monoclonal IgG antibodies to the particles gave inferior results as compared to indirect coupling of MoAb bound to cells and corresponding PAb bound to particles. Also Kemshead demonstrated the necessity of applying a cocktail of MoAbs to ensure an effective binding to all tumor cells and that the method only resulted in a small decrease in the proliferation capacity of the progenitor cells in the bone marrow.

The use of the M-450 beads for treatment of patients with neuroblastoma has, since Kemsheads first attempts in 1983, been started at a number of hospitals in Europe and U.S.A. It has been claimed[33] that on purging the bone marrow for neuroblastoma cells with magnetic beads much more effective removal of cells is achieved by a two times purging.

In model experiments cultured tumor cells premarked with the supravital DNA binding fluorochrome Hoechst 33342 were seeded into normal bone marrow. As few as 1 viable tumor cell could be detected among one million normal cells. When the normal bone marrow cells were mixed with 10% premarked neuroblastoma cells, one and two times purging resulted in a 4, and 6 log depletion of tumor cells, respectively[34].

6.3. Removal of B-Cells From Bone Marrow

Quite recently it has been shown that the M-450 beads may be used for a rapid and efficient removal of B-lymphoma cells from bone marrow[35]. In this case the B-cell specific monoclonal IgM antibodies were coupled chemically directly to the M-450 particles and incubated with Rael Burkitt lymphoma cells admixed to fresh mononuclear human bone marrow cells. After incubation and removal of particle coated cells and excess beads by cobalt samarium magnets, the remaining clonogenic tumor cells were assayed by a soft agar procedure.

These experiments showed for the first time that with the large IgM antibodies one could couple the MoAb directly to the particles with good results.

The bone marrow was seeded with 10% tumor cells. The ratio of beads to total mononuclear cells was 7:1. With a two times purging procedure there were no detectable tumor cells found in the bone marrow by the colony assay method, showing a tumor cell depletion of at least 6 logs. These experiments seem to support that two times purging is better than one, as described above for neuroblastoma cells. One time purging only resulted in a 4 log depletion of tumor cells. Clonogenic capacity of the bone marrow progenitor cells was reduced only by about 30% as compared with the bone marrow before treatment.

Binding of the M-450 beads to a Burkitt lymphoma cell is illustrated in Figure 5. The monoclonal IgM antibodies applied were AB-1, a pan-B cell antibody (CD19), and AB-4, a HLA-DR monoclonal antibody that will bind to most non-Hodgkins lymphomas and leukemias.

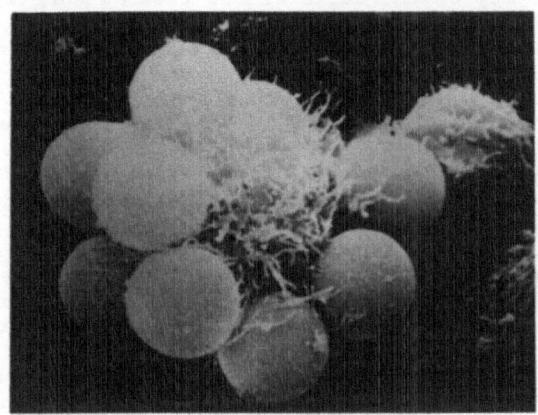

Figure 5: Burkitt-lymphoma cell to which are attached M-450 beads.

6.4. Depletion of T-Cells from Peripheral Blood and Bone Marrow

Graft versus host disease (GVHD) frequently observed after allogeneic bone marrow transplantation is initiated by alloreactive donor T-cells present in the transplant. To avoid this complication attempts have been made to deplete T-cells from the marrow inoculum.

The methods used so far tend to be either cumbersome, time consuming or not always effective in removing T-cells or may cause considerable non-specific loss of non T-cells. The first attempts to apply the M-450 beads to remove T-cells were carried out by Lea et al[36] on mononuclear cells isolated from human peripheral blood. Depletion experiments were carried out by use of a mixture of monoclonal antibodies directed against the T3 (CD3), T4 (CD4) and T8 (CD8) antigens. The lack of T-cells in the cell mixture after depletion was shown by indirect immunofluorescence and from the lack of rosette formation when reacted with AET - treated SRBC. These results were verified by studies of mitogenic, antigenic and allogeneic induction of T-cell proliferation before and after T-cell depletion.

Selective removal of T-lymphocytes from bone marrow by the M-450 beads has been carried out at a number of medical research institutes. The work of Platsoucas et al[37] is described in detail in this volume. Common procedures have involved incubation with mouse MoAbs against different T-antigens followed by treatment with magnetic beads with sheep anti-mouse PAb on them.

Quite recently removal of T-cells from bone marrow was attempted by Vartdal et al[38], using the method of interaction, denoted method 3 above, with the secondary PAb and the mouse MoAbs attached to the beads in subsequent steps. In this case the particles were coated with rabbit anti-mouse Ig. The microspheres were subsequently coated with a secondary layer of T-cell specific MoAbs by incubation with MoAbs specific for T11 (CD2) and T3 (CD3). The ratio of beads to T-cells was 10:1. Untreated bone marrow cells contained 22 - 31% AET- SRBC rosette forming cells, whereas 0.025% was detectable in the depleted marrow of the seven donors investigated. The T-cell depletion efficacy (the ratio between the percentage of T-cells after and before depletion) was thus about 10^{-3}. This result was supported by a limiting dilution T-cell clonogenic assay and by flow cytometric studies applying indirect immunofluorescence. A median of 58% non-AET-SRBC rosette forming cells were recovered after depletion. The viability of the cells of the treated bone marrow was 99% as tested by trypan blue exclusion.

The authors found that the spontaneous proliferation of non-mature bone marrow progenitor cells in bone marrow made the use of mitogenic, antigenic and allogeneic induction of T-cell proliferation as a measure of T-cell depletion uncertain. A more clear cut evidence for the abrogation of T-cell associated functions in the depleted bone marrow was obtained by examination of IL2-production in PHA stimulated BMC cultures and BMC cultures depleted of T-cells. After immunomagnetic T-cell depletion PHA induced IL2-production was reduced to background control levels. The T-cell depletion of BMC did not appear to cause any damage to the stem cells, as revealed in an assay where T-cell depleted bone marrow was assayed for GEMM and GM progenitors. After depletion of T-cells unchanged or increased numbers of CFU-GEMM and increased numbers of CFU-GM were observed.

Recent data suggest that GVHD does not occur in patients receiving less than 10^5 - 10^6 T-cells per kg body weight. Although the minimum number of viable progenitor cells required for stable engraftment is not known, engraftment is regularly observed after transplantation of at least 10^7 - 10^8 nucleated cells per kg body weight. Taken together these data indicate that the marrow inoculum should contain less than 0.1 - 1% T-cells while a majority of stem cells should be preserved during the depletion procedure. The immunomagnetic T-cell depletion technique described above[38], seems to fulfill both requirements. It should be noted that the process involves that the antibodies, the goat anti-mouse Ab and the mouse MoAbs, are attached to the beads in subsequent steps before incubation with the BMC. This does as described above greatly facilitate the ease and speed of the washing procedures involved in the process.

6.5. Positive Selection of Cells by use of Magnetic Beads

The application of beads described above have all involved the use of beads for negative cell separation, i.e. the magnetic beads have been applied for removal of "unwanted" cells. The procedure applied for incubation of the cell mixture with the beads may lead to a strong interaction between the beads and the target cells. Moreover the ratio of beads to target cells has normally been so high that each target cell during the incubation becomes surrounded by a relatively large number of beads.

No methods have so far been developed by which one can remove the beads from the cells after separation. Quite recently[39,40] it has been demonstrated that the M-450 beads may be used for positive selection of cells when a low ratio of beads to target cells is applied. Even with a low ratio of beads to target cells, in the order of 5:1, one may get a very effective isolation of target cells from the cell mixture. In this case the isolated cells will have fewer beads attached to them. An IgM

monoclonal antibody of T8 (CD8) specificity was attached to the M-450
beads by physical adsorption and used for positive selection of T8 cells
from PBMC and BMC. No attempts were made to remove the beads from the
isolated cells. Their presence did not interfere with the effector
functions of the T8 subset as cytotoxic and supressor activity. That the
beads do not affect the functionality of the cells is perhaps not
surprising as one in the present case with 1-2 beads per cell only will
have a very small fraction of the cells covered with beads. This in
connection with the fact that the MoAbs are bound directly to the beads
results in minimal occupancy of surface determinants reactive with MoAb.
This may be of paramount importance for the subsequent functional
analysis of the isolated cells after positive selection.

Positive selection of T-cells and B-cells from PBMC has also been
demonstrated by Lea et al[41]. Also in this case a low ratio of beads to
target cells was applied. However, the method of linking of beads to
cells was the method 2 described above, where the MoAbs were added to the
cell mixture in a separate step before addition of M-450 with the
attached secondary PAb.

It is concluded that provided the MoAbs do not react with membrane
structures involved in the transduction of activating signals highly
purified quiescent cell populations may be recovered in a single
fractionation step. In most instances it was found that the particles
detached from the isolated cells by overnight culture and then could be
removed from the system by a suitable magnet. T-lymphocytes,
subpopulations of T-lymphocytes and B-lymphocytes were isolated and
studied in a variety of functional assays. Comparison with cells obtained
after negative selection demonstrated that the positive selection was
useful especially if the membrane markers selected are not directly
engaged in the activation processes.

The M-450 beads may be used for fast isolation of cells for HLA
typing[42]. The beads were coated with monoclonal antibodies specific for
CD8 T-cells or for HLA class II monomorphic epitopes and then used to
obtain HLA class I or class II positive cells directly from blood. The
cells (attached to the beads) were subsequently applied in microcytotoxic
HLA typing.

It was found that the immunomagnetic HLA typing was specific and had
a sensitivity superior to that observed with conventional assays. The
method is fast and technically simple. Only small samples of blood ~0.5
ml are necessary. The high selectivity and yield of viable cells in the
separation step is especially advantageous in the typing of uraemic and
leukaemic patients where HLA typing often is difficult due to low
percentage of HLA II$^+$ cells.

7. MAGNETIC SEPARATION DEVICES IN CELL SEPARATION OF BONE MARROW

For analytical purposes one may use a very simple magnetic equipment
for achieving the separation of cells with attached beads from the
system. For depletion of cells from bone marrow which is to be reinfused
into the patient the requirement to the magnetic system is obviously much
more severe. The system should ensure complete sterility. An incomplete
removal of target cells with beads as well as excess beads may result in
microembolisms.

Two different systems have been applied. The one originally
constructed by Kemshead involved that the bone marrow mixture after
incubation with the beads was continuously pumped by a peristaltic pump
through two chambers made from small polycarbonate bags connected with a
silicon tubing. The chambers were placed upon an assembly of cobalt-

samarium magnets arranged on a frame of soft iron. The pretreated bone marrow was pumped through the system at a speed of 1.5 ml per minute and the chambers flushed free of haematopoetic cells with saline and 5% PPF.

This system has been improved by R.B. Freeman (private communication). The bond between the cell/MoAb/Pab and bead may be relatively weak and subject to disruption by excess jarring, eddy currents or pulsative flow in the sample. In the magnetic device constructed by Freeman the magnetic separation takes place in a chamber with smooth walls and no sharp corners and with a steady even flow of the sample. Most important Freeman applies a non-uniform magnetic field. A weak magnetic field at the inlet of the separation chamber provides a "soft" attractive force to prevent breaking the bond of conjugated beads at this point in the chamber. As one passes through the separation chamber the magnetic field is intensified to ensure removal of all remaining magnetic particles. Freemans apparatus has been shown to work excellently in removal of tumor cells from bone marrow.

Another system which has been applied involves that the bone marrow sample is mixed with the beads in a normal freezing bag of about 200 ml (C.P. Reynolds, private communication). The discontinuous system for depletion of target cells involves that one after incubation places the bag in an assembly so constructed that the cobalt samarium magnets become clamped to the sides of the bag. The rotation is stopped and the bag placed in a position where it may be emptied for bone marrow through a tube. In order to ensure a complete removal of the excess of beads the bone marrow is passed through a tube with cobalt samarium magnets on the outer tube periphery.

8. OTHER APPLICATIONS
8.1. Isolation of viruses and organelles

The M-450 beads and modification of these have proven to be very effective for selective separation of organelles and viruses[43,44]. The separation takes place in a magnetic flow system. The magnetic beads with bound specific antibodies are suspended in a 15 ml isolation chamber, maintained at 4°C and placed within the magnetic field generated by an electromagnet. The shape of the magnetic field is designed to permit the beads to be maintained in a disperse suspension and to be contained within the chamber while the washing buffer flows through the chamber to remove the unbound and non specifically bound components. The gentle washing conditions prevent losses due to vesicularization of the specifically bound vesicles. Improved results have recently been obtained by use of a M-450 bead modified in the way that it contains less magnetic oxide and an outer layer of non-magnetic material.

8.2. Magnetic particles for immunoassays

As stated above a number of different magnetic particles have been introduced for use in different immunoassays. The advantage of magnetic particles is obviously that they are much more easily isolated after incubation than non-magnetic particles where one has to use centrifugation.

The weak point for most of the magnetic beads used up to now has been their tendency to agglomerate. As stated above our particles as for example M-450 do not show this behaviour. However, these particles turned out to have two main disadvantages in immunoassays. They were too heavy and therefore had a tendency to settle too fast. Secondly, while they showed very low non-specific binding to large molecules they showed a too high non-specific binding of small molecules like triiodothyronin. To overcome these difficulties, we have recently prepared magnetic particles which are considerably smaller and have a lower density than the M-450 beads. At the same time the use of another polymer for filling up the pores has reduced the non-specific binding of triiodothyronin to practically zero.

9. Acknowledgement:

This work was supported by grants from the Norwegian Cancer Society and from Royal Norwegian Council for Scientific and Industrial Research.

LIST OF ABBREVIATIONS

AET,	2-aminoethyl isothioroniumbromide
BMC,	bone marrow cells
BSA,	bovine serum albumin
CFU,	colony forming unit
FPLC,	fast protein liquid chromatography
GEMM,	granulocyte erythroid monocyte megakaryocyte (multipotential haematopoetic progenitor cells)
GM,	granulocyte monocyte
GVHD,	graft-versus-host disease
HLA,	human leukocyte antigen
Ig,	immunoglobulin
IL2,	interleukine 2
MoAb,	monoclonal antibody
PAb,	polyclonal antibody
PBMC,	peripheral blood mononuclear cells
PHA,	phytohaemagglutinin
PPF,	purified plasma protein fraction
SRBC,	sheep red blood cells

REFERENCES

1. Ugelstad J; Mork PC; Kaggerud HK; Ellingsen T and Berge A: *Adv. Colloid Interface Sci.* **13**, 101 (1980).
2. Ugelstad J; Mfutakamba HR; Mork PC; Ellingsen T; Berge A; Schmid R; Holm L; Jorgedal A; Hansen FK and Nustad K: *J. Polym. Sci.* **72**, 225 (1985).
3. Ugelstad J; Mork PC; Nordhuus I; Mfutakamba H; Soleimany E; Berge A; Ellingsen T and Khan AA: *Makromol. Chem. Suppl.* **10/11**, 215 (1985).
4. Ugelstad J; Soderberg L; Berge A and Bergstrom J: *Nature* **303**, 95 (1983).
5. Nustad K; Johansen L; Ugelstad J; Ellingsen T and Berge A: *Eur. Surg. Res.* **16**, (Suppl. 2), 80 (1984).
6. Millan JL; Nustad K and Norgaard Pedersen B: *Clin. Chim.* **31**, 54 (1985).
7. Johansen L; Nustad K; Orstavik TE; Ugelstad J; Berge A and Ellingsen T: *J. Immunol. Methods* **59**, 225 (1983).
8. Nustad K; Closs O and Ugelstad J: in *Developments in Biological Standardization* **57** Monoclonal Antibodies: Standardization of their Characterization and Use, p.321 Basel (1984).
9. *Biomag*™ *Technical Bulletins* Advanced Magnetics Inc., Cambridge, Massachusetts U.S.A. (1984)
10. Rembaum A; Yen SPS and Molday RS: *Macromol. Sci. Chem.* **A13**, 603 (1979).
11. Margel S; Beitler U and Ofarim M: *J. Cell. Sci.* **56**, 157 (1982).
12. Daniel JC; Schuppiser JL and Tricot M: U.S. Patent 4,358,388 (1982).
13. Buske N and Goetze T: *Acta Polymerica* **34**, 184 (1983).
14. Mosbach K and Schroder U: *FEBS Lett.* **102**, 112 (1979).
15. Schroder U; Stahl A and Salford LG: in "Microspheres and Drug Therapy", Davis SS(eds). Illum L; McVie JG and Tomlinson E: Elsevier Science Publ. p.427 (1984).
16. Schroder U: *Methods in Enzymology* **112**, 116 (1986).
17. Widder KJ and Senyei AE: In reference 15, p.393.
18. Senyei AE; Driscoll CF and Widder KJ: *Methods in Enzymology* **112**, 56 (1986).
19. Poyn CH; Dicke KA; Culbert S; Frankel LS; Jagannath S and Read CL: *Lancet* **524** (1983).
20. Kronick P and Gilpin RW: *J. of Biochem. and Biophys. Methods* **12**, 73 (1980).
21. Ugelstad J; Ellingsen T; Berge A; Helgee og B: PCT Int. Appl. WO83/03920 (1983).
22. Morup S: DTH, D-2800 Lyngby Denmark, private report.
23. Fjellvag H and Skjeltorp A: IFE, N-2007 Kjeller Norway. Private report.
24. Cullity RD: *Introduction to Magnetic Materials*, Addison Wesley Co. Inc., U.S.A. 1972.
25. Bagchi P and Birnbaum SM: *J. Coll. Interface Sci.* **83**, 460 (1981).
26. Nilsson K and Mosbach K: *Biochem. Biophys. Res. Commun.* **102**, 449 (1981).
27. Danielsen H; Funderud S; Nustad K; Reith A and Ugelstad J: *Scand. J. Immunol.*, **24**, 179 (1986).
28. Treleaven JG; Gibson FM; Ugelstad J; Rembaum A and Kemshead J: *Magnetic Separation News* **1**, 103, (1984).
29. Treleaven JC; Gibson FM; Ugelstad J; Rembaum A; Philip T; Caine GD and Kemshead JT: *Lancet* **1**, 70 (1984).

30. Kemshead J; Treleaven JG; Gibson FM; Ugelstad J; Rembaum A and Philip T: in "Advances in Neuroblastoma Research, Progress in Clinical and Biological Research", 175, p.413. Evans AE; D'Angir GJ; Seeger RC and Liss AK(eds), New York (1985).

31. Ugelstad J; Rembaum A; Kemshead JT; Nustad K; Funderud S and Schmid R: In reference 15, p.365.

32. Nustad K; Danielsen H; Reith A; Funderud S; Lea T; Vartdal F and Ugelstad J: in "Microspheres: Medical and Biological Applications". Rembaum A(ed): CRD Press Inc. in press.

33. Seeger R; Reynolds CP; Vo DD; Ugelstad J and Wells J: In reference 30, p.443.

34. Reynolds CP; Seeger R; Vo DD; Black AT; Wells J and Ugelstad J: Cancer Research, 46, 5882 (1986).

35. Kvalheim G; Fodstad O; Pihl A; Nustad K; Pharo A; Ugelstad J and Funderud S: Cancer Research, 47, 32 (1987).

36. Lea T; Vartdal F; Davies C and Ugelstad J: Scand. J. Immunol. 22, 207 (1985).

37. Platsoucas C; Chae FH; Collins N; Kerner N; Laver J; Ellingsen T; Stenstad P; Bjorgum J; Rembaum A; O'Reilly R and Ugelstad J: In reference 32.

38. Vartdal F; Kvalheim G; Lea TE; Bosnes V; Gaudernack G; Ugelstad J and Albrechtsen D: Transplantation, in press.

39. Albrechtsen D; Gaudernack G; Kvalheim G; Lea T; Ugelstad J and Vartdal F: Transplantation, Vol. I, 108 (1986).

40. Gaudernack G; Leivestad T; Ugelstad J and Thorsby E: J. Immunol. Methods 90, 179 (1986).

41. Lea T; Smeland E; Funderud S; Vartdal F; Davies C; Beiske K and Ugelstad J: Scand. J. Immunol., 23, 509 (1986).

42. Vartdal F; Gaudernack G; Funderud S; Bratlie A; Lea T; Ugelstad J and Thorsby E: Tissue Antigens 28, 301 (1986).

43. Howell K; Ansorge W and Grunberg J: In ref. 15, p.443.

44. Howell K: EMBL Research Reports p.24 (1984)

THE USE OF POLYSTYRENE LATEXES IN MEDICINE

J.M. Singer, M.D.

Montefiore Medical Center
Department of Pathology
Division of Microbiology and Immunology
111 East 210th Street
Bronx, New York 10467

1. INTRODUCTION

Latexes are colloidal dispersions of submicroscopic polymer spheres and thus have a dual character: colloidal and polymeric. The properties of particles of such a small size are determined predominantly by their surface characteristics. As polymers prepared from various synthetic monomers, latex particles have the properties inherent to that particular polymer species. Most latexes belong to the lyophobic classification, i.e. their particles are comprised of hydrophobic polymers (in contrast, the lyophilic colloids are exemplified by polymer molecules in solution e.g. methylcellulose and polyvinyl alcohol. Colloidal sols are considered to derive their stability from one of two mechanism: 1) double layer repulsion and 2) steric stabilization. The Verwey-Overbeek theory of double layer repulsion describes the interaction of colloidal particles as the additive combination of two types of forces: 1) electrostatic forces of repulsion arising from the double layer and 2) London-van der Waals forces of attraction. A second mechanism of stabilization is based on the steric effect of adsorbed polymer molecules. Such adsorption can have one of two effects: 1) it can enhance the stability of the colloidal sol by preventing the mutual approach of colloidal particle to a distance at which the London-van der Waals forces of attraction are predominant or 2) it can cause flocculation of the colloidal particles by the formation of polymeric bridges between adjacent particles. Most of the commercial particles are polymerized in the presence of a monomer, an initiator, a buffer and an emulsifier. Latex particles can also be prepared in the absence of emulsifier (Kotera et al., 1970). The emulsifier serves to stabilize primary particles and prevents flocculation by giving the hydrophobic particles a negative charge. The adsorbed emulsifier ions introduce the particle surface persulfate ions (SO_4-) which are used to initiate polymerization and contribute to the particle charge. The negative electrical charge of the latex particle may vary from 2 to -15uc (microcoulomb) cm-2. Negatively charged particles have their highest charge at an alkaline pH which results in increased stability. Thus glycine buffer of pH 8.2 is used in the latex system. Latexes devoid of emulsifier would flocculate when subjected to dialysis or centrifugation. The emulsifier can be removed without flocculation of the latex by the use of ion exchange resins (Dowex 50 and Dowex 1) and conductometric technique can be used to characterize the surface characteristics of latexes (Hull and Vanderhoff, 1968). A soap titration method described by Maron et al. is frequently used to measure particle size and specific surface area of a synthetic latex. The application of the seeded emulsion polymerization technique has allowed Bradford and Vanderhoff (1950) and

Vanderhoff et al (1956) the preparation of a wide range of particle sizes. A series of ten different sizes ranging from 880 to 11,710Å were initially prepared. Particle size is a key factor in colloidal behavior of latexes. Performance properties such as stability, viscosity, rheology, application properties and film properties can all be influenced by particle size and particle size distribution. In the latex fixation test of Singer and Plotz (1956), the first application of this technology, latex particles of 0.8μ were utilized. The number of latex particles in 0.10gm of latex is 3.56×10^{11} particles per ml. The total surface area per ml of latex is 7.15×10^{19}Å^2. Particles with a size larger than 0.8μ tend to settle on standing but easily resuspend on shaking.

The electrophoretic behavior of polystyrene latexes measured at various soap and salt concentration is explained on the basis of the Stern theory of the double layer (Sieglaff and Mazur, 1960) and in terms of general concept associated with hydrophobic colloids (Kruyt, 1946). Latex particles have been shown to follow the Schultze and Hardy law in that the effectiveness of an electrolyte is determined primarily by the nature of the ion opposite in charge to colloidal particles and secondly as the valency of the ion increases the effectiveness of the electrolyte increases markedly (Halberstam et al., 1965). Latex particle suspensions become unstable in the presence of polyvalent cations but not anions.

The influence of anionic, cationic and nonionic detergents was studied in a simplified Tisselius apparatus. Particles with an initial electrophoretic mobility toward the anode of 1.91×10^{-4} cm/sec/V/cm achieved an electrophoretic mobility of 3.20×10^{-4} toward the anode, 1.31×10^{-4} toward the anode in sodium lauryl sulfate (anionic), benzalkonium chloride (cationic) or Tween 80 (nonionic) respectively (van Oss and Singer, 1966).

When considering the use of latex particles in immunological procedures, one must examine their protein adsorbing properties. The adsorption of protein on polystyrene surfaces is an irreversible process. It could be reversible only when dioxane in large amounts is added to the solid. The first test developed with latex particles was the latex fixation test of Singer and Plotz (1956) in which latex particles were coated with human gammaglobulin and reacted with antibody to human gammaglobulin present in sera of patients with rheumatoid arthritis (RA). When 0.8μ latex particles are coated with human globulin their colloidal stability characteristics are that of the protein coat. At pH 5 (net charge positive of HGG) in the presence of polyvalent anions the results is instability of the coated particles while at pH 9 (net charge negative) the latex particles are stable (Halberstam et al., 1965).

When gammaglobulin is added to 0.8μ latex particles the initial electrophoretic mobility of 1.91 to 10^{-4} cm/sec/V/cm is reduced to 0.14×10^{-4} and was raised to 0.57×10^{-4} when sodium lauryl sulfate was added (van Oss and Singer, 1965). The binding of human gammaglobulin and antibodies is wholly pH independent. HGG binds to negative latex particles of size 2200A° solely by the van der Waals London type of hydrophobic bonds. The binding of human albumin and rabbit hemoglobin is pH dependent, however, electrostatic forces also play a role, since the two proteins are more hydrated than human gammaglobulin.

The binding of proteins decreases in the presence of detergents; in the case of NaLS this decrease in binding seems mainly due to competition between detergent and proteins (van Oss and Singer, 1966). When polystyrene latex plates manufactured without detergent (size 2 x 20cm, Almac Plastics, Inc.) were immersed in radiolabeled protein the

adsorption of antibodies (HGG) was the same as for polystyrene latex
particles of uniform size (van Oss and Singer, 1966). Catt and Tregear
(1967) proposed the use of solid support, such as polystyrene tubes, to
bind antibody in radioimmunoassay. Engvall and Perlman (1972) have shown
that polystyrene binding tubes and plates can be used for quantitation in
immunoenzymatic assays. Fourteen commercial preparations of human
gammaglobulin varied widely in their ability to sensitize latex particles
for agglutination with RA serum (Oreskes et al., 1963). While
gammaglobulin destablizes latex particles easily, albumin and whole human
or animal sera can stabilize them. Addition of small amounts of HGG to
0.2 or 0.8μ size latex particles (about 10μg HGG per mg latex) will cause
latex particles to flocculate, whereas larger amounts increase particle
stability. The kinetics of flocculation of latex particles by HGG was
studied by a photometric technique. Flocculation rate was studied by
measuring the change in turbidity after addition of various amounts of
human gammaglobulin. The rate of flocculation was dependent on the
fraction of latex particle surface that is covered with protein with a
maximum rate of 50% coverage and no reaction at all (protection) at 100%
coverage. Flocculation was a bimolecular reaction with the rate determined by the surface coverage of the latex particles (Singer et al.,
1973). An earlier study of antibody adsorption of latex particles using a
modified Langmuir equation showed that the adsorbance was determined as a
function of latex particle size (0.2 and 0.8μ) and pH. The results
indicated that adsorption apparently occurred in two steps characterized
by different binding constants. It appeared that either multilayer
adsorption of IgG occurred or that adsorption of proteins just prior to
full surface coverage produced a change in the orientation of the
adsorbing molecule to occupy the remaining available spaces. At pH 9 the
maximum binding of IgG occurred showing 75,000 molecules for 0.8μ latex
particles and 6700 molecules per 0.2μ latex particle (Oreskes and Singer,
1961).

In another study using 0.2μ Monsanto Lytron particles, which are
different from the 0.2μ Dow Chemical Co. latex particles, 2500 molecules
of IgG adsorbed onto one latex particle. The 0.2μ size Monsanto latexes
contain more than one emulsifier in the dispersed medium (Kochwa et al.,
1967). The binding of individual proteins is decreased in the presence of
mixed proteins in solution and is dependent on both the protein and
polymer used (Lee et al., 1974). If latex particles are first covered by
albumin very little gammaglobulin is adsorbed. When both gammaglobulin
and albumin in mixture are adsorbed onto the surface of latex particles
they will produce agglutination with both RF and antialbumin (Singer et
al., 1961). The adsorptive characteristics of polystyrene tubes for seven
proteins of different molecular weight and ionic charges have been
studied by Cantarevo et al., (1980). At given concentrations the amount
of protein bound varied for different proteins. For all the proteins
studied there was an input concentration of up to 1000ng per 6.5cm^2
tubed surface wherein the proportion bound was independent of the amount
added. The total number of moles bound was inversely proportional to
protein size, suggesting that the upper limit of the region of
independence adsorption, a protein layer of one molecule thick covers the
effective surface of polystyrene. Working in higher concentration of
proteins, there is a protein-protein association, possibly multilayers
are formed that may be antigenically non-effective due to steric
hindrance. Once adsorbed protein remains stable bound throughout the
assay step. As a practical application Cantarevo et al., (1980) suggested
that each protein should be considered specifically, that it is desirable

to work with a concentration in the region of independence since about this point protein input is wasteful (about 500ng of protein per 6.5cm² tube) and when working with mixtures the total amount of protein added should be below the upper limit of the region of independent binding to insure that all of the protein in a mixture will be adsorbed onto the polystyrene.

Pesce et al. (1976) showed that the maximal adsorption of 0.2ml of rat albumin or rabbit IgG to a 12 x1 75mm polystyrene tube was about 5μg after 24 hours of incubation. The binding characteristics of the competitive and ELISA techniques are different. The ELISA technique has an optimal time during which the adsorbed IgG is immunoreactive while the competitive assay approaches maximal bonding in an asymptotic manner. Adsorption of I¹²⁵ labelled rabbit IgG to plastics (cellulose nitrate, polystyrene and polyvinyl) was primarily independent on initial antibody concentration and to a lesser extent, on time allowed for adsorption. All of the plastics tested adsorbed IgG to approximately the same degree with the exception of cellulose condition. The results have also shown that at 18h of adsorption in polystyrene tubes, 55% of the total amount adsorbed can be obtained by the use of only 10% of the input IgG (100μg/ml sample)(Hermann and Collins, 1975). The variability in adsorption of polystyrene (Chesum and Denmark, 1978; Krick et al., 1980), evaluation of various microtiter plates (Shekarki et al. 1984; McCullough and Parkinson, 1984) and their thermal characteristics (Burt et al., 1979) have been investigated. The strategies for their use for antigen mixture (Kenny and Dunomor, 1983). Quality control and standards to be utilized has been proposed (Denmark et al., 1978; Wall, 1980; and Yolken, 1982).

The binding of human IgG by latex particles becomes stronger with increasing molecular weight of the gammaglobulin. Monomeric 7SIgG would not react with RF in a precipitation reaction when both are in solution. However, when 7S is adsorbed onto the surface of the latex particles they will sensitize them and react with RF in an agglutination reaction. When 7SIgG is aggregated by heating it will react in a precipitation reaction, thus only denatured HGG will react with RF. The presence of surfactants on the surface of a latex particle by itself may cause denaturation of the adsorbed 7SHGG and so react with RF (Oreskes et al., 1962). The molecular unfolding and required immunogenicity of IgG molecule concomitant with adsorption to a series of polystyrene Lytron latex particles (DuPont Co.) has been studied by Kochwa et al. (1967) using a potentiometric titration. Conformational changes accompanied by the rupture of intramolecular bonds and exposure of additional titrable groups should result in an increased acid uptake per adsorbed molecule. Adsorption under conditions of low surface coverage apparently leads to a molecular unfolding. The conformational changes of IgG were reinforced by the finding that IgG adsorbed at a low concentration behave immunogenetically like heat denatured material, but when adsorbed at high concentration behave normally.

The development of latexes that give optimal and colloidal properties to a coating composition has led to the inclusion in the polymerization recipe of a so-called functional monomer. The seeded polymerization, developed by Smith in 1948 was a method used particularly to produce uniform particles or particles with larger size from small latex particles (Vanderhoff et al., 1956). This technique, which is still widely used today, has been extended to a method of incorporating the desired functional group in the latex particle by using a common feed containing monomers of desired functionality. These chemically bound functional groups on the particle surface then can be used as reactive sites for further chemical reactions to other functional molecules through a wide variety of chemical reactions. The list of monomers frequently used for industrial purposes belong to the following groups: carboxyl, sulfonate, hydroxy, amide, quaternary amine, polyoxyethylene; the most common basic monomers are styrene, vinyl acetate, methylacrylic acid, itaconic and methacrylamide (Young, 1964). In addition to polystyrene important styrene containing polymers includes styrene acrylonitrites, unsaturated polyster resins and styrene-butadiene rubbers. However, for medical applications considering the production of latex particles of uniform size only latexes with functional groups have been produced and only in response to the demands of the few researchers in this field.

The most common procedure used in the past and still prevalent today for adsorbing antigen or antibodies for use in diagnostic tests has been through physical adsorption. In this method proteins, glycoprotein antigens or antibodies are coated onto the hydrophobic latex particles through van der Waal forces. A second method is available for coating latex particles is used, covalent coupling, less for clinical diagnostic tests but much more for cell labelling and separations which will react in a highly specific way with target cells, viruses and other antigenic material.

Covalent coupling methods are available for binding most species to latex particles. In both methodologies protein materials are irreversibly adsorbed. Attachment of antibodies, antigen or other materials should be carried out through available functional groups on the proteins which are non-essential for their biological activity under conditions which do not denature the ligand. The most important consideration which cannot be overemphasized, is to keep the latex suspension stable at all times. One of the first covalent bindings has been used by Hans Hager (Roche Laboratories, 1974) in the development of carbodiimide techniques for coupling antibody to choriogonadotropic hormone to latex particles.

Molday et al. (1975) copolymerized metacrylic spheres 300-3, 500Å diameter containing carboxy and hydroxy groups. These latexes were activated by cyanogen bromide and tagged with dansyl epsilon lysine or tritiated glycine to obtain a fluorescent or radioactive particle.

It is possible to add a spacer arm with a terminal amino or carboxyl group between the latex particle and the ligand. This provides a flexible linkage of ligand to the latex, protecting the antigen sites of the ligand and preventing excess linking of the ligand in solution. The spacer arm used in this technology were 1,6 diaminohexane or 1,7 diaminoheptane, hydrazine, gluteraldehyde, succinic anhydride, paratoluene sulfonyl chloride (tosylchloride), 2,2,2-trifluoroethane sulfonyl chloride, carbonyl-di-imidazole, etc. There is a three to four step sequence to covalently couple this particle. Diaminoheptone is added to the carboxylated latex and with carbodiimide (CDI) as coupling agent, the carboxyl group is converted to an amino group. Thereafter coupling with antibody is performed in the presence of glutardehyde. Carboxylate latex can also be coupled by the use of CDI with epsilon amino caproic acid. The carboxyl group at the end of epsilon amino caproic is coupled with the ligand by CDI.

Dorman (1977) reacted the amide modified latexes with hydrazine converting them to hydrazide modified latexes. This hydrazide group can be coupled directly to the ligand by the use of glutardehyde. Succinic anhydride may be added to the activated hydrazide latex particles. The end results are particles with a carboxylated group on the end of the spacer arm which can be further coupled with the ligand by CDI. The hydrazide group on the latex particles can be converted to an azide group if reacted with nitrous acid.

Some proteins are liable to lose biological activity when coupled by their amino acid residue. However, if these proteins also contain carbohydrate chains which do not contribute to their reactivity, these carbohydrate groups may be used to couple proteins to insoluble supports (Quash et al., 1978). Methods for coupling glycoproteins such as immunoglobulin G to the latex particles is described in detail by Quash et al. (1978). Carboxylated latex particles hexamethylenediamine are added to create an amine covered surface (HMD). To HMD latexes hydrazinobenzoic acid is added in the presence of CDI to obtain an aromatic hydrazine substituted group. These particles can be coupled with the ligand via CDI. To other HMD latexes, adipic dihydrazide is added in the presence of glutardehyde to obtain an aliphatic hydrazine substituted group. The gammaglobulin is oxidized in the presence of sodium periodate. The oxidized gammaglobulin can then be added either directly to the HMD latex particles or to the hydrazine latex particles in the presence of sodium borchydride.

A sulfhydryl group at the end of a spacer arm may also be coupled to an SH group of a protein by means of an ammoniacal potassium ferricyanide (Epton and Thomas, 1981).

Latex spheres containing aromatic amines can be activated by diazotization and covalently coupled to immunoglobulin (Gonda et al., 1978). Tosyl chloride (paratoluene sulfonyl chloride) can be used to activate the hydroxyl groups on latex particle surfaces which in turn will couple with amine groups on the ligand (Nilson and Mosbach, 1980). Sulfate groups on latex particles may be spontaneously hydrolyzed to hydroxyl groups on old latex particles.

Monodisperse polymer particles of the shell and core type were prepared by copolymerization of methylmetacrylate, hydroxyethyl methacrylate, and ethylene glycol dimethylacrylate in the presence of particles of methylstyrene divinylbenzene 2.9μ in diameter (Ugelstad et al., 1980). Final particles are of 3.3μ diameter density $1.07 g/cm^3$, and are commercially available from Dyno Industries A/S, Oslo, Norway.

Proteins can be covalently bound to polystyrene glycidyl methacrylate latexes through the reaction of the free amino group of the protein with the epoxide group of the polymer at pH between 7-9 (U.S. patent 4,210,723,980).

Poly (4-vinyl pyridine) acrylamide (PVP) and poly (2 hydroxyethyl methacrylate) latexes were synthesized by copolymerization of the monomer in the presence of bisacrylamide as described by Rembaum et al. (1976, 1978). These microspheres varied in diameter from 0.05 to 10μ. They were coupled with hydrazine to yield hydrazide groups on the surface. Chemical binding was carried out with the ligand in the presence of polyglutardehyde dissolved in dimethylsulfoxide. This method can be used for labelling cell surface receptors. These particles are distinguished from the polystyrene latex particle by having a more hydrophilic surface characteristic.

Polyacrolein microspheres were lately prepared by BioScience Inc., capable of undergoing typical reaction of carboxyl group as oxime formation, reaction with Fehling's reagent, and the formation of silver mirrors on the microspheres with silver nitrate-ammonia complex. This particle exhibited low to no nonspecific bindings to protein and did not require further activation to bind proteins. There are many general references on chemistry of coupling liquids to latex particles and the subject was reviewed recently by Bangs (1984).

Polybutadiene polystyrene latex particles (Seragen Diagnostics) can be radioiodinated (Singer et al., 1969) by labelling butadiene latex particles with a combination of bromine and iodine (Vanderhoff and Tarnowski, 1972). If a radiomonomer is chosen, the radiotag can be incorporated directly into the particle during polymerization (Bogen, 1970; Szende and Udvarhelyi, 1975). Latexes can also be labelled with a variety of radioactive isotopes: Gr, Mn, Ni, Cu, Zn, Fe, Co, Ru, Rh, Au, Yt and V (Hinrich, 1975). Gamma radiation from a ^{60}Co source may bind ^{131}I onto latex particles (Huh et al., 1974). Large particles and small particles can also be dyed by a technique developed by Dow Chemical Co. as well (Vanderhoff, 1974; Bangs, 1984). Tritiated glycine and dansyl-epsilon lysine were coupled to copolymers acrylic spheres by the cyanogen bromide (Molday et al., 1975) to obtain radioactive and fluorescent particles.

The number of commercial firms producing polystyrene or styrene divinyl benzene or other copolymers are numerous. However, many of the particles are not of uniform size, the additives and detergents are not specified, thus many of these particles do not lend themselves to coupling with proteins. A limited listing of these commercial latexes is: Airco, Celanese, Bordon, AMSCO, GAF, UVIREZ, Catalin, Rhom and Haas, and Goodrich. Occasionally some of these latexes may be used in serological reactions such as the Lytron series from Monsanto. The most used latex particles are manufactured by Dow Chemical Co., Rhone-Poulenc, Aubervilliers, France and also manufactured and distributed by Seragen Diagnostics, Indianapolis for Dow Co., Polyscience Inc., Warrington, Pa; and Covalent Technology Corp., Ann Arbor, Michigan.

1.1. Covasphere Technology Corp., Ann Arbor, Michigan.

1.1.1.   Polymers. Covasphere CX particles, polystyrene spheres with carboxylic sites; Covasphere FX particles with amino group and Covasphere MX with a more hydrophilic surface.

1.1.2   Particle Types. Diameters - 0, 5, 0.7 and 1μ. Fluorescent particle - blue, green or red. Also MY-1 Covasphere coated with monoclonal antibodies and a fluorescent label, diameter 0.7μ.

1.2. Rhone-Poulenc, Aubervilliers, Paris, France.

1.2.1   Polymers. Polystyrene, polyvinyl toluene, styrene butadiene, styrene/MHA and acrylic ester.

1.2.2   Particle Types. Diameter from 0.15 to 3μ.

1.2.3   Reactive Particles. Carboxylated, amino, amino and carboxylic and hydroxylated.

1.2.4   Dyed Particles. Available.

1.2.5   Fluorescent Particles. Rodhamine B, alpha NPO, Dansyl chloride, BBOT, DHPOPOP and fluorescein.

1.2.6   Magnetic Particles. Available.

1.3. Seragen Diagnostics, Indianapolis, Indiana.

1.3.1   Polymers. Styrene butadiene, styrene divinyl benzene and vinyl toluene butylstyrene.

1.3.2   Particle Types. Diameter from 0.038 to 3.01μm. Large particles - 6.4-90.7μm. Uniform sizes - 2,5,7,10,15 and 20μm.

1.3.3   Reactive Particles. Carboxylated, amide and amino modified, activated particle.

1.3.4   Magnetic Particles. Diameter 0.7 to 1.75μ.

1.3.5　　　Dyed Particles. Pink, blue, yellow and black.
1.3.6　　　Fluorescent Particles. Pink and rhodamine B.

1.4　Polysciences. Inc.. Warrington. PA.
    1.4.1　　　Polymers. Polystyrene, epoxymicrospheres - poly (styrene-glycidyl methacrylate) polyacrolein, polymethyl-met-acrylate and chloromethylstyrene.
    1.4.2　　　Particle Types. Diameter from 0.05 to 5μ. Large particles.
    1.4.3　　　Reactive Particles. Carboxylated, amino and hydroxylated.
    1.4.4　　　Dyed Particles. Red, blue and yellow.
    1.4.5　　　Fluorescent Particles. Yellowish green, yellowish-orange and bright blue.
    1.4.6　　　Magnetic Particles. Available.

1.5　Interfacial Dynamic Corp.. Portland Oregon.
    1.5.1　　　Polymers. Polystyrene latexes prepared in the absence of surfactants.
    1.5.2　　　Particle Types. Diameter from 0.03 to 4μ.
    1.5.3　　　Reactive Particles. Sulfate, carboxyl and amidine.

1.6　DYNO Industrier A/S. Norway. Various sizes, highly porous, compact particle, magnetic particle.

2. LATEX AGGLUTINTION TESTS

There were four techniques introducing the latex particle as a solid support for antibody in serological procedures (Singer and Plotz, 1956). The latex fixation is a test for the laboratory diagnosis of rheumatoid factors present in sera of patients with rheumatoid arthritis. The original paper recommended the use of 0.8μ diameter polystyrene latex particles and a solution of human gammaglobulin as antigen. The HGG was adsorbed physically onto the latex particles. In a tube technique and particles in uniform suspension with a glycine buffer of pH 8.2 latex particles would agglutinate in the presence of patient sera. The highest serum dilution with 1+ agglutination was reported as titer. Agglutination at a dilution of 1:80 or greater is considered a positive test for the 19SIgM antihuman gammaglobulin antibodies (RF). The control tube should not show any granularity (Singer and Plotz, 1956).

In a modified latex fixation text, latex particles were precoated with human gammaglobulin and the excess removed by repeated washing (Singer and Plotz, 1961). Most of the commercial latex particle kits are patterned on this technique.

A technique was developed by which latex particles are precoated with an antibody for the detection of an antigen. The antibody was a rabbit antiserum for C reactive protein and the antigen C reactive protein present in patient sera (Singer et al., 1957).

In this procedure uncoated latex particles are agglutinated by RA sera. Latex particles are coated by the patient's own gammaglobulin present in the serum and reacted with the patient's own rheumatoid factor (Singer and Plotz, 1958). This was a slide technique using a drop of serum and a drop of latex suspension. Agglutination occurred when RF was present in the serum.

There are many modifications of these techniques and the latex fixation test has served as a tool for the development of other serological antigen-antibody reactions in medicine.

Attempts have been made for the standardization of the latex fixation test (Singer et al., 1973, 1979) and a rheumatoid arthritis

serum proposed by a subcommittee of the National Committee for Clinical
Laboratory Standards (NCCLS) is available at the Centers for Disease
Control, Biological Products Division, Atlanta, Ga. The latex fixation
test for RA has been reviewed by Singer (1961) and the biomedical use of
latex particles by Vanderhoff (1964) by Kenny (1976,1978), Hechemy and
Michaelson (1984) and Bangs (1984,1985).

Latex particles in the biomedical field are used in 1) medical
serological testing, 2) latex particles as first or second antibody in
the immunoassay ELISA techniques, 3) latex particles in immunoassays
utilizing photometric techniques, 4) latex as immunosphere probes for
surface cell receptor recognition, 5) studies of phagocytosis in vitro or
in vivo, 6) studies of the reticuloendothelial system, 7) used as an
adjuvant for the enhancement of antibody production, 8) other
miscellaneous uses in physical biotechnology. Table I represents the
medical applications of the serological tests for the detection of
antigen or antibody in the area of microbiology, virology, mycology,
parasitology and endocrinology in therapeutic drug monitoring and for
other miscellaneous uses.

There are many commercial kits available in the USA for diagnostic
use. They can be classified under two headings: use in emergency
procedures for the etiologic diagnosis of acute meningitis and for
routine laboratory assays. Since it is crucial that physicians initiate
appropriate animicrobial therapy in the shortest possible time (within
minutes preferably) (Coonrod et al., 1983) the following kits are used:
for detection of antigens in cerobrospinal fluid, Neisseria meningitidis
group A, B, C, X, Y, WR-135, Streptococcus group B, Hemophilus influenzae
group B, Streptococcus pneumonie and for the detection of antibody to
Cryptococcus neoformans.

3. POLYSTYRENE OR POLYVINYL TOLUENE AS A SOLID SUPPORT FOR ANTIGEN AND
    ANTIBODY REACTIONS.
    3.1 Latex Particles as an Emulsion.
    Polystyrene as a Solid Support on Tubes, Plates or Membrane.
    Factors affecting the two forms of support would depend on a) the
chemistry of their surface, b) material to be absorbed on their surfaces,
and c) reagents used in the preparation or performance of the tests. The
material to be adsorbed on these surfaces is serum or other fluids, a
mixture of proteinaceous materials, carbohydrates and minerals present in
all fluids. Both systems are similarly affected. There are many
nonspecific factors which may inhibit the specific antigen-antibody
reaction and detection.
    3.1.1. Nonspecific Factors Which May Interfere in the Test System.
    A. Lipemic (high fat content), hemolytic (bloody specimen),
bacterial contaminants, aged serum, drugs present.
    B. CIQ or Cl are components of the complement system, present in all
sera, may attach to the solid phase, preventing the attachment of
antigens with their specific antibody. The complement components may be
readily inactivated by heating sera at 56°C for 30'.
    C. Pre-existing antibodies in patient sera, even in small amounts
may bind to the antigen added and form immune complexes, thus inhibiting
antigen detection. A variety of dissociating methods for AgAb complexes
have been described which permit the detection of antigens despite pre-
existing antibodies (Coonrod and Leach, 1978; O'Reilly et al., 1975;
Weiner and Stephen, 1979; Meckstroth et al., 1981). Several of these
techniques are based on the thermodissociation of complexes (Sugarman and
Hart, 1973; Wheateal, 1979, 1983).

D. The rheumatoid factor represents a major interference factor in all immunoassay techniques. It can react with all gammaglobulin preparations used as antibodies in the test system. RF is present in 0.5 to 2% of normal sera, in 25-30% sera from people over 70 years of age and in many diseases other than rheumatoid arthritis. The interference created by RF can be eliminated by a number of procedures: heating sera and use of S-S reduced agents such as 2 mercaptoethanol, N-acetylcystine or dithioerythiol (Yolken and Stopa, 1972; Gordon and Lapa); by EDTA and heat extraction (Eng and Peron, 1981); by the use of pronase-CB protease (Stockman and Roberts, 1982); by adsorption of sera with IgG coated latex particles (Aranjo and Remington, 1980); by prior testing of serum with an RF kit; comparing the effect of immune verses nonimmune serum added to the assay; use as a control as latex particles coated with nonimmune antibody used.

An alternative procedure commonly used today in most of the immunoassays is the replacement of the antibody used with the $F(ab^1)_2$ fragment of the same antibody. RF reacts only with the Fc fragment of the immunoglobulin but not with the $F(ab^1)_2$ fragment. $F(ab^1)_2$ carries the antibody site of the Ig molecule. The $F(ab^1)_2$ fragment can be prepared by the technique of Mandy et al. (1961), Nisonoff et al. (1961), Ishikawa et al (1981).

E. Spinal fluid may have a nonspecific component which may be destroyed by either a reducing agent or heating 100°C 15' and using the supernatant by using pronase. RF may interfere in the test if the specimen is bloody.

F. The prozone phenomenon has been described for all immunoassays. Excess of antigen or antibody in the procedure would interfere with the test. Using dilution of the serum and/or titration may eliminate this phenomenon.

G. Cross reactivity is a common interference feature for all serological tests and has to be recognized. The use of highly avid monoclonal antibody in the assay may be one answer in the future concerning the specificity of the test.

3.1.2 Latex as an Emulsion.

A dominating and unifying characteristic of the latex system is that latexes are a colloidal suspension. Control of the colloidal properties is a fundamental rule in the preparation of a test procedure utilizing latexes in suspension.

A lyophobic sol such as latex particles does not constitute a stable system from the thermodynamic point of view. In order to maintain such a system stable the particle must be kept separated and kept unflocculated at all times. The colloidal properties would vary dependent on chemical composition of the latexes, their surface characteristics, the components of the disperse medium, the uniformity of the latex suspension in sizes, displacement of surfactant from their surface, the pH of the dispersed medium, or that used in the test procedure, the flocculating effect of salt, protein in small amounts may destabilize the system or when in large amounts colloidally protect them, chemical coupling procedure would destabilize them, centrifugation, dialysis, freezing, drying particle, bacterial contamination, keeping suspension in metal containers. All would effect colloidal stability.

The adsorption on plates or tubes is also effected by their chemical composition. Adsorption of protein on polystyrene, nylon, lucite and polysulfane vary from one plate to another and even from well to well (Shekarchi et al., 1984). Chemical formulation and production procedures were directed primarily toward improving optical clarity and uniformity. Wide lot to lot, plate to plate and within plates variation may be encountered. It is recommended that each ELISA test procedure as a matter of record should include a description of plates used, giving manufacturers, type of plastic, and a lot number. Well to well variation should also be considered. Titers determined by serial dilution are less susceptible to well to well variation. Polystyrene as a hydrocarbon has a low surface energy and is classified as a nonspecific adsorbent (Kiselev, 1968). This leads to its poor wetability by water and low adsorptive properties. In order to improve the properties of polystyrene surfaces polar sites are created at the surface of the polymer by chemical treatment. Pretreatment of plates, cleaning agents used in order to improve antigen binding, especially detergent, use of chemical coupling procedures for binding of protein to plates all would interfere with the binding of the ligand but also favor nonspecific adsorption of serum components.

Ability of the plate to absorb proteins or other compounds by noncovalent bond may cause physical adsorption of the biologically active protein during the coupling reaction. Using nonimmune sera or fetal calf serum after the solid phase is coated with the antibody minimizes the nonspecific binding. As an emulsion the adsorption of protein is irreversible, however, detachment from the plastic during washing has been reported. Competition between proteins and enzyme for adsorption sites on plastic was studied (Kenny and Durrsmor, 1983). It has been shown that IgG is adsorbed well on all types of plates but albumin may differ. Factors affecting physical adsorption and chemical binding are very similar for particles, beads, plates and tubes made of polystyrene or polyvinyl toluene.

4. PARTICLES - IMMUNOPHOTOMETRIC ASSAY

Optical measurement based on light scattering has been recognized as a means of determining particle size in ranges where the light microscopy is difficult. Scattering by aggregation of very small particles is readily described by Raleigh's theory. For somewhat larger particles Raleigh-Gans-Debye approximation have been used. For particles up to 10 micron the Mie theory may be applied. The most important impediment in the past was the lack of suitable monodisperse particles for experimental verification. Monodisperse particles developed by Bradford and Vanderhoff (1955) have been shown to be one of the most accurate systems for investigation of light scattering. They are used as particle models in light scattering. They are used as particle models in light scattering, turbidity, low angle neutron scattering, reflectance light, aerosol spectrometry, standardizing, calibrating or checking electronic light microscope, particle counter, flow cytometry (Bangs, 1984). The measurements of light scattering in an antigen-antibody reaction may give direct information on the rate of reaction, being also a sensitive indicator measuring differences in the reaction rate between different Ag-Ab (Tangerdy and Small, 1966). Many methodologies have been used to quantify the agglutination of suspension of latex particles: 1) turbidimetric, 2) measurement of light scattering at certain angles, 3) counting of the unagglutinated latex particles and 4) liquid scintillation counter. These may be further divided and used for direct agglutination or for inhibition techniques.

5. SPECTROPHOTOMETRY, NEPHELOMETRY AND LIGHT SCATTERING

Mathies, 1966; Dezelic et al., 1971; Singer and Chu, 1971; Singer et al., 1973; Blume and Greenberg, 1975; Bonefay and Grange, 1976; von Schultess, 1976;. Cambiaso, et al., 1977, 1978; Grange et al., 1977, 1978; Virella et al., 1978; Hallegren, 1979; Finlay et al., 1979; Limet et al., 1979; Cambiaso and Masson, 1979; Leflor, 1980; Leek et al., 1980; Ripoll et al., 1980; Kimura, 1980; Sado et al., 1980; Cassart et al., 1980; Bernard and Lawyers, 1981; Husman et al., 1981; Crane, 1981; Magnuson et al., 1981; Sindic et al., 1981; Bernard et al., 1982; Babson et al., 1981; Ito et al., 1983; Makojime, 1983; Magnuson et al., 1983; Bernard and Lawyers, 1983; Merlini et al., 1983; Bernard and Lawyers, 1984; Kent and Blackmore, 1985; Fan et al., 1983; Caffo et al., 1983; Yamagashi et al., 1983; Kubota et al., 1983; Oka et al., 1983; Kuroso et al., 1983; Suzuki et al., 1983; Ikawa et al., 1983; Becker and Kapmeyer, 1983; Lida et al., 1983; Nakamura, 1983; Kariya et al., 1983; Hashimoto et al., 1983; Suzuki et al., 1983; Tsubota, 1983; Mukojima, 1983.

6. QUASI ELASTIC LIGHT SCATTERING IMMUNOASSAY

Described by Cohen and Benedek (1975) and McConnell (1981), it is a light spectroscopy, a helium-neon laser technique for following the Brownian motion of a particle in a medium. The Brownian motion is related to the diffusion constant (D) of the particle which in turn is inversely related to the particle size. As the particle size increased due to the antigen-antibody agglutination, the spectral line width of the scattered laser light is broadened. The sensitivity of the measurement may be increased by using anisotropic light measurement. Angular anisotropy measure scattered light at two angles usually greater and less than 90° e.g. 45° and 135°. The ratio of forward-backward scattered light is indicative of degree of agglutination (van Schulthess, 1980).

7. FULLY AUTOMATED SPECTROPHOTOMETER

Developed in Japan - LPIA 1, LIPA-300 by Mitsubishi Co., EL-100 system by Kyowa Co., and LA-200 by Eiken Chemical Co. These are fully automated latex immunoassay systems, 20-160 tests per hour, sequential multianalyzer, fully computerized, sensitivity of order 10^{-9} to 10^{-12}g. A number of kits are available in Japan such as carcinoembryonic antigen, alpha fetoprotein, haptoglobulins, complement C_3, C_4, beta 2 microglobulins, human placental lactogen, IgG, IgA, IgM, fibronectin and phenobarbital, IgE, estriol 16-glucuronide (E_3-IgG), tobramycin and theophylline.

8. PARTICLE COUNTING

Described by Cambiaso et al. (1977) as the PACIA system technique applied in a fully automated instrument - IMPACT, Arcada SA, Brussels, Belgium. It measures the agglutination of 0.8μ latex particle coated with antigens, antibodies and haptens. Includes a sample processor, multichannel peristaltic pump, and particle counter. The number of unagglutinated latex particles are counted by an optical particle counter and a computer collates the standard curves, compares simple values and print-out results.

Cambiaso et al., 1977, 1978; Masson et al., 1979, 1981; Collet-Cassart et al., 1981, 1982, 1983; Magnuson et al., 1982, 1984; Lindic et al., 1979, 1980, 1981, 1982, 1983, 1984; Limet et al., 1979, 1982; Bernard and Lawyers, 1981, 1982, 1983, 1986; Holy and Reicher, 1981, 1982; de Steenwinkel, 1982; Djurup et al., 1983; Castracane et al., 1984; Elby and Berck, 1982.

Kits available in Belgium with the Impact System - Alpha fetoprotein (AFP), ferritin, human placental lactogen, C-reactive protein, thyroxine

(T4), triiodothyronine (T3), thyroid stimulating hormone (TSH) and thyroxine binding globulin (TBG).

9. LIQUID SCINTILLATION
9.1. Scintillation Proximity Assay.
In this method titrated particles are used with scintillant dyed particles. Both kinds of particles are coated with Ag. When the two particles are mixed with AB agglutination of the two kinds of particle results in scintillation which is measured in a liquid scintillation counter. Nanogram sensitivities can be obtained (Hart and Greenwald, 1979).

10. ENZYME IMMUNOASSAY
Measurement of substances in human fields by immunoassay is of great importance in clinical chemistry and medicine. Most immunoassay methods depend on completition of sample antigen and labelled antigen for a limited number of antibody binding sites. After determination of the fraction of labelled antigen bound by antibody, the concentration of sample antigen can be calculated by comparison with results for appropriate standards. The sensitivity of the technique depends upon the label. This label can be a radioisotope (Yallow and Benson, 1960), and enzyme (Englwall and Perlman, 1981), fluorescence (Sokol et al., 1962 and Dandleker and Feigen, 1964), chemiluminescence substances (Cerlan and Halmann, 1978), organometallic material (Cais et al., 1977), coenzymes (Carrico et al., 1976), and enzyme inhibitors (Howley and Tonkes, 1977).

Enzyme immunoassay is a type of binding assay that depends on the antigen-antibody reaction as a base and the enzyme reaction as a marker. The various EIA or ELISA that have been used can be divided into several groups as follows (Windon, 1978; Voller et al., 1977 and van Weener and Schurr, 1976):
1. ELISA in which the substances to be assayed are antigen or haptens or antibodies.
2. EIA in which the reactions are competitive or noncompetitive.
3. EIA in which the bound or free forms are separated (heterogeneous) or nonseparated (homogeneous).

The assays which do not require a separation step are called homogeneous. Those immunoassays that include a separation step to distinguish the complexed (bound) labelled component from the uncomplicated (free) labelled compound are said to be heterogeneous.

A subclass of the heterogeneous immunoassays coming into widespread use involved enzyme labelled antibodies. In the majority of heterogeneous and some homogeneous assays the solid phase may be a polystyrene or polyvinyl toluene tube, microtitration plate, beads or latex suspension. The separation step in heterogeneous assays are performed by the use of ammonium sulfate, polyethylene glycol or by the second antibody technique.

The antigen or antibody can be immobilized on the solid phase. In the sandwich method the analyte reacts with the matrix antibody and the matrix is washed and reacted with an enzyme antibody conjugate. After a second wash the enzyme reaction is quantitated.

The EIA technique has been improved by:
1. Using two particle reagents, one particle carrying the first substance and a second particulate reagent carrying the second substance. The first substance may be an antigen and a second substance an antibody. These two reagents are such that they will agglutinate when mixed together but remain unagglutinated prior to mixing (Cambiaso et al., U.S.

patent 4,184,849). Many PACIA techniques are based on this principle.

2. Three types of latex particles, small 0.3μ - density 1.03, intermediate 0.93μ - D 1.14 and very large particle D 1.15 are used in the procedure. The agglutinated particles are separated from the free particles by centrifugation in a sucrose gradient procedure. This has been·used in assays for insulin, human choriogonadotropic hormone and the C_3 component of the complement (Kimura, 1980). A copolymer of styrene-TBPA (tribromophenylacrylate) of particle size 0.9μ and D 1.4 for the microtiter used has been developed by Kariya et al., 1983.

3. The use of polystyrene balls in mm diameters (Precision Plastic balls, Chicago, Ill.) as a solid phase for antigen or antibodies; insulin determination (Ishikawa et al., 1981), α_2 microglobulin test (Kawai and Takagi, 1981), beta$_2$ microglobulins (Ferrua et al., 1981), carcioembrionic antigen (CEA) (Masseyeff et al., 1981), antibody to thromboglobulin (Tanaka and Koto, 1982), coated beads by glutardehyde treated insulin antibody (Shinkai et al., 1980; Hashimoto and Kawaoi, 1981; Tanaka and Kato, 1982; Galleti, 1982; Yamamoto and Ywata, 1982; Sata and Yamamoto, 1982; Matsuoka, et al., 1983; Nagelkerken et al., 1983; Estevenon and Figarella, 1983; Kata et al., 1983; Kawamura et al., 1984; Fieve et al., 1984; Chang et al., 1984;. Johns and Kumar, 1984; Ferrua and Masseyeff, 1985). In a CEA assay two different polystyrene balls are coated with two different antibodies (Luis et al., 1984).

4. Combination of polystyrene plates and IgG coated silicone rubber (Othtaki and Endo,1 1981), glars beads and polystyrene plates. In a sandwich enzyme immunoassay of HBS Ag, glass beads (5 mm diameter) coated with IgG were compared with silicone discs (5 mm diameter). The ELISA with glass beads appeared more accurate than the EIA with silicone disc (Adachi et al., 1981).

5. EIA have been reported utilizing substances other than polystyrene; nylon in various forms (Pillai and Bachhawat, 1977); gelatin and microporous membrane (Millipore Co.,), combined polystyrene plastics (Hashimoto and Kawaoi, 1977); sheep red blood cells (Hashimoto and Kawaoi, 1977); nitrocellulose membrane (Hawkes et al., 1982); Protein A colloidal gold particles (E and Y Laboratories, Inc., San Mateo, CA.).

6. In an agglutination test using Staphylococcus aureus containing protein A (a component of the Staphylococcus cell wall), the antibody to be coated is attached via its Fc portion to protein A, the Fab portion of the antibody is then available for reaction with antigen (Edwards et al., 1980). The coagglutination kits for the detection of various antigens are available from Pharmacia. The coagglutination test has been compared with latex particle agglutination in many publications reviewed by Coonrod et al. (1983).

7. Fluorescent immunochemical label introduced by Coons et al. in 1941 has been developed in a wide variety of both heterogeneous and homogeneous fluorescence immunoassays in the last several years. The sensitivity of fluorometric method can be refined to detect analytes at concentrations of 10^{-15}M (Nakamura, 1983). Fluorecent measurement capabilities, fluorescent immunoassay with magnetizable solid phase antibody, fluorescent protection immunoassay, substrate labelled fluorescent immunoassay, fluorescent excitation transfer immunoassay has been reviewed by Nakamura et al., 1980, 1983.

Chemiluminescent reactions have been used instead of radioisotopes to monitor the immunoassay (Schroeder et al., 1980; Bogusloski, 1983). The peroxidase-antiperoxidase technique of Stemberger et al. (1978), peroxidase-avidin biotin complex technique of Hsu et al. (1981) and glucose-oxidase-avidin biotin technique of Robb (1981) are new enzyme immunolectin assays (Robb, 1983).

8. The use of monoclonal antibodies of higher avidity is the ultimate goal in the replacement of polyclonal antibody in the immunoassay procedure.

9. Covalent binding of alkalin phosphatase to the surface of the carboxylated polystyrene latex particles was achieved by using a carbodiimide technique (Michael, 1985). A continuous stirred tank reactor was developed with the latex AKP conjugate. Long operational time of 2 to 16 hours were achieved in all reactors with covalently bound AKP (Michael, 1985).

10. Liposomes bearing a second ligand added to latex antigen suspension has shown to produce a marked enhancement of agglutination reaction. To the latex covalently conjugated with human gammaglobulin liposomes covalently coupled with antihuman IgG Fab[1] fragment were added: This technique has been applied in a commercial kit for the detection of RF.

11. LATEX IMMUNOMARKERS

A large variety of active molecules have been shown to bind with a high affinity to specific type of cells. These include antibodies, lectins, toxins, enzymes, inhibitor drugs, chemical transmitters. They will bind to the cell surface markers. Latex particles may serve as a visual marker in microscopy, scanning microscope, electron microscope and fluorescent microscope as a reagent for qualitative and quantitative studies of cell surfaces and their receptors. They may be synthesized to incorporate radioactive, electromagnetic, opaque, highly colored, fluorescent markers. Proteins can be physically adsorbed, covalent chemically bound to the surface, protein made fluorescent. Fluorescent immunospheres can also be coupled with Protein A. In many investigations, marker tests are associated with cytological staining techniques by the use of enzymes, lectins, and protein A. Many of these microspheres are coated with highly specific monoclonal antibodies. Many instruments are using these microspheres - the Fluorescence automated cell sorting (FACS) (Parks et al., 1979); particle concentration fluorescence immunoassay (PCFIA) (Jolly et al., 1984; JacGrindle et al., 1985); time resolved fluoroimmunoassay (Meurman et al., 1982; Esiola et al., 1982), particle enhanced turbidimetric inhibition immunoassay (Petinia, Schifreen et al., 1985; Polliak and Gordon, 1975). In addition to polystyrene latex particles other particles have been used: polystyrene latex particles (Le Bugilo et al.); polymethyl methacrylate (Fuchs and Bachi, 1975); hydroxyethylmethacrylate (Rembaum et al., 1976, 78); polyvinyl pyridine (Smolka et al., 1979); polyacrylamide (Rembaum et al., 1976, 1978; Ljungsted et al., 1978); polyacrolein (Kempner et al., 1982); methylmethacrylate with hydroxyethyl metacrylate and ethylene glycol dimethylacrylate (Ugelstad et al., 1980), styrene divinylenzene with epoxy and hydroxyl groups (Ugelstad et al., 1979, 1984).

For the separation of cell two other techniques have been used: density perturbation which is used for preparative scale separation of cells and subcellular organells using routine differential and/or gradient density centrifugation (Lungsted et al., 1978); also this particle was used for altering the electrophoretic motility of the cells (Smolka et al., 1979; Kempner et al., 1982).

Many papers on this subject have been described: Le Buglio et al., 1970; Rinehart et al., 1971; Molday et al., 1974, 1975, 1976, 1977, 1980, 1981, 1982, 1984; Rembaum et al., 1976, 1978, 1979, 1980, 1986; Rembaum and Dryer, 1980; Gonda et al., 1977, 1978; Yel et al., 1976; Bracke and Markovetz, 1978; Israel et al., 1979; Sjoberg and Ingan, 1979; Capoet et al., 1979; Parks, 1979; Margi et al., 1979; Sjoberg, 1980; Mason et al., 1980; Bast et al., 1980; Bartlett et al., 1980; Passwell et al., 1980; Mirro et al., 1981; Higgins et al., 1981; Civin et al., 1981; Paul, 1981; Kormet et al., 1981; Horan, 1981; Ross et al., 1982; Pretlow and Psetlow, 1982; Waymouth, 1982; Bernardt et al., 1982; Parish and Higgins, 1982; Jonak et al., 1982; Trevisian et al., 1982;. Lambris and Ross, 1982; Sheetz and Spudich, 1983; Strauss et al., 1983; Cupp et al., 1984; Slifkin et al., 1984; Stanley et al., 1984; Wahlin et al., 1984; Beckerle, 1984; Stanley et al., 1984; Wahlin et al., 1984; Beckerle, 1984; Katz et al., 1984; Travassoli, 1984; Brower and Fraser, 1984; Loda and Travassoli, 1984; Katakoa and Travassoli, 1984; Mac Grindle et al., 1985; Mirro et al., 1985.

12. MAGNETIC MONOSPHERES

These monospheres can serve both as electron dense markers for transmission electron microscopy and as reagent for cell separation. Antigen and/or antibodies are physically adsorbed or coupled to the bead. A magnet is used to separate the beads from the liquid phase and thus the time to carry out the procedure is simplified and shortened.

Ferromagnetic iron monospheres used in magnetic cell separation were developed by Guesdon and Avrameas in 1977 in a ELISA technique; hydroxyethylmethacrylate by Kronick et al., 1978; copolymeric methacrylate by Molday et al., 1977, in cell labelling and separation; polyacrylamideagarose beads: Antoine et al., 1978; magnetic protein A microsphere: Wider et al., 1979; Kronick et al., 1979 developed magnetic hydrogel monosphere to label neuroblastoma cell containing surface ganglioside GMI. In addition magnetic albumin microsphere: Wider et al., 1981; polyglutardehyde: Margel et al., 1979; polyacrolein/polystyrene: Rembaum et al., 1982; dextran microsphere: Molday and MacKenzie 1982; iron dextran: Molday and MacKenzie, 1982, Molday, 1984; cobalt magnetic albumin immunocolloid; Poynton et al., 1983, 85; protein magnetite; Owens and Sykes, 1984; Owen, 1986; and lately highly porous monosized polystyrene particle styrene divinyl benzene magnetite, Ugelstad et al., 1983. Styrene divinylbenzene were coated with epoxy hydroxyl group, Vartdal et al., 1985.

These magnetic microspheres coupled with antibody have been used for:

1. Removal of neuroblastoma cells from bone marrow for therapeutic purposes, Kronick et al., 1978, 1980; Treleaven et al., 1984; Kembshead et al., 1985; Seeger et al., 1985; Reynold et al., 1985.

2. Removal of T cell from bone marrow, Antoine et al., 1978; Kernan et al., 1985; Goudernack et al., 1985; Platsoucas et al., 1986; Platsoucas see this workshop; bone marrow transplantation (bone marrow purging) Kembshead et al., 1986; Poynton et al., 1985; Treleaven et al., 1984; Nustad et al., 1986; cell separation for mixture of various cells and in various immunoassays, Molday et al., 1977; Margel et al., 1979, 1982; Rembaum et al., 1982; Antoine et al., 1978; Molday and MacKenzie, 1982; Molday, 1984; Poynton et al., 1983, 1985; Owen and Syles, 1984; Ohman et al., 1985; Owen, 1978; Owen et al., 1983, 1984, 1986; Guadermack et al., 1985, Vartdal et al., 1985; Johansen et al., 1983, 1984; Nustad et al., 1982; Bormer, 1982.

Hydrophilic monodisperse shell and core particles of $2.5\mu m$ prepared by copolymerization of methylmethacrylate with hydromethylmethacrylate according to Ugelstad et al., (1980). These particles are commercially available as Dynospheres XP4001 from Dyno Industry A/S Oslo, Norway. These particles can be coupled with antibody by the method of Nilsson and

Mosbach (1980) using a sulfonyl chloride reagent. These particles have been investigated by Nustad et al., (1986) (CRC in Press) and were used in an ELISA technique (Bormer et al., Nustad et al., 1984; Johansen et al., 1983; Johansen et al., 1984; Nustad et al., 1982; Bormer, 1982). Antibody and Protein A can also be coupled to tosylated particles, Nustad et al., in CRC Press Inc., to be published.

13. PHAGOCYTOSIS
    In the study of phagocytosis latex particles have been used for: 1) sequential steps of the process of phagocytosis, 2) metabolic responses to the phagocytosis of latex particles, kinetic study, rates of ingestion, effect of temperature and metabolic inhibitors, 3) mechanism of adhesion to membrane, 4) recognition of various cells based on surface markers, 5) isolation of the phagocytic vacuole, 6) effect of various drugs on phagocytosis, 7) phagocytosis by platelets and 8) phagocytosis in disease (Sbara and Karnowski, 1959; Kvarstein, 1969; Weiss, 1974; Werb and Cohen, 1972; Atkinson et al., 1975; Goodal, 1972; Rabinovitch and DeStephano, 1973; Baehner and Natan, 1967; Segaland Levi, 1975; Homan Muller et al., 1975; Weening et al., 1975; Root, 1975; Baehner, 1966; Weening et al., 1976; Falke et al., 1975; Imanishi et al., 1975; Cooper et al., 1972; Leffell and Spitznagel, 1975; Schiffer et al., 1975; Werb and Gordon, 1975; Bleier et al., 1975; White, 1972; Lee and Outterridge, 1976; Molday et al., 1976; Bubentick and Molkovski, 1977; But and Nuttal, 1978; Le Ferrectal, 1978; Fimbel and Polliak, 1979; Bubentick et al., 1979; Soman and Kaplow, 1979; Spector et al., 1979; Ibrahim et al., 1978; Kaiser, 1979; Mayorg et al., 1979; Estrada et al., 1979; Hamvell et al., 1979; Moering and Shuteck, 1979; Forrester and Balazi, 1980; Gudewicz et al., 1980; Lewis and Andre, 1980; Rimland and Hand, 1980; Birmingham and Jeska, 1980; Hutchinson et al., 1980; Sank, 1980; Sjoberg, 1980; Hung, 1981; Levine et al., 1981; Cron and Jansa, 1981; Villiger et al., 1981; Levine et al., 1981; Adamson and Bowden, 1982; Parish and Higgins, 11982; Flyn et al., 1982; Bernhard et al., 1982; Finozzi et al., 1982; Doll et al., 1982; Kasai et al., 1982; Stock et al., 1982; Bassu et al., 1982; Kelly and Grand, 11982; Christman and Schwartz, 1982; Moske et al., 1982; Kawaguchi, 1982; Fase et al., 1982; Wagner and Hynes, 1982; Gill et al., 1982; Pucci et al., 1982; Geisler et al., 1982; Ageller and Werb, 1982; Shuter et al., 1983; Leibinsk et al., 1983; Pnezanski et al., 1982; Hard and Rischell, 1983; Shirakawa et al., 1983; Matschumara et al., 1983; Matsuda et al., 1983; Talstad et al., 1983; Idebska et al., 1983; Parod and Brian, 1983; Lundin et al., 1983, Onozaki et al., 1983; Foket et al., 1983; Hand et al., 1983; Cofano et al., 1983; Sabloniere et al., 1983; Dunn et al., 1983; Seyfried et al., 1983; Abee and Garinell, 1983; Martin et al., 1983; Rhee et al., 1983; Schroeder and Kier, 1983; Agnew et al., 1983; McLaughlin et al., 1983; Schwartz and Juliano, 1984; Ilum and Davies,1 1984; Greenspan and Marrow, 1984; Pesanti and Shanky, 1984; Mayernik et al., 1984; Edwards et al., 1984; Obeso and Auerbach, 1984; McCusky et al., 1984; Mosseson, 1984; Holt et al., 1984; Donald et al., 1984; Mittelman et al., 1984; Bernard et al., 1984; Lyod, 1984; Kanai et al., 1984; Asano, 1984; Shatrov, 1984; Yashida et al., 1984; Pesianini et al., 1984; Boulton et al., 1984; Grinell, 1984; Suss et al., 1984; Diazo et al., 1984; Mattews et al., 1985; Oliver and Veir, 1985; Roberts et al., 1985; Lemert and Tech, 1985; Soland and Nouze, 1985; Aoucasstal, 1985; Lima and Kierzenbaum, 1985; Lochner et al., 1985; Souland and Nouza, 1985; Dan and Waki, 1985).

14. STUDY OF PHAGOCYTOSIS IN VIVO
    Radioiodinated latex particles and combined radioisotope histologic and electronic microscopic studies traced the fate of the particles in RES of mice (Singer et al., 1966, 1967; Adlersberg et al., 1967, 1968). Other studies of in vivo phagocytosis include Heinant et al., 1978; Mohrig and Schittek, 1979; Slinter et al., 1983; Asano, 1984; Samuelson et al., 1984; Klum and Davis, 1984; Lugitt et al., 1985.

15. ADJUVANCY FOR ANTIBODY PRODUCTION
    Antigens adsorbed onto latex particles can be used as immunizing agents in animals for the enhancement of antibody production (Singer, 1983; Litwin and Singer, 1966; Kochwa et al., 1967).

16. STUDY OF THE VASCULARIZATION OF ORGANS
    The effect of microsphere emobolization (occlusion) of larger vessels, precapillaries and capillaries has been extensively investigated (Fassol et al., 1977; Kurarstein and Dale, 1978; Le Fevre et al., 1978; Tissim, 1979; Updike and Diesem, 1980; Christie, 1980; Kendel and Eissman, 1980; Evans et al., 1981; Zbrodowski, 1981; Bisaillon, 1981; Erth et al., 1982; Delathoe et al., 1982; Grunteroth et al., 1982; Lee and Yeh, 1983; McCuskey et al., 1983; Key et al., 1983; Gelberman and Mastensen, 1983; Ducasse et al., 1984; Cinca et al., 1984; Scala et al., 1984; Kogan et al., 1984; Evemeev and Dorokhova, 1984; Goncharevskaia and Szemeredi, 1984; Stridbeck et al., 1984; Massobrio et al., 1985; Mirvis et al., 1985.

17. MISCELLANEOUS USES
    1. The use of latex particles for calibration of measuring instruments:
        A. Electron microscope modification, shadow angle and thickness of shadowing material.
        B. Various light scattering instruments and techniques.
        C. Various electronic particle counting instruments, e.g. the Coulter Counter.
        D. Optical microscopes.
        E. Various aerosol counting instruments.
        F. Ultracentrifuges.
        G. X-ray diffraction.
    2. Counting of virus particles.
    3. Determination of pore size.
    4. Analysis of flagellate locomotion mechanism.
    5. Aerosolysing particles.
    6. Drying particles.
    7. Dispensing particles in other liquids.
    8. Dissolving particles.
    9. Particle as model for diffusion, acoustics, rheology, electrophoresis.
    10. Particles as replacement for blood cells, gas molecule, oil mist and cosmic dust have been reviewed by Vanderhoff (1964) and Bangs (1984).

TABLE I

LATEX AGGLUTINATION TESTS

MICROBIOLOGY

Aeromonas salmonicida	Ag
Antrax	Ab
Bordetella pertussis	Ab
Brucella abortus	Ab
Brucella polysaccharide	Ag
Campylobacter fetus	Ab
Candida albicans	Ag*
Candida albicans	Ab*
Candida mannan	Ag
Clostridium botulinism	Ab
Clostridium difficile toxin	Ag*
Clostridium perfringens enterotoxin	Ag*
Corynebacterium diphtheriae	Ab
Corynebacterium diphtheriae toxin	Ag
Edwardsella	Ag
Escherichia coli	Ag
Escherichia coli	Ab
Escherichia coli enterotoxin	Ag*
Haemophilus influenzae	Ab
Klebsiella (6 types)	Ab*
Legionella pneumophilia	Ab
Leptospira	Ab***
Mycobacterium phospholipid	Ag
Mycobacterium tuberculosis	Ag
Old tuberculin	Ab
Moraxella	Ag
Neisseria gonorrhoeae	Ag
Neisseria meningitidis	
Group A,B,C,D,X,Y,Z, W135, 29E	*
Pasteurella tularensis	Ab
Pasteurella pestis	Ab
Proteu.	Ab
Pseudomonas aeruginosa	Ag
Reiter protein	Ab
Salmonella	Ag*
Shigella	Ab*
Staphylococcus aureus	Ab*
Staphylococcus aureus	
Clumping factor & Protein A	Ag*
Staphylococcus enterotoxin	Ag*
Staphylococcus toxin	Ag*
Streptococcus mucopeptide	Ab
Streptococcus mutans	Ag
Streptococcus M Protein	Ab
Streptococcus pneumoniae	Ag*
Streptococcus group A	Ag*

TABLE I (Continued)

Streptococcus
  Group A,B,C,D,E,F,G,Q   Ag*
Streptolysin O   Ab*
Syphilis   Ab
Tetanus toxoid   Ag
Tularemia   Ab
Vibrio cholerae enterotoxin   Ag*
Whooping cough   Ab
Yaws   Ab
Mycoplasma hyosynoviae   Ab
Mycoplasma pneumoniae   Ab
Mycoplasma swine   Ab
Rickettsia prowazekii   Ab
Rickettsia rickettsii   Ab
Rickettsia typhi   Ab

VIRAL INFECTION

Avian encephalomyelitis   Ab
Arbovirus   Ab
Bovine leukosis
Cytomegalic virus   *
HTLV-III (AIDS)   Ab*
Hepatitis   Ag*
Hepatitis   Ab
Herpes   Ag*
Infectious mononucleosis   Ab*
Measles   Ab
Mumps   Ab
Poliomyelitis   Ab
Pseudorabies in swine   Ab*
Rotavirus   Ag*
Rubella   Ab*

PARASITOLOGY

Amebiasis   Ab*
Angiostrongyloides   Ab
Ascaris   Ab
Babesia argentina   Ab
Canine heart worm   Ab*
Chagas disease   Ab
Cysticercosis   Ab
Echinococcus granulosum   Ab
Fascioliasis   Ab
Filariasis   Ab
Kalaazar   Ab
Malaria   Ab

TABLE I (Continued)

Pneumocystis carinii	Ab
Toxoplasma gondii	Ab*
Trichinella spiralis	Ab*
Trypanosoma congolensis	Ab
Trypanosoma rhodensis	Ab

MYCOLOGY

Aspergillus fischeri	Ab
Candida albicans	Ab*
Candida albicans	Ag*
Candida albicans mannan	Ag
Coccidioides immitis	*
Cryptococcus neoformans	Ag*
Farmers lung	Ab
Histoplasma capsulatum	Ab*
Sporotrix schenkii	Ab*

ENDOCRINOLOGY

Chorionic gonadotropin HCG	Ag*
Estriol	Ag**
Estrogen	Ag
Human placental lactogen	Ag**
Progesterone	Ag
Somatotropin	Ag
Thyroid stimulating hormone	Ag**
Thyroxin (T4)	Ag**
Triiodothyronine	Ag**

DRUG ASSAY

Amikacin	Ag*
Barbiturate	Ag
Digoxin	Ag***
Gentamycin	Ag*
Metamphetamine	Ag
Morphine	Ag
Netilmycin	Ag*
Phenobarbital	Ag
Phenytoin	*
Theophyline	Ag*
Tobramycin	Ag*

TABLE I (Continued)

<u>MISCELLANEOUS</u>

Albumin	Ag
Alpha fetoprotein (AFP)	Ag**
Alpha antitrypsin	Ag*
Anti NaDase	Ab**
Antinuclear antibodies	Ab*
Antistreptokinase	Ab
Antithrombin III	Ag**
Antizona pellucida	Ag
Apoprotein	Ag
Beta 1 glycoprotein	Ag
Beta 2 microglobulin	Ag**
Beta thromboglobulin	Ab
C reactive protein	Ag*
Canine fibrin degradation product	Ag*
Carcinoembrionic antigens (CEA)	Ag**
Circulating immunocomplexes	Ag
Collagen	Ag
Complement C3 and C4	Ag**
DNP-lysine	Ag
Deoxynucleoprotein	Ab*
Ferritin	Ag***
Fibrin fibrinogen degrad. product (FDP)	Ag*
Fibrinogen	Ag*
Fibronectin	Ag***
Haptoglobin	Ag*
Histamin	Ag
Human semen	Ag
Immunoglobulin E	Ag**
Immunoglobulin G,A,M	Ag**
Lectins	Ag
Mixovirus neuraminidase	Ag
Multitumor	Ab*
Myoglobin	Ag
Myelin basic protein	Ag
Plasmin	Ag
Plasminogen	Ag***
Polyamine	Ag
Polyamine	Ab
Protein quantitation	Ag
Rat b_2 microglobulin	Ag
Retinol binding protein	Ag**
Rheumatoid factor	Ab*
S-100 protein	Ag
Sarcoidosis	Ag
Snake venom	Ag

TABLE I (Continued)

Sperm	Ab
Vitamin K	Ag
von Willebrand factor VIII antigen	Ag*

Kits available:

*	USA
**	Europe
***	Japan

REFERENCE

1. Singer JM; Plotz CM: The latex fixation test. 1. Application to the serological diagnosis of rheumatoid arthritis. *Am. J. Med.* 21: 888, (1956).
2. Singer JM: The Latex Fixation Test in Rheumatic Disease. A review. *Am. J. Med.* 31: 766, (1961).
3. Vanderhoff JW: The Use of Monodisperse Latex Particles in Medical Research. Preprint. *Div. Organic Coating and Plastic Chem.* 24:223, (1964).
4. *Singer JM: Standardization of the Latex Test for Rheumatoid Arthritis Serology. Bull. Arthr. Rheum.* 6-7: 762, (1973-74).
5. Weetal HH: *Immobilized Enzymes, Antigens, Antibodies and Peptides.* Marcel Dekker, Inc. New York, (1975).
6. Nakamura RM; Ditto RW; and Tucker III ES: *Immunoassays, Clinical Laboratory Techniques for the 1980's.* Alan R. Liss Inc., N.Y., (1980).
7. Ishikawa E; Kawai T and Miyai K: *Enzyme Immunoassay.* IgaKu-Shoim, Tokyo, Japan, (1981).
8. Stites DP; Stobs JD; Fudenberg HH and Weeks JV: *Basic and Clinical Immunology*, Lang Medical Publication, Los Altos, Ca., (1982).
9. Coonrod JD; Jung JL and Ferraro JM: *The Direct Detection of Microorganisms in Clinical Specimens.* Academic Press, Inc., (1983).
10. Nakamura RM; Ditto RW; and Tucker III ES: *Clinical Laboratory Assay. New Technology and Future Direction*, Masson Publishing, Inc., (1983).
11. Stevens WR: Agglutination tests in diagnosis of bacterial infections, *Laboratory Medicine*, 25 (1983).
12. Hechemy EK; Michaelson EF: Latex particle assays. *Laboratory Medicine*, 22(6): 27 and 22(7): 26.

NOTE: Literature reviewed in this paper can be obtained from: Medline (Date Base), Bethesda, Md., National Library of Medicine.

INDEX

acceptor 279, 284
acid 375
acoustic spectroscopy 186
acoustics 388
acoustophoresis 250
acoustophoretic fractionation 13
acoustophoretic mobility 16
acrylamide 7
acrylate ester 5, 23
acrylic acid 7, 80
acrylonitrile 34, 36, 37, 67, 69, 80, 99
acrylonitrile (AN) 66
acrylonitrile-butadiene 9
adhesion 6, 7, 11
admittance 293, 295
adsorption chromatography 248
adsorption isotherm 270
aerosol spectrometry 381
affinity chromatography 311
agglutinates 309, 310
agglutination 309, 310, 315, 318, 321, 373, 374, 378, 382, 384, 389
albumin 308, 309, 323, 324, 325, 326, 328, 330, 331, 334, 335, 336, 372, 373, 386
aldehyde groups lactones 357
algorithms 248
allophycocyanin 311
allyl methacrylate 49
alpha value 302
amide 375
amino acids 317
analysis 332
angular velocity 154
anhydride groups 357
anionic 316
anisometric particles 113
anisotropic particles 113
anisotropy 121, 382
anthracene 284
antibodies 307, 308, 309, 311, 312, 315, 316, 318, 322, 323, 326, 330, 331, 332, 334, 336, 337, 339, 360, 361, 362, 364, 366, 373, 375, 377, 378, 379, 380, 383, 384, 385, 386
antibody V, 308, 309, 310, 315, 316, 317, 318, 319, 322, 326, 330, 334, 336, 337, 338, 339, 355, 356, 359, 360, 361, 363, 365, 368, 372, 373, 374, 378, 379, 380, 383, 384, 385, 386, 387, 388
anti-cancer drug V
antigen 309, 315, 316, 317, 318, 326, 330, 361, 368, 375, 378, 383, 384, 385, 386
antigen antibody 319, 321, 379
antigen antibody agglutination 382
antigenic 311
antigens 318, 321, 322, 388
antisera 316

aplastic anemia 338
area per molecule 271
azide 376
2,2'-azobis (2,4-dimethyl-valeronitrile) 40

B cells V, 362, 365
B lymphocytes 365
batch 8, 23, 24, 28, 29, 34
batch copolymerization 32, 73
batch emulsion copolymerization 67
batch polymerization 5, 8, 23, 24, 29, 31, 32, 33, 35, 38
batch semi-continuous polymerizations 23
Bingham stress 146
binomial distribution 230
bioluminescent lectins 315
biomedical applications 243, 307, 321, 355
biomedical field 305, 307, 312, 321, 379
biotechnology 307, 312, 379
bladder 325
bleeding 312
blood 318, 363, 368
blood cells 322, 388
blood flow 321
blood vessels 113
body channels 113
body fluids 321
bone 326, 338
bone marrow V, 307, 308, 328, 329, 332, 333, 334, 335, 336, 337, 338, 339, 340, 361, 362, 363, 364, 365, 366, 368, 386
branching 11, 12, 16
Brinkman's equations 143
Broglie's relation 254
Brownian forces 133
Brownian motion 115, 120, 132, 163, 172, 244, 382
bulk modulus 139
bulk polymerization 3, 226
butadiene 24, 25, 26, 38, 49, 62
butadiene-styrene 5, 9
butyl acrylate 28, 72, 74
butyl acrylate-methyl methacrylate 24
butyl acrylate-methyl methacrylate systems 23
butyl acrylate-rich 28
butyl methacrylate 30
butyl methrylate 30

calorimetric measurements 249
cancer V, 322, 356, 362
cancers 307, 336
capacitance 168, 292, 293
capacitors 289
capillary chromatography 245
carbonylation VI
carboxy 375
carboxyl 375

396

carcinoma 324, 325, 336
catalysts VI, 246
cationic nonionic detergents 372
centrifugal fields 108
centrifugal forces 121
centrifugation 316, 332, 380, 384, 385
charcoal 309, 322
charge densities 300
charge density 122, 202, 291, 295, 297, 299, 301, 302, 316
chelating agents 322
chemi 315
chemical potential 72, 169, 200
chemical potentials 75
chemical shift 245
chemiluminescence 310, 383
chlorinated hydrocarbons 211
chromatography 38, 310, 311, 315
chromatography (GPC) 12
chromophore 277, 281
chromophores 278, 284
Coacervation 188
coagulation rate constant 82
coagulation rate constants 101
coefficients 75
coherent neutron scattering length 257, 258
coherent scattering length 256, 257, 263
coherent scattering length density 257
coil radius 229
collisions 274
colloidal stability 6,7, 8, 36
comb structures 194
combustion analysis 248
commodity polymers IX
complex modulus 112, 170, 171
computer simulation 115, 131
computer simulations 116
concentrated dispersions X
conductivity 137, 138, 139, 140, 142
conductivity titration 297
conductometric 12
contact angle 250
continuous polymerization 5, 8, 23, 24, 31
copolymerization 93
core-shell latexes 9
correlation 229
correlation function 271
correlation length 233
Couette 152, 154
Couette flow 151, 157
Coulombic forces 119, 122
crosslinker 49, 50, 53, 56, 57, 59, 60
crosslinking 11, 12, 13, 14, 16
cyclohexane 283, 285

Debye length 203
Debye radius 122
Debye-Huckel parameter 272
decay 286
decay time 281

decay times 286
density 120, 195, 231, 234, 257, 309, 384
density profile 234
depleted layer 189
depletion flocculation 124, 169, 170, 179
depletion layer 169, 185, 201
depletion layers 183, 185, 200
detergent 316, 372
detergents 316, 377
dextran 214, 324, 328, 334, 335, 356, 386
di-2-ethylhexyl phthalate 151, 159, 160
dialysis 12, 264, 316, 380
diameter 264, 291, 309, 315, 325, 331, 355, 357, 377
diameters 244, 281, 377
dielectric 289, 293, 295, 297, 299, 301
dielectric loss 293, 300
dielectric measurements 302
dielectric spectra 293
dielectric spectrometer 290
dielectric spectroscopy 110, 111, 250, 290, 302
differential scanning calorimetry 54
differential scattering calorimetry 62
diffraction 109, 122, 158, 160, 161, 162
diffuse layer 289
diffusion 75, 76, 114, 211, 230, 264, 281, 282, 289, 388
diffusion coefficient 75, 101, 198, 244, 281
diffusion coefficients 10, 11, 75, 76, 264
diffusion constant 382
diffusivity 132, 279
dimensional analysis 120
dipole 278, 284
dipoles 289
dispersion polymerization 209, 210, 212, 217, 218, 220, 226
distribution 295
distribution functions 232
disulphide bridges 318
divinylbenzene 39, 49, 72
DLVO 115, 128, 290
DLVO theory 76, 289
DNA 312
DNA/RNA 312
dodecane 169
donor 279
double layer 122, 296, 302
drug 307, 308, 312, 323, 324, 325
drugs 308, 321, 322, 323, 324
drying 316
dynamic mechanical 38
dynamic mechanical spectroscopy 56, 62, 250
dynamic yield stress 112, 113

education 19
Edwards formalism 230
effective medium 138, 139
effective medium approximation 137
effective medium theory 137, 139
eigenfunctions 232, 233
eigenvalues 233
Einstein value 122

elastic moduli 138
elastic modulus 169
elastomer 47
electric magnetic fields 109
electrical double layers 272, 289, 297, 302
electrical field 289
electrokinetic 198
electrokinetic methods 198
electron 278
electron micrograph 357, 358
electron microscope 385
electron microscopy 25, 49, 54, 62, 169, 247, 262, 322, 361, 386
electron spin resonance (ESR) spectroscopy 12
electrophoresis 198, 250, 388
electrophoretic 13, 16
electrophoretic mobilities 293
electrophoretic mobility 13, 250, 272, 290, 293, 294, 372
electrophoretic motility 385
electrostatic potential 111
electroviscous effect 122, 123, 192
elongation 57, 58, 59, 60, 61, 62
elongational strain 113
elongational stress 113
emission intensity 277
emulsifier 32, 66, 67, 72, 74
emulsion 93, 209, 371
emulsion copolymerization X, 1, 3, 10, 11, 12, 20, 29, 65, 70, 79, 80, 90, 92, 99
emulsion polymerization IX, 3, 4, 5, 6, 7, 8, 9, 10, 20, 23, 48, 49, 75, 79, 80, 90
endocrinology 379
energy transfer 284, 285, 286
enthalpies 249
entropy 75, 169, 244
enzyme 310, 215, 321, 324, 383
enzyme immunoassay 383
epoxy groups 357
erythrocyte 308
erythrocyte antibody 332
erythrocytes 332, 334, 338, 339
ESR 186, 196, 198
ester groups 357
ethyl acrylate 69
2-ethyl-hexyl acrylate 90
ethyl methacrylate 43
ethylene glycol dimethacrylate 49
exchange 296, 322
excimer 278
exciplex 277
exotherm 23

fast protein liquid chromatography 311, 355, 368
ferrimagnetism 357
field 324
filtration 316
first-generation latexes 7
flame ionization detection (TLC-FID) 12, 35, 38
flocculation point 178

Flory-Huggins equation 84
Flory Huggins interaction 231
Flory-Huggins interaction parameter 72, 102, 200
Flory-Huggins lattice 84
Flory-Huggins lattice theory 83
Flory-Huggins model 231
Flory-Huggins theory 71
Flory-Krigbaum theory 167
flow cytometry 311, 312, 381
fluorescein 311
fluorescence 15, 186, 247, 248, 249, 277, 281, 282, 283, 286, 287, 308, 310, 321, 383
fluorescence immunoassays 384, 385
fluorescence quenching 277, 282
fluorescent 245, 279, 282, 286, 322
fluorescent immunoassay 384
fluorine 312
fluorine-19 312
fluorochrome 362
fluorophore 311, 315
Fourier transform 196, 197, 273
fourier transform infrared spectroscopy (FTIR) 12
Fourier transformation 268
free energy 168, 229, 232, 233, 235
free radical-initiated 3
free radical-initiated polymerization 3
free volume 237
freeze fracture 247
freezing 380
frequencies 293, 302
frequency 171, 172, 174, 250, 289, 292, 295, 299, 300
Frequency Response Analyzer 290
fructose 317
FTIR 247

galactose 317
gamma radiation V
gamma scintillation 311
gammaglobulin 372, 373, 374, 376, 378, 380, 385
gas chromatography 97
Gaussian distribution 232
gel filtration 311
gel permeation 12
gel permeation chromatography 12, 13, 16, 19, 249
gelling 124, 126
genes 118
glass transition range 54
glass transition temperatures 54, 55, 62, 248, 262
globular 282
globulin 317, 338, 372, 383
glucose 163
glycoproteins 316, 318
graft copolymer 9
granulocytes 332, 368
gravitational 121
gravitational centrifugal fields 108
gravity fields 108
Green's functions 138
Guinier analysis 270
Guinier equation 260

Hamaker constant 168, 191, 193
haptens 382, 383
hemoglobin 372
hemoperfusion 312, 322
Henry's law 80, 249
hepatocytes 361
heptane 168
hexane 214
high frequency modulus 136
high-performance liquid chromatography (HPLC) 13
hormones 321, 375, 383, 384
HPLC 248
hydrodynamic chromatography 245
hydrodynamic chromatography (particle size
distribution) 19
hydrodynamic methods 198
hydrodynamics 132, 134
hydroformylation VI
hydromethylmethacrylate 386
hydrosilylation VI
hydroxy 375
hydroxyethyl methacrylate 321, 376, 385
hydroxypropylcellulose 217

immunoassay 310, 315, 318, 379, 380, 383, 384, 385
immunoassays 307, 309, 310, 312, 315, 318, 321, 355,
356, 359, 366, 380, 383, 386
immunoenzymatic assays 373
immunofluorescence 332, 337, 363, 364
immunoglobulin 315, 317, 318, 326, 330, 331, 334,
335, 337, 339, 340, 368, 376, 380
immunoglobulin antibody 332
immuno-labeling V
immunonephelometer assay 319
immunophotometric assay 381
immunotoxins 308
impedance 289, 292, 295
infinite cluster 137, 140, 142
infrared spectroscopy 30
initiation 3, 6
integration algorithm 87
intensities 281
intensity 196, 311
interfacial energy 83
interfacial forces 71
interfacial tension 32, 102
interpenetrating polymer network (IPN) 47
interpenetrating polymer network latexes 47
ion 296, 322
ion exchange 12, 268, 309, 311
ion exchange resin 290
ion-exchangers VI
ionic exchange resins 311
ion-selective electrodes 12
IPN 47, 48, 49, 50, 51, 52, 53, 54, 55, 56, 57, 58, 59,
60, 61, 62
is poly(methyl methacrylate) 203
isocyanate 357

ketones 279

kidney 325
kinetic 221
kinetics 183, 220, 226, 290, 373

Langevin's equations 116
Langmuir equation 373
Langmuir trough 250
laser 243
laser-Raman spectroscopy 186
lasers 311
latexes 9
layer 289
lectin agglutination 308, 338
lectins 315, 331
length 229
leukemia 324, 335, 336, 338
leukemias 363
leukocytes 339
lifetime 281, 282, 286
ligand 317, 330, 375, 376, 381, 385
ligands 322, 326
light 253
light reflectance 168
light scattering 12, 244, 248, 253, 254, 259, 272, 290,
291, 381, 388
light-neutron-scattering 290
light scattering immunoassay 382
liposomes 312
liver 325
London-Keesom-Debye 315
London-van der Waals forces 371
loss modulus 29
low volume fractions 274
luminescence 277, 279
lung 325
lymphocyte 307, 308, 312, 321, 322, 332, 340
lymphocytes 322, 330, 332, 335
lymphoid cells 322
lymphoma 336, 362, 363

macro-emulsions 79, 187
magnetic 324
magnetic field 324, 325, 326, 355, 366
magnetic moment 254
magnetic properties 310
magnetic susceptibility 341
mannose 317
marrow 326
mass spectrometers 247
mass spectrometry 246
mass transfer coefficient 89, 90, 98, 101
material balance 86
mechanical properties 57, 58
mechanism 220
meningitis 379
methacrylic acid 321
methyl acrylate 69, 80, 327
methyl acrylate (MeA) 69, 70
methyl methacrylate 38, 121, 210, 212, 214, 216, 218,
219, 221, 222, 321, 330, 376, 386

methyl methacrylate (MMA)-styrene 66
methylacrylic 375
methylstyrene divinylbenzene 376
micelles 4, 94, 209, 277
micellization 224
micro-emulsions 187
microbes 318
microbiology 318, 379, 389
microdomains 48, 53, 54, 58, 248, 277
microemulsions 79, 277
microgel 191, 192
microphases 283, 284
microscopes 388
microscopy 381, 385
microspheres 311
microtitration 383
Mie theory 381
Miller indices 158
miniemulsion 79, 80, 87, 88, 90, 91, 92, 93, 95, 96, 97, 100
mobilities 281
mobility 75
modulus 37
molecular weight 4, 12, 13, 14, 16, 18, 19, 94, 123, 124, 169, 173, 175, 179, 212, 213, 214, 220, 221, 222, 234, 245, 248, 249, 265, 322, 323, 373, 374
molecular weights 12, 66, 126
momentum 144
monomer/initiator VI
morphologies 43, 47, 56, 65
morphology 36, 38, 43, 44, 47, 48, 49, 56, 62, 72, 73, 76, 224, 225, 243, 247, 248, 253, 268, 279, 283, 286, 318
Mossbauer spectroscopy 357
multiple scattering 109
mycology 379

naphthalene 284, 286
n-butyl acrylate 49, 80
negative staining 247
nephelometry 381
neuroblastoma 327, 336, 337, 362, 363, 386
neutron 244, 255, 256
neutrons 244, 253, 254, 258, 272
neutron scattering 109, 157, 184, 186, 194, 196, 198, 254, 255, 259, 290
neutron scattering length 259, 265
neutron scattering length density 259, 265
NMR 13, 186, 196, 198, 245, 248, 312
non-Newtonian fluids 127
non-Newtonian viscosities 123
Nondraining Flocs 146
nonionic 316
nonionic surfactants 124
nuclear magnetic resonance (NMR) 12
nuclear magnetic resonance spectroscopy 15, 16, 32, 38
nuclear magnetic resonance spectroscopy (NMR) 12
nucleation 4, 5, 8, 16, 23, 65, 66, 68, 71, 74, 82, 88, 223, 224, 226
nuclei 272

number concentration 273

optical properties 183
oscillatory stress 112
osmotic pressure 170, 201, 238
osmotic underpressure 238
ovarian cancers 336

pair potential 194, 199, 202, 229, 272
pair-potentials 185, 188
parameter 231
parasites 318
parasitology 379
particle counting 382
particle diameter 216, 217, 219
particle morphologies 11
particle morphology IX, X, 1, 3, 6, 8, 10, 11, 14, 15, 17, 65, 66, 71, 72, 80
particle number 68
particle radius 126
particle size 36, 44, 214, 217, 218, 224, 243, 253, 259, 289, 295, 299, 316, 371, 372, 373, 382
particle size distribution 9, 66, 209, 215, 243, 253, 261, 302
particle sizes 372
particle volume 217
partition coefficients 66, 71, 72, 221
peclet number 120
peclet numbers 115
PEO 70, 169, 172, 173, 174, 175, 176, 177, 178, 179, 193, 194, 200, 203
percolation 138, 140
percolation theories 134
percolation theory 137
perfusion rates 321
permeability 145, 283
permittivity 293, 300
persulfate ion initiator 25
perturbation potential 199
phagocytosis 379, 387, 388
phase diagram 122
phase separation 48
phase transition 122
phase transitions 186
phenanthrene (Phe) 279
phosphorescence 277
photon correlation light scattering 198
photon correlation spectroscopy 13, 169, 253, 272
phycoerythrin 311
pigments 119
pili 318
plasma 322
plasma etching 247
plasma protein 368
plastic viscosity 146
plasticizers 164
platelets 113, 387
(PMMA) 279
Poisson's ratio 138, 139, 143
poly (12-hydroxy stearic acid) 211, 215

400

poly(2-ethylhexyl methacrylate) 279
poly(α-methylstyrene sulphonate) 203
poly-(butyl-2-cyanoacrylate) 214
poly(ethylene oxide) 175, 193
poly(methyl methacrylate) 123, 175, 211, 215, 216, 225, 247, 279
poly (styrene-b-dimethylsiloxane) 211, 214
poly(vinyl acetate) 10, 279
'poly(vinyl alcohol) 193
poly(vinylchloride) 123
polyacrylamide 5, 321
polyacrylate seed latex 48
polybutadiene 9, 24, 38, 39, 377
polybutyl acrylate 32, 33, 34, 35
polybutyl acrylate seed latex 32
polychloroprene 5
polydispersity parameter 291
polyethyl methacrylate 43, 44
polygluteraldehyde 321, 322, 327, 331
polyketones 188
polymer microstructure 11
polymerization 3, 371
polymethyl methacrylate 38, 39, 40, 41, 42, 257, 274, 378, 385
polyoxyethylene 375
polypeptides 318
polysaccharides 193, 316, 318
polystyrene 35
polystyrene-polybutyl acrylate 32
polytetrafluoroethylene 5
polyurethanes, polyimides, polyamides, polyesters, polysulphones, aromatic polyethers 188
polyvinyl acetate 5, 6, 28
polyvinyl alcohol 371
polyvinyl chloride 5, 38, 159, 160
poropidium iodide 311
positive staining 247
poststabilization 7
potential 235, 238, 290, 292, 293
potential difference 290
potential energy 185, 232
potentiometric conductometric titrations 111
potentiometric titration 12
pressure 198
probabilities 230, 236
probability 230, 231
probability density 132
probability distribution 197
propagation 3, 6
propagation rate constants 68, 69, 70, 82, 101
prostate 325
protein 282, 309, 310, 315, 316, 317, 319, 322, 328, 335, 355, 359, 372, 373, 374, 376, 379, 380, 381, 385
protein antigen 316
protein liquid chromatography 311, 355
proteins 282, 312, 316, 317, 318, 321, 360, 372, 375, 376, 381, 385
pus 318
PVAc 279
pyrene groups 249

quasi-elastic light scattering 253
quaternary amine 375
quenched 282, 283, 284
quencher 247, 279, 281, 282
quenching 277, 278, 281, 282, 283, 284, 286, 287

radial distribution function 133, 273
radial number density 273
radiation 109
radii 261
radioactive isotopes 325, 377
radioimmunoassay 310, 321, 373
radioisotopes 315, 384
radiomonomer 377
radiotheraphy 361, 362
radius 195, 250, 258, 261, 265, 268, 295, 299
radius of gyration 74, 170, 178, 249, 261
rate constant 281
reaction rate 381
reactivity ratios 11
reactivity ratios (r_{ij}) 67
reciprocal space 272
red blood cells 312, 327, 328, 331, 335, 368, 384
reduced density 121
reflective infrared spectroscopy 12
refractive index 245
regression procedure 69
relative density 120
relative permittivity 272
relative viscosity 120, 172
relaxation 293
relaxation frequency 299
relaxation processes 198
remanence 310, 355, 359
resistance 292
resistors 289
Reynold's number 113, 120, 121
rheological modifiers IX
rheological properties 107, 109, 111, 302
rheology X, 11, 13, 105, 107, 112, 114, 116, 117, 119, 120, 127, 131, 142, 144, 146, 164, 206, 244, 372, 388
rheumatoid 378
rheumatoid arthritis 378, 380
rhodamine 311
rigid spheres 119
rods 113
rotating anode 244

S-butadiene₄ 66
salicylic acid 317
SANS 206, 244, 248, 267
sarcoma 324
sarcomas 325
scanning microscope 385
scattered intensity 271
scattering angle 256, 259
scattering cross-section 256
scattering length density 260, 266
scattering vector 255
Scheutjens-Fleer theory 200

Schiff base 334
Schroedinger equation 232
scintillation 383
second-generation latex 7
second-generation latexes 7
secondary ion mass spectrometry (SIMS) 246
sectioning 247
sedimentation field flow fractionation 245
seed latex 5, 8, 9, 23, 24, 25, 29, 34, 37, 38, 74
seed latexes 34, 43, 49, 187
segment density 229, 233, 234
selective dissolution 247
semi-batch polymerization 5, 23
semi-continuous 8, 23, 35
semi-continuous copolymerization 11, 30, 32
semi-continuous copolymers 30
semi-continuous emulsion polymerization 9
semi-continuous polymerization 8, 11, 23, 24, 25, 28,
    29, 30, 31, 32, 33, 34, 38
serological identification 318
serology 318
SERS (Surface Enhanced Raman Scattering) 246
serum 324, 325, 330, 378, 379, 380
serum albumin 315, 368
serum antibodies 318
shear 151, 152, 153, 155, 156, 157, 158, 159, 161, 162,
    163, 170, 172, 176, 178, 183, 192, 250
shear moduli 140
shear modulus 56, 140, 141, 142, 143, 145, 146
shear rate 156, 157, 159, 160, 162, 170, 172, 175, 176
shear stress 172
shear thickening 128, 129
shear thinning 128, 129, 131, 132, 134, 148
silicone 124, 126, 163
silicone rubber 199
silicones 121, 124
singlet triplet 278
site fraction 138, 139, 140
size 295, 309
size distribution 9, 220, 312
small angle neutron scattering 244, 261, 262, 268
Small Angle Neutron Scattering (SANS) 248
small-angle X-ray scattering 13, 251, 254
Smoluchowski's equation 80
SNMS (Secondary Neutral Mass Spectrometry) 246
sodium dodecyldiphenyl oxide sulfonate 49
sodium styrene sulfonate 7
solubility 279, 282
solution polymerization 3
specialty polymers IX
spectrometry (SIMS) 246
spectrophotometry 319, 381, 382
spectroscopy 289, 295
spinal fluid 380
stabiliser sheath 192, 193, 194
statistical mechanics 148
steam-stripping 7
steric stabilization 191, 267, 371
steric stabilizers 212, 213, 217
sterically-stabilized 199

Stern 289
Stern layer 202, 297, 299, 302
Stern-Volmer 281
Stern-Volmer equation 281
stochastic matrix 230
Stokes equations 133, 143
stool 318
storage modulus 29
strain amplitude 112
stress relaxation modulus 108
structure factor, S(Q) 271
structured 9
styrene 34, 36, 37, 38, 99, 219
styrene divinyl benzene 386
styrene divinylenzene 385
styrene (s) emulsion copolymerization 69
styrene-2-ethylhexyl acrylate-methacrylic acid 25
styrene-acrylonitrile 9, 25, 34, 35
styrene-butadiene 5, 6, 47
styrene-divinylbenzene 310
styrenebutadiene 10
sulfhydryl groups 317, 376
sulfonate 375
surface 293
surface active agent 270
surface area 38, 41, 42, 74
surface charge 111, 290
surface charge density 243, 253
surface domains 34
surface electric charge 72
surface electric field 71
surface energy 381
surface potential 243, 250, 253, 272, 289
surface tensiometry 19
surface tension 12, 17, 71, 72
surfactants 245
suspension polymerization 3, 4, 209
synchrotron 244
synthesis 47, 48
synthetic rubber IX

T antigens 363
T B cells 327
T B lymphocytes 327, 328, 330, 331
T cell V, 307, 336, 339, 340, 364, 386
T cells 307, 328, 332, 333, 335, 338, 363, 364, 365
T and B lymphocytes 328
T lymphocyte 329, 330, 332, 334, 336, 338, 339, 340,
    363, 365
target cells 360, 361, 364, 365, 375
Taylor series 197
tensile strengths 30, 57, 58, 59, 60, 61
termination 3, 6, 7
testicular 336
tetraphenyl lead 284
Texas red 311
thermal properties 55
thermodynamical potential 235
thin-layer 38
thin-layer chromatographic 13

thin-layer chromatography 12, 32, 35, 248, 250
thin-layer chromatography-flame ionization detection 32
third-generation latex 7
thixotropy 109
throat 318
titration 290, 296, 297, 298, 299, 301, 380
toroidal rings 155, 156
torque 154
torsional flows 152, 154
transfer 3, 6
transfer area 88, 90
transfer coefficient 81, 82
transference 289
transition temperature Tg 9
transition zone 278
transmission electron microscopy 13, 16, 32, 38, 261
2,2,4-trimethylpentane 220, 221, 222
tumors 307, 308, 323, 324, 325, 326, 336, 337, 340, 341, 361, 362, 363, 366
turbidity 19

ultracentifugation 12, 13, 198
ultracentrifuges 388
ultramicroscopic methods 184
ultrasonic potential 111
ultrasonication 316
urine 318
UV 311, 333
UV fluorescence spectroscopy 196

valence 295
van der Waals attraction 168, 191, 193
van der Waals forces 121, 131, 148, 184, 202, 317, 375
van der Waals interaction 202, 206, 315
velocities 145
velocity 152, 261
Verwey-Overbeek theory 371
video enhanced image analysis 247
vinyl acetate 5, 69, 72, 74, 80
vinyl acetate-butyl acrylate 11, 23, 24, 28, 29, 30
vinyl chloride 5

vinylidene chloride 9, 30
vinylidene chloride-butyl methacrylate 30, 31
vinylidene chloride-butyl methacrylate copolymers 31
viral antibodies 318
viral infection 318
virology 318, 379
virus 388
virus genomes 318
viruses 375
viscoelastic properties 80, 107, 167
viscoelasticity 172, 176
viscometry 16
viscosities 125
viscosity 8, 12, 120, 122, 126, 127, 128, 131, 132, 133, 134, 136, 143, 159, 160, 171, 172, 173, 198, 211, 249, 372
volume 263
volume fraction 120, 134, 137, 140, 142, 145, 147, 156, 160, 162, 169, 172, 173, 174, 175, 177, 192, 198, 200, 204, 205, 231, 258, 259, 272, 273, 274
volume restriction 167
volumes 263
von Mises criterion 143, 145

wave function 232
wave vector 255
wetability 381

Xe 245, 246
X-ray photoelectron spectroscopy (ESCA) 12
X-ray diffraction 30, 388
X-ray induced Ion Mass Spectrometry 246
X-ray photoelectron (ESCA) 12
X-ray photoelectron spectroscopy (XPS) 246
X-ray scattering 284, 285
X-rays 244, 246, 253, 254

Young's moduli 30, 113
Young's modulus 32

zeta potential 13, 122, 195, 250, 272